Advances in Industrial Control

For further volumes:
http://www.springer.com/series/1412

Dong-Ping Song

Optimal Control and Optimization of Stochastic Supply Chain Systems

 Springer

Dong-Ping Song
School of Management
University of Plymouth
Drake Circus
PL4 8AA Plymouth
United Kingdom

ISSN 1430-9491   ISSN 2193-1577 (electronic)
ISBN 978-1-4471-4723-7   ISBN 978-1-4471-4724-4 (eBook)
DOI 10.1007/978-1-4471-4724-4
Springer London Heidelberg New York Dordrecht

Library of Congress Control Number: 2012954539

Springer is part of Springer Science+Business Media (www.springer.com)

To my wife, Li Jin, and my son, Tianyi

Series Editors' Foreword

The series *Advances in Industrial Control* aims to report and encourage technology transfer in control engineering. The rapid development of control technology has an impact on all areas of the control discipline: new theory, new controllers, actuators, sensors, new industrial processes, computer methods, new applications, new philosophies, and new challenges. Much of this development work resides in industrial reports, feasibility study papers, and the reports of advanced collaborative projects. The series offers an opportunity for researchers to present an extended exposition of such new work in all aspects of industrial control for wider and rapid dissemination.

For many readers from the control community, the notion of a "supply chain" probably conjures up the advanced information technologies used for the demand and supply of goods and products for commercial enterprises such as retail and supermarket chains. However, as society infrastructure becomes ever larger and more complex, supply chains exist across a wide variety of industrial and commercial activities. For example, effective supply chains are important in bringing together the components in aircraft manufacture and in automobile production lines.

Common features of a supply chain include a focal "consumer" point at the top of a supply chain that is issuing "orders" or demands for finished goods and processed materials. These demands create information flows down to the suppliers (manufacturers) who in turn attempt to meet these demands and transport the required goods and products along the chain to the consuming focal point. Thus, supply chains have an information–material flow duality. In many supply chains, the suppliers themselves are also a link in a supply chain of their own, so it is easily seen how the structure of a supply chain can quickly become very complex. Other complicating factors might be "competition" between the supplier vendors at various points of the supply chain or, alternatively, policies where the activities in the whole supply chain are coordinated or integrated using cooperation strategies to try to make efficiency gains.

Another aspect of the supply chain field is the "people and organization" dimension that does not usually exist in more conventional engineering process or industrial control system studies. Consequently, there is an extensive literature on "business management" approaches and philosophies for the operation of an efficient, reliable, and cost-effective supply chain. To enhance and improve supply chain performance, there are also a number of professional organizations and institutes that provide training and other services to supply chain operatives and company personnel.

Given the importance of the field, the editors of the *Advances in Industrial Control* monograph series welcome this very first monograph in the series on supply chain control. Professor Dong-Ping Song's monograph *Optimal Control and Optimization of Stochastic Supply Chain Systems* studies these systems using an analytical framework and the techniques of control systems including system modeling, optimal control derivation, system simulation, and the formulation of the best suboptimal control solutions. Supply chains rapidly become very complex, and Prof. Song's monograph reports work that is focused on bridging the gap between modeling complexity and solution simplicity in stochastic supply chains. The control solutions presented use optimal control methods, and these are investigated and explored to indicate what can be learnt from the structure of these optimal policies. Subsequently, Prof. Song derives easy-to-implement suboptimal solution policies and reports on solutions for situations with multiple inventory and production decisions in supply chain systems in the presence of uncertainty and stochastic effects. Overall, the aim is to emphasize the global integration of the supply chain rather than hierarchical decision-making.

Other entries in the *Advances in Industrial Control* series that have some relation to the supply chain topic include:

1. Palit, A.K., Popovic, D.: Computational Intelligence in Time Series Forecasting. (2005). ISBN 978-1-85233-948-7
2. Bogdan, S., Lewis, F.L., Kovacic, Z., Mireles, J.: Manufacturing Systems Control Design. (2006). ISBN 978-1-85233-982-1

Professor Song's monograph makes an invaluable addition to this small subset of monographs on these important enterprise and manufacturing subjects.

Industrial Control Centre   M.J. Grimble
Glasgow, Scotland, UK   M.A. Johnson

Preface

Subject of the Book

In the last two decades, supply chain systems have attracted a huge amount of attention from both the industrial and academic communities. Leading companies now see their supply chains as an important source to gain competitive advantages. The success of Wal-Mart in 1990s was partly attributed to the application of innovative supply chain management strategies, for example, the continuous replenishment program or the vendor-managed inventory strategy which coordinates the inventory management across the retailer and its suppliers. Business organizations are increasingly recognizing the importance of breaking down the barriers between functions and entities and tend to control and optimize their supply chains as an integrated system.

Although many companies have been involved in the analysis of their supply chain systems to seek performance improvement, in most cases, this analysis is performed based on experience and intuition. Very few analytical models have been used in this process (Simchi-Levi et al. 2009). This gap may be explained from two perspectives. The first perspective is the difficulty in obtaining solutions from the analytical models. Supply chain management emphasizes the integration and optimization of the entire supply chain system. A number of factors contribute to the complexity of modeling supply chain systems. Firstly, a supply chain may involve different functions and entities, a number of tasks and resources, and many different types of decisions. Secondly, supply chain components interact with each other in a variety of relationship formats. Such relationships influence the level of information flow, the degree of uncertainty, and the responsibility of decision-making. The modeling and solution are therefore usually problem dependent. Thirdly, supply chain systems are dynamic and subject to various uncertainties, which may exist in external environment and in internal activities. As a stochastic system, it is not guaranteed that the system will be stable if not well designed or controlled. The complexity of the supply chain systems obviously makes analytical tools difficult to solve many supply chain management problems.

The second perspective is the difficulty in implementing the solutions in practice. The solutions from analytical models, even if they exist, are often not robust and flexible enough to allow industry to use them effectively. From an industrial perspective, the operationability and simplicity of control policies are vital to their successful implementation and execution. Therefore, there is a need to bridge the gap between the complexity of the models due to the requirement of optimization and the simplicity of the solutions due to the requirement of implementation.

In this book, we intend to fill this gap by formulating analytical models for various typical stochastic supply chain systems, investigating the structural characteristics of the optimal control policies, constructing easy-to-operate suboptimal policies, establishing the system stability conditions, and addressing the optimization of suboptimal policies.

How the Book Is Structured

Chapter 1 provides a general introduction to stochastic supply chain systems. Various types of uncertainties in supply chain systems and different channel relationships between supply chain entities are discussed. The manufacturer is regarded as a focal company in the supply chain. The aim of this book is to tackle the optimal control and optimization problem for stochastic supply chain systems. More specifically, the main objective is to seek the optimal production control policies and the optimal ordering policies in the supply chain by taking into account a variety of uncertainties such as random customer demands, stochastic processing times, unreliable machines, and stochastic material lead times. The basic assumptions are stated, and the structure of the book is outlined.

In Chaps. 2, 3, 4, 5, 6, and 7, several typical stochastic supply chain systems will be studied. Analytical models are formulated and analyzed in detail. The purpose is to investigate and establish the structural characteristics of the optimal policies.

In Chaps. 8, 9, 10, 11, and 12, the structural knowledge of the optimal control policies obtained in earlier chapters will be utilized to construct easy-to-operate suboptimal control policies for various stochastic supply chain systems accordingly. Here the focus is to achieve the trade-off between the closeness to the optimality of the constructed policies and the degree of simplicity in terms of their implementation. Extensive numerical examples are provided to demonstrate the effectiveness of the proposed threshold-type policies. In addition, the system stability issues will also be addressed, which is essential when steady-state performance measures are concerned.

In Chaps. 13, 14, and 15, the optimization of threshold-type control policies and their robustness are addressed. The value iteration-based method and the stationary distribution-based method are first introduced to optimize the threshold parameters. Then, simulation-based optimization methods including genetic algorithm, simulated annealing, and ordinal optimization are presented. A range of numerical examples are given to demonstrate their efficiency. Finally, the main conclusions and limitations are summarized, and further research directions are discussed.

Reference

Simchi-Levi, D., Kaminsky, P., Simchi-Levi, E.: Designing and Managing the Supply Chain: Concepts, Strategies and Case Studies, 3rd edn. McGraw-Hill, Irwin (2009)

Acknowledgements

I would like to thank the following people for joint works and/or insightful discussions on topics that are related to the materials covered in this book in various periods of time: F.S. Tu, Y.X. Sun, W. Xing, Q. Zhang, C. Hicks, C.F. Earl, and J. Dinwoodie. I also thank the University of Plymouth for providing great support and an excellent academic environment during the past years.

Contents

Chapter 1
Stochastic Supply Chain Systems

1.1 Introduction

Globalization brings many opportunities to manufacturers, but it also gives rise to problems in planning procurement, inventory, production, and distribution. For example, the bullwhip effect, which represents that a little uncertainty in customer orders may result in the instability of the supply chain due to information distortion, could leave manufacturers and other members in supply chains with excessive inventories or severe shortages that incur very high costs (Lee et al. 1997a, b; Chatfield et al. 2004). There has been a clear trend that many individual firms no longer compete as independent entities but as an integral part of supply chain (Christopher 1992). This reflects the importance of supply chain integration and coordination.

An integrated supply chain may be defined as "*a network of connected and interdependent organizations mutually and co-operatively working together to control, manage and improve the flow of material and information from suppliers to end users*" (Christopher 2010). More specifically, an integrated supply chain system should coordinate a series of interrelated business processes and activities in order to (Min and Zhou 2002) (1) acquire raw materials (including parts and components), (2) add form value by transforming raw materials into finished goods, (3) add time and place values to finished goods by inventory and transportation, and (4) arrange information exchange between channel members (e.g., suppliers, manufacturers, distributors/retailers, third-party logistics providers, and customers).

With the manufacturer as the focal firm, a supply chain system is often regarded as consisting of two business processes: physical supply (inbound logistics) and physical distribution (outbound logistics), as shown in Fig. 1.1, in which the dotted lines indicate information flows and solid lines indicate material flows.

The objective of supply chain management (SCM) is to achieve efficient and cost-effective flow and storage of materials and information across the entire supply

D.-P. Song, *Optimal Control and Optimization of Stochastic Supply Chain Systems*, Advances in Industrial Control, DOI 10.1007/978-1-4471-4724-4_1, © Springer-Verlag London 2013

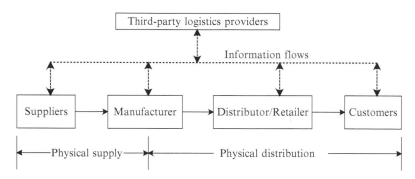

Fig. 1.1 A generic supply chain

chain system to meet customer requirements and minimize total system-wide costs incurred in the physical supply and physical distribution processes. With this in mind, supply chain management problem may be regarded as an optimal control and optimization problem for the underlying supply chain system subject to a set of constraints.

However, the optimal control and optimization of supply chain systems is difficult. The main challenges are related to three aspects: the desire to pursue global optimization, the presence of multiple uncertainties, and the involvement of relationship management.

The global optimization refers to the process of finding the optimal system-wide strategy. On the one hand, it is desirable to consider the supply chain as an integrated system since entities are interdependent and may have conflicting goals. On the other hand, contributions to the literature have demonstrated that even for a single entity, minimizing the cost while maintaining a certain customer service level could be difficult, for example, in situations with failure-prone machines and/or multiple product types. The difficulty can increase exponentially when an entire supply chain with more entities is optimized (Simchi-Levi et al. 2009).

Uncertainty is an inherent characteristic in every supply chain (Simchi-Levi et al. 2009). There are various types of uncertainties that may exist in a supply chain system, internally and externally, which makes the supply chain optimization challenging. The common types of uncertainties in supply chains will be addressed in detail in the next section.

The involvement of relationship management is another important characteristic of supply chain systems. From the cyclic perspective, a supply chain consists of a series of cycles, and each cycle represents an interface between two entities in which one entity places orders and the other fulfills orders. By nature, the human factor, the technology, and other factors will influence the relationships between entities in the supply chain. Different types of relationships may lead to significantly different supply chain management problems. Channel relationship will be further addressed in Sect. 1.3.

1.2   Uncertainties in Supply Chain Systems

Uncertainty can be defined as *"the unknown future events that cannot be predicted quantitatively within useful limits"* (Cox and Blackstone 1998). According to the degree of predictability, uncertainty may be classified into four levels: unpredictable events, unknown type of distribution with known moments (e.g., mean and variance), known type of distribution with unknown parameters, and known distribution. The last one, that is, using known probability distribution to describe uncertainty, is widely adopted in the literature.

There are many sources of uncertainties in a supply chain. The three most important categories of uncertainties are probably (e.g., Buzacott and Shanthikumar 1993; Davis 1993) (1) supply uncertainty such as material supply or subcontract order, (2) manufacturing uncertainty such as processing time and resource availability, and (3) customer uncertainty such as customer orders or demand arrivals.

Taking the manufacturer as the focal company, the first category is an external uncertainty, which occurs during raw material supply or outsourcing. The common measurement is suppliers' on-time delivery, average lateness, and degree of inconsistency. The second category is an internal uncertainty, which may be caused by set-up times, machining times, transfer times, machine failures and repairs, routine maintenance, human errors and absenteeism, and other events inside the manufacturer. The third category is an external uncertainty, which is caused by unpredictable market environments and customer requirement changes.

Throughout this book, the uncertainties in raw material (RM) supply, finished goods (FG) manufacturing process, and customer demand arrivals will be considered explicitly in the majority of the chapters. We will not address the issues on how to eliminate or reduce the uncertainties in supply chains. Instead, our focus is on how to achieve optimal control and optimization of the supply chain systems in the presence of multiple stochastic factors.

1.3   Channel Relationships in Supply Chain Systems

There are many options of relationship between channel members in supply chains. The different relationship options can be viewed in the form of a continuum ranging from arm's length to partnership, strategic alliance, joint venture, and vertical integration (Cooper and Gardner 1993). The arm's length relationship is the traditional channel relationship, which is conducted through the marketplace with price as its foundation, whereas vertical integration represents the common ownership (or fully integration acting as if a single organization). Each of these relationship types has motivating factors that drive its development, which then influences the operating environment and business processes. The duration, breadth, strength, and closeness of the relationship varies from type to type, from case to case, and from time to time.

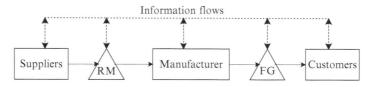

Fig. 1.2 Basic integrated supply chain with information flow

There has been a trend to move toward integration in supply chain management. The basic step to achieve supply chain integration is the information sharing among entities. Lee et al. (1997b) stated that information sharing is one of the most important mechanisms to counteract the bullwhip effect. There are various pieces of information that could be shared in the supply chain. For example, upstream entities could share the information such as inventory level, production schedule and capacity, and delivery lead times with downstream entities; downstream entities would avoid over-ordering products and handle the out-of-stock situations better. On the other hand, if downstream entities could share the information such as demand and inventory with upstream entities, upstream entities would be able to better manage the production and inventory of products. In reality, the implementation of information sharing mechanism between entities also relies on the applications of technologies such as electronic data interchange (EDI) and Internet.

A further step to achieve supply chain integration is the channel alignment or coordinated management. Instead of each entity in the supply chain making ordering decisions by itself, such decisions could be coordinated among the channel members in the entire supply chain. The vendor managed inventory (VMI) practice is a typical example of coordinated management, in which the upstream entity (e.g., supplier, vendor) is responsible for managing the inventory levels and availability for the customer based on the demand and inventory information provided by the customer (Song and Dinwoodie 2008).

In this book, we would normally assume that the information is shared among channel members and the management in terms of order placing and order fulfillment is coordinated to reflect the trend of supply chain integration in the recent decade. We will focus on the optimal control and optimization for integrated stochastic supply chain systems.

From the manufacturer's viewpoint, distributors or retailers are its customers. Moreover, in many supply chains, the manufacturers actually sell products directly to customers. The concept of "disintermediary" reflects such practices. Therefore, the generic supply chain in Fig. 1.1 may be simplified into a three-entity supply chain as shown in Fig. 1.2, in which raw material (RM) procurement and finished goods (FG) production are to be managed appropriately to meet customer demands. We call the supply in Fig. 1.2 the basic integrated supply chain or the basic supply chain. Most of the chapters in this book will use the basic supply chain as a reference point and extend it in different aspects. Due to the relative simple channel structure, it enables us to analytically explore the optimal control structure of the stochastic supply chain systems in various contexts. The knowledge developed may then be extended and applied to more complicated supply chain systems.

1.4   Optimal Control and Optimization in Stochastic Supply Chains

Although many companies have been involved in the analysis of their supply chain systems to seek performance improvement, in most cases this analysis is performed based on experience and intuition. Very few analytical models have been used in this process (Simchi-Levi et al. 2009). Two main factors may explain the difficulty. First, the complexity of the supply chain systems makes analytical tools difficult to solve many supply chain management problems. The complexity may be understood from three aspects: (1) the involvement of different functions and entities, tasks and resources, and many different types of decisions; (2) the desire for global optimality and integration; and (3) the presence of a variety of stochastic factors which may exist in the external environment and in the internal activities.

Second, the optimal solutions from analytical models, even if they exist, are often not robust and flexible enough to allow industry to use them effectively. From an industrial perspective, the operationability and simplicity of control policies are vital to their successful implementation and execution. Therefore, there is a need to bridge the gap between the complexity of the models due to the requirement of optimization and the simplicity of the solutions due to the requirement of implementation.

This book aims to fill this gap by formulating analytical models for typical stochastic supply chain systems using the stochastic dynamic programming approach, investigating the structural characteristics of the optimal control policies, constructing easy-to-operate suboptimal policies, establishing the system stability conditions, and addressing the optimization of suboptimal policies using numerical optimization methods and simulation-based meta-heuristics methods.

More specifically, the following stochastic supply chain systems will be examined in this book: basic supplier–manufacturer–customer supply chains, multistage serial supply chains, supply chain systems with backordering decisions, supply chain systems with preventive maintenance decisions, supply chain systems with assembly operations, and supply chain systems with multiple products.

We will make two basic assumptions to clarify the research context of this book. Firstly, in the literature, inventory information is assumed to be available either continuously or periodically. Accordingly, the inventory replenishment decisions can be made continuously or periodically. Both situations are reasonable in practice depending on the business operational context and management preference.

With the emphasis on supply chain visibility and integration facilitated by the development of technologies such as radio frequency identification (RFID) and global position system (GPS), continuous or more frequent inventory counts and reviews become commoner than before. We therefore assume that inventory information is continuously reviewed and decisions are made continuously.

Secondly, the material flow and finished goods production can be treated as continuous variables or discrete variables. The continuous material flow model, often termed as fluid flow model (e.g., Gershwin 1994; Sethi and Zhang 1994), is a good approximation for production of sufficient volume. Mathematically, such treatment makes the state variables and control variables continuous and enables the application of some sophisticated mathematical techniques. However, due to the discrete nature of manufacturing systems, it is probably more reasonable to model the manufacturing supply chains as discrete-state dynamic systems. Therefore, we assume raw materials and finished goods are represented by discrete variables. The above two basic assumptions are stated explicitly as follows:

Assumption 1.1. Inventory information is continuously reviewed and management decisions are made continuously.

Assumption 1.2. Inventory positions (including raw material, finished goods, and backlogs) are represented by discrete variables.

Therefore, the context that we are dealing with is based on continuous-time and discrete-state supply chains. In addition, the majority of the chapters assume that unmet customer demands are completely backordered, which makes the state space of the finished goods inventory position infinite.

1.5 Structure of the Book

This book consists of four parts. In Part I (Chap. 1), a general introduction to stochastic supply chain systems is provided. The manufacturer is regarded as a focal company in the supply chain. It interacts with other functions or entities such as customers, suppliers, and transport companies in many different ways. The main objective is to seek the optimal production control policies and the optimal ordering policies in the supply chain by taking into account a variety of uncertainties such as random customer demands, stochastic processing times, unreliable machines, and stochastic raw material lead times.

In Part II (Chaps. 2, 3, 4, 5, 6, and 7), several typical stochastic supply chain systems are studied, including the basic integrated supply chains, generalized serial supply chain, supply chains with backordering decisions, supply chains with preventive maintenance decisions, supply chains with assembly operations, and

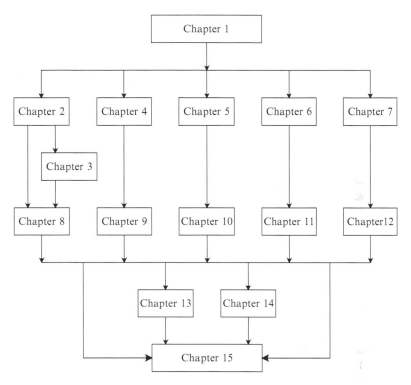

Fig. 1.3 Flowchart of chapter relationships

supply chains with multiple products. Analytical models for these stochastic supply chain systems are formulated and analyzed in detail. By applying the stochastic dynamic programming approach, the optimal control policies will be investigated. More importantly, the structural characteristics of the optimal policies will be established through the investigation of the properties of value functions. In some special cases, the explicit forms of the optimal control policies can be derived analytically, whereas in general cases, the switching structure of the optimal state-feedback control policies will be illustrated either rigorously or numerically.

In Part III (Chaps. 8, 9, 10, 11, and 12), the structural knowledge of the optimal control policies obtained in Part II will be utilized to construct easy-to-operate suboptimal control policies for the stochastic supply chain systems accordingly. Here, the focus is to achieve the trade-off between the closeness to the optimality of the constructed policies and the degree of simplicity in terms of their implementation. We propose threshold-type control policies to approximate the optimal policies. Threshold-type control policies can be characterized by a set of threshold parameters or a set of switching manifolds. A range of numerical examples will be provided to demonstrate the effectiveness of the proposed control policies. The linkage of the proposed policies to traditional base-stock and Kanban control policies will be discussed.

This part will also address the system stability issues. As the supply chains are subject to many random factors, poorly designed and controlled supply chains can lead to instability. For example, poor control of production rate or an inappropriate replenishment policy could result in exceeding warehouse capacity or accumulating unsatisfied customer demands increasingly.

Part IV (Chaps. 13, 14, and 15) discusses the optimization of threshold-type control policies and their robustness. Although threshold-type control policies are fairly simple and intuitive, their implementation in practice still requires the determination of the threshold parameters. It is therefore necessary to investigate the optimization of the threshold parameters and the robustness of the policy with respect to these parameters. In Chap. 13, numerical methods based on the value iteration algorithm or steady-state probability distribution are presented to optimize threshold parameters, which also make use of some analytical results in terms of the bounds and relationships of those threshold parameters. In Chap. 14, simulation-based optimization methods are presented. This includes the common meta-heuristics such as evolutionary strategy/genetic algorithms and simulated annealing, which are applicable to generic parameter optimization problems. In addition, the ordinal optimization method is introduced and combined with the genetic algorithms to speed up the threshold parameter optimization in stochastic supply chain systems. Chapter 15 summarizes the monograph with some conclusions and indicates the directions for further research.

We conclude this chapter with a flowchart to illustrate the relationships of different chapters in Fig. 1.3.

References

APICS: In: Cox, J.F., Blackstone, J.H. (eds.) APICS Dictionary – The Educational Society for Resource Management, 9th edn. APICS, Falls Church (1998). ISBN 1-55822-162-X

Buzacott, J.A., Shanthikumar, J.G.: Stochastic Models of Manufacturing Systems. Prentice Hall, Englewood Cliffs (1993)

Chatfield, D.C., Kim, J.G., Harrison, T.P., Hayya, J.C.: The bullwhip effect – impact of stochastic lead time, information quality, and information sharing: a simulation study. Prod. Oper. Manage. 13(4), 340–353 (2004)

Christopher, M.: Logistics and Supply Chain Management: Strategies for Reducing Costs and Improving Services. Pitman Publishing, London (1992)

Christopher, M.: Logistics and Supply Chain Management, 4th edn. Financial Times/Prentice Hall, London (2010). ISBN ISBN9780273731122

Cooper, M., Gardner, J.: Building good business relationships – more than just partnering or strategic alliances? Int. J. Phys. Distrib. Logist. Manage. 23(6), 14–26 (1993)

Davis, T.: Effective supply chain management. Sloan Manage. Rev. 34(4), 35–46 (1993)

Gershwin, S.B.: Manufacturing Systems Engineering. Prentice-Hall, Englewood Cliffs (1994)

Lee, H.L., Padmanabhan, P., Whang, S.: Information distortion in a supply chain: the bullwhip effect. Manage. Sci. 43, 546–558 (1997a)

Lee, H.L., Padmanabhan, P., Whang, S.: The bullwhip effect in a supply chain. Sloan Manage. Rev. 38, 93–102 (1997b)

Min, H., Zhou, G.: Supply chain modeling: past, present and future. Comput. Ind. Eng. **43**, 231–249 (2002)

Sethi, S., Zhang, Q.: Hierarchical Decision Making in Stochastic Manufacturing Systems. Birkhause, Boston (1994)

Simchi-Levi, D., Kaminsky, P., Simchi-Levi, E.: Designing and Managing the Supply Chain: Concepts, Strategies and Case Studies, 3rd edn. McGraw-Hill, Irwin (2009)

Song, D.P., Dinwoodie, J.: Quantifying the effectiveness of VMI and integrated inventory management in a supply chain with uncertain lead-times and uncertain demands. Prod. Plan. Control **19**(6), 590–600 (2008)

Chapter 2
Optimal Control of Basic Integrated Supply Chains

2.1 Introduction

The supply chain system under consideration in this chapter consists of three entities (i.e., supplier, manufacturer, and customer) and two finite capacitated warehouses (raw material warehouse and finished goods warehouse). The manufacturer produces discrete parts. The system is subject to multiple types of uncertainties, for example, uncertain replenishment lead times, stochastic production times, and random demands. The objective is to minimize the expected total cost consisting of material inventory cost, product holding cost, and demand backordering cost.

To achieve the global optimization of the supply chain, the decision activities for raw materials, finished goods, and customer satisfaction should be considered in an integrated mode. For example, the delay and uncertainty in raw material (RM) replenishment, the RM inventory level, and RM warehouse capacity may affect the production decision. On the other hand, the processing time, uncertainties in production and demand, the finished goods (FG) inventory level, and FG warehouse capacity may also affect the raw material ordering decision. These complex interactions necessitate the investigation of integrated ordering and production policy.

It is assumed that the supply chain is integrated in the sense that the supplier and the manufacturer have an agreement that the manufacturer can change or cancel the order at any future decision point before it arrives. This may be regarded as a type of partnership between the supplier and the manufacturer, in which both parties aim to meet the final customers as closely as possible and reduce the downstream inventories. Note that downstream warehouses usually incur higher inventory costs. In addition, the costs associated with placing orders and changing order sizes are not considered explicitly because they become less important when trading parties have formed a close partnership, adopted advanced technologies (e.g., Internet, EDI), and focused on the overall performance of the supply chain. The assumption that the outstanding order can be modified at any time together with the exponential lead-time assumption implies that there is at most one order outstanding at any time.

D.-P. Song, *Optimal Control and Optimization of Stochastic Supply Chain Systems*, Advances in Industrial Control, DOI 10.1007/978-1-4471-4724-4_2, © Springer-Verlag London 2013

The manufacturer can place an order to the supplier with any quantity subject to the constraint of the raw material warehouse capacity, which triggers the raw material delivery activity from the supplier. The raw material order replenishment lead time is exponentially distributed. The production is done in discrete mode with lot size one. The processing time of each part follows an exponential distribution with a fixed production rate. Customer demands arrive as a Poisson process. They can be satisfied immediately if products are available in the finished goods warehouse; otherwise, unmet demands are backordered. The problem for the integrated supply chain is to determine when and how much quantity of raw material to order and whether or not to produce finished goods in order to minimize the expected sum of raw material inventory cost, finished goods holding costs, and demand backordering costs.

The unique aspects of this model compared to the literature are the integration of ordering policy and production policy and the consideration of finite warehouse capacities for both raw materials and finished goods in stochastic situations. In addition, because unmet demands are completely backordered and the warehouses have finite capacities, the system stability requires careful examination.

The rest of this chapter is organized as follows: The problem is formulated into an event-driven Markov decision process in the next section. In Sect. 2.3, we focus on the discounted cost case and obtain the optimal control policy. In Sect. 2.4, the structural properties of the value function are established. In Sect. 2.5, we demonstrate that the optimal integrated ordering and production policy can be characterized by two monotonic switching curves. The asymptotic behaviors of the switching curves are then analyzed. In Sect. 2.6, other issues related to practical supply chain management are discussed. Section 2.7 notes the relevant literature and others' contributions.

The stability conditions of the basic supply chain and the approximation to the optimal policy will be given in Chap. 8. The generalization of the model will be given in Chap. 3.

2.2 Problem Formulation

Consider a supply chain that consists of three entities (supplier, manufacturer, and customer) and two warehouses (a raw material (RM) warehouse and a finished goods (FG) warehouse) as shown in Fig. 2.1. Both warehouses are capacitated. Customer demands are exogenous inputs. The supplier and the manufacturer form

Fig. 2.1 The basic integrated supply chain

a partnership in order to effectively meet the customer demands and reduce the inventories in two warehouses.

The manufacturer produces one product (or part) at a time, and the part-processing time is exponentially distributed with a rate u. Physically, u represents the production rate. It is a control variable that takes 0 or r, which represents an action "not produce" or "produce at a speed r," respectively (the model can be extended easily to the case of allowing u taking more values between 0 and r, but the results remain the same). Customer demands arrive one at a time to the FG warehouse according to a Poisson process with arriving rate μ. A demand is satisfied immediately if there are finished goods stored in the FG warehouse; otherwise, unmet demands are backordered. The manufacturer also makes decisions on the quantity of orders placed to the supplier for raw materials. The quantity of an order is denoted as q. The raw material replenishment lead time refers to the total time that elapses between an order's placement/modification and its receipt, which is assumed to follow an exponential distribution with the mean $1/\lambda$, where λ is termed as the lead-time rate. The lead time is assumed to be independent on actual order quantity, which may be justified by the fact that the dispatching equipment/vehicle usually can deliver a range of order quantity in a single trip. It is assumed that one unit of raw material is used for producing one product and unused raw materials will be stored in the RM warehouse. Let M and N denote the warehouse capacities for raw materials and finished goods, respectively. In Fig. 2.1, two decision variables u and q are denoted as $u(t)$ and $q(t)$ to represent their dependence on time t dynamically.

We summarize the assumptions as follows:

Assumption 2.1. There is at most one outstanding order at any time with its size changeable without incurring any cost.

Assumption 2.2. The lead time for procuring raw materials is exponentially distributed, which is independent of the size of the order.

Assumption 2.3. The manufacturing times follow an exponential distribution.

Assumption 2.4. The customer demands arrive as a Poisson process.

Assumption 2.5. Both raw material and finished goods warehouses have finite capacities.

In the literature, the assumption of at most one outstanding order at any time was used in Berman and Kim (2001, 2004) and Kim (2005). The exponential order replenishment lead time was also assumed in Berman and Kim (2004) and Kim (2005). They employed the Markov decision process theory and characterized the optimal replenishment policy on whether or not to place a fixed order but did not consider the service rate control and the multiple choices of the order size. The exponential manufacturing time was a common assumption that has been adopted in more literature, for example, Veatch and Wein (1992, 1994), Ching et al. (1997), Song and Sun (1998, 1999), Feng and Yan (2000), He et al. (2002a), and Feng and Xiao (2002). These studies mainly focused on the structural properties of the optimal production control, in which material ordering decisions were not explicitly considered. The Poisson demand arrival is one of the most common assumptions in the related literature (cf. Buzacott and Shanthikumar 1993). The majority of the above literature assumed infinite capacity of warehouses or internal buffers. However, in practice, it is more realistic to assume finite capacity of warehouses, which essentially includes the case of infinite capacity since we can let the capacity go to infinity.

Clearly, the decision of order quantity for raw materials is constrained by the RM warehouse capacity M and the current RM inventory level, while the production rate is constrained by the capacity r, the available raw materials, and the available FG warehouse space. Let $x_1(t)$ denote the on-hand inventory of raw materials at time t and $x_2(t)$ denote the on-hand inventory of finished goods at time t. Here $x_2(t)$ could be negative, which represents the number of backordered demands. We have $x_1(t) \in [0, M]$ and $x_2(t) \in (-\infty, N]$. The system state space can be described by $X = \{\mathbf{x} := (x_1, x_2) \,|\, x_1 \in [0, M] \text{ and } x_2 \in (-\infty, N)\}$.

More specifically, the manufacturer needs to make two types of decisions: the production rate $u(t) \in \{0, r\}$ and the raw material order quantity $q(t) \in [0, M - x_1(t)]$. Define an admissible control set $\Omega = \{(u(t), q(t)) | u(t) \in \{0, r\}$ if $x_1(t) > 0$ and $x_2(t) < N, u(t) = 0$ if $x_1(t) \leq 0$ or $x_2(t) \geq N; q(t) \in [0, M - x_1(t)], t \geq 0\}$. The problem is to find the optimal integrated policy $(u, q) \in \Omega$ by minimizing the infinite horizon expected discounted cost (depending on the initial state):

$$J(\mathbf{x}_0) = \min_{u,q} E\left[\int_0^\infty e^{-\beta t} g(\mathbf{x}(t), u(t), q(t)) \, dt \,\Big|\, \mathbf{x}(0) = \mathbf{x}_0 \right] \qquad (2.1)$$

where $0 < \beta < 1$ is a discount factor and $g(.)$ is a cost function to penalize raw material inventories, finished goods inventories, and backordered demands. For example, $g(.)$ can be defined as

$$g(\mathbf{x}, u, q) = c_1 x_1 + c_2^+ x_2^+ + c_2^- x_2^- \qquad (2.2)$$

where c_1, c_2^+, and c_2^- are the unit costs of raw material inventory, finished goods inventory, and backordered demands, respectively, and $x_2^+ := \max(x_2, 0)$ and $x_2^- := \max(-x_2, 0)$.

There are three types of events in the system, that is, demand arrivals, production completions, and placed order arrivals. Due to the memoryless properties of Poisson process and exponential distribution, unfinished production or unarrived order interrupted by an event is statistically equivalent to that of restarting. Three types of events incur three different directions of state transition. The system state changes if and only if one of the above events occurs. The event occurring epochs comprise the set of decision epochs. At each decision epoch, a policy (u, q) specifies whether or not a replenishment order is placed, how much of the order size is, and whether or not a product is produced. The state transition map at a state (x_1, x_2) under the control (u, q) is depicted in Fig. 2.2.

Fig. 2.2 System state
transition map in a serial
supply chain

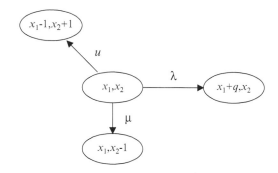

The original problem is a continuous time Markov decision problem. Following the uniformization technique (Bertsekas 1987), the problem can be formulated into an equivalent discrete-time Markov chain problem with positive but unbounded cost per step and infinite countable state space. Let $v = \mu + r + \lambda$ be the uniform transition rate. Under a stationary control policy $(u, q) \in \Omega$, the one-step transition probability functions from (x_1, x_2) are given by

$$\text{Prob}\{(x_1 + q, x_2) \,|\, (x_1, x_2), (u, q)\} = \lambda \cdot I\{q > 0\}/v, \tag{2.3}$$

$$\text{Prob}\{(x_1 - 1, x_2 + 1) \,|\, (x_1, x_2), (u, q)\} = r \cdot I\{u > 0\}/v, \tag{2.4}$$

$$\text{Prob}\{(x_1, x_2 - 1) \,|\, (x_1, x_2), (u, q)\} = \mu/v, \tag{2.5}$$

$$\text{Prob}\{(x_1, x_2) \,|\, (x_1, x_2), (u, q)\} = [\lambda - \lambda \cdot I\{q > 0\} + r - r \cdot I\{u > 0\}]/v, \tag{2.6}$$

where $I\{.\}$ is an indicator function, which takes 1 if the condition is true and takes 0 otherwise.

To simplify the narrative, we denote $\mathbf{u} := (u, q)$. Let $0 = t_0 < t_1 < t_2 < \ldots < t_n < \ldots$ be the potential state transition epochs. Let $\mathbf{x}_k := \mathbf{x}(t_k)$ be the destination state of the kth transition and $\mathbf{u}_k := \mathbf{u}(t_k)$ be the control decision of the kth transition. Thus, $\mathbf{x}(t) \equiv \mathbf{x}_k$ and $\mathbf{u}(t) \equiv \mathbf{u}_k$ if $t \in [t_k, t_{k+1})$.

Computing the cost function defined in (2.1) for a given initial condition $\mathbf{x}(0) = \mathbf{x}_0$ and control policy $\mathbf{u}(t)$, we have

$$E\left[\int_0^\infty e^{-\beta t} g(\mathbf{x}(t), \mathbf{u}(t))\, dt \,|\, \mathbf{x}(0) = \mathbf{x}_0\right]$$

$$= E \sum_{k=0}^\infty \int_{t_k}^{t_{k+1}} e^{-\beta t} g(\mathbf{x}_k, \mathbf{u}_k)\, dt$$

$$= E \sum_{k=0}^\infty \frac{1}{\beta} e^{-\beta t_k} \left[1 - e^{-\beta(t_{k+1} - t_k)}\right] \cdot g(\mathbf{x}_k, \mathbf{u}_k). \tag{2.7}$$

Note that $t_k = \sum_{j=1}^{k}(t_j - t_{j-1})$. Random variables $(t_j - t_{j-1})$ are independent for any $j > 0$ and follow the same exponential distribution with the "uniform transition rate" v. Due to the independence of the three terms on the right-hand side of the above equation and exchanging the mathematical expectation with the sum operator, the cost function can be further simplified using

$$E e^{-\beta t_k} = \left(\int_0^{\infty} e^{-\beta \tau} \cdot v e^{-v\tau} d\tau \right)^k = \left(\frac{v}{\beta + v} \right)^k, \tag{2.8}$$

$$E \left(1 - e^{-\beta(t_{k+1} - t_k)} \right) = \frac{\beta}{\beta + v}. \tag{2.9}$$

Hence, we have

$$E \int_0^{\infty} e^{-\beta t} g(\mathbf{x}(t), \mathbf{u}(t)) \, dt = \frac{1}{\beta + v} \sum_{k=0}^{\infty} \left(\frac{v}{\beta + v} \right)^k E g(\mathbf{x}_k, \mathbf{u}_k)$$

$$= \frac{1}{\beta + v} \sum_{k=0}^{\infty} \theta^k E g(\mathbf{x}_k, \mathbf{u}_k) \tag{2.10}$$

where $\theta = v/(\beta + v)$. Now the equivalent discrete-time Markov chain can be formulated, and it is an infinite state Markov decision process with unbounded cost and infinite action set.

2.3 Optimal Control Policy

To further simplify the narrative, we define:

Definition 2.1.

- e_i be a unit vector in the state space whose ith element is one.
- $D := \{d_1, d_2, d_3\}$ be the transition set, where $d_1 = e_1$, $d_2 = e_2 - e_1$, and $d_3 = -e_2$. Namely, d_1 represents replenishment of one item of raw materials, d_2 represents production, and d_3 represents a demand.
- $D(\mathbf{x}) := \{d \in D \,|\, \mathbf{x} + d \in X\}$ be the feasible transition directions from state $\mathbf{x}$.

Defining $V(\mathbf{x}) := \min \sum_{k=0}^{\infty} \theta^k E g(\mathbf{x}_k, \mathbf{u}_k)$ and following the standard stochastic dynamic programming approach (e.g., Bertsekas 1987), we have the Bellman optimality equation for this discrete dynamic programming problem below:

$$V(\mathbf{x}) = \min_{u,q} \{ g(\mathbf{x}) + \theta \cdot [\mu/v \cdot V(\mathbf{x} + d_3) + r/v \cdot V(\mathbf{x}) \cdot I\{\mathbf{x} + d_2 \notin X\}$$

$$+ (u/v \cdot V(\mathbf{x} + d_2) + (r - u)/v \cdot V(\mathbf{x})) \cdot I\{\mathbf{x} + d_2 \notin X\}$$

$$+ \lambda/v \cdot V(\mathbf{x} + q \cdot d_1)]\}.$$

Note that $V(\mathbf{x}) = (\beta + v)J(\mathbf{x})$ and $\theta = v/(\beta + v)$; the above equation can be rewritten as

$$(\beta + v) J(\mathbf{x}) = \min_{u,q} \{g(\mathbf{x}) + \mu \cdot J(\mathbf{x} + d_3) + r \cdot J(\mathbf{x}) \cdot I\{\mathbf{x} + d_2 \notin X\}$$

$$+ (u \cdot J(\mathbf{x} + d_2) + (r - u) \cdot J(\mathbf{x})) \cdot I\{\mathbf{x} + d_2 \notin X\}$$

$$+ \lambda \cdot J(\mathbf{x} + q \cdot d_1)\}.$$

In addition, note that the ordering decision and the production decision are not coupled; the above equation can lead to the following simplified optimality equation for our original problem:

$$J(\mathbf{x}) = \frac{1}{\beta + v} [g(\mathbf{x}) + \mu \cdot J(\mathbf{x} + d_3) + r \cdot J(\mathbf{x}) \cdot I\{\mathbf{x} + d_2 \notin X\}$$

$$+ \min_{u=0,r} \{u \cdot J(\mathbf{x} + d_2) + (r - u) \cdot J(\mathbf{x})\} \cdot I\{\mathbf{x} + d_2 \in X\}$$

$$+ \lambda \cdot \min \{J(\mathbf{x} + q \cdot d_1) \,|\, q \in [0, M - x_1]\}]. \tag{2.11}$$

The fourth term on the right-hand side (RHS) of (2.11) represents the production decision, which takes an action "produce" or "not produce" and is subject to the constraints of raw material availability and FG warehouse capacity. The last term on RHS of (2.11) represents the ordering decision, which is subject to the current available RM warehouse space. From Eq. (2.11), the optimal integrated ordering and production policy $(u^*(\mathbf{x}), q^*(\mathbf{x}))$ is given by

$$u^*(\mathbf{x}) = \begin{cases} r & J(\mathbf{x} + d_2) \le J(\mathbf{x}), \quad \text{if } \mathbf{x} + d_2 \in X \\ 0 & \text{otherwise} \end{cases} \tag{2.12}$$

$$q^*(\mathbf{x}) = j, \quad \text{if } J(\mathbf{x} + j d_1) = \min \{J(\mathbf{x}), J(\mathbf{x} + d_1), \ldots, J(\mathbf{x} + (M - x_1) d_1)\}. \tag{2.13}$$

The optimal discounted cost function can be approximated numerically using the value iteration algorithm (cf. Chap. 13), that is, if we let $J_0(\mathbf{x}) \equiv 0$ for any $\mathbf{x} \in X$, and

$$J_{k+1}(\mathbf{x}) = \frac{1}{\beta + v} [g(\mathbf{x}) + \mu \cdot J_k(\mathbf{x} + d_3) + r \cdot J_k(\mathbf{x}) \cdot I\{\mathbf{x} + d_2 \notin X\}$$

$$+ \min_{u=0,r} \{u \cdot J_k(\mathbf{x} + d_2) + (r - u) \cdot J_k(\mathbf{x})\} \cdot I\{\mathbf{x} + d_2 \in X\}$$

$$+ \lambda \cdot \min \{J_k(\mathbf{x} + q \cdot d_1) \,|\, q \in [0, M - x_1]\}] \tag{2.14}$$

for $k > 0$ and $\mathbf{x} \in X$, where $J_k(\mathbf{x})$ is the k-stage cost function for state x, then

$$\lim_{k \to +\infty} J_k(\mathbf{x}) = J(\mathbf{x}), \quad \text{for } \mathbf{x} \in X. \tag{2.15}$$

The convergence of the k-stage policy and cost function to the infinite-horizon optimal policy and cost follows from the fact that only finitely many controls are considered at each state (Bertsekas 1987). The optimal integrated policy given in (2.12) and (2.13) is implicit. Although (2.14) and (2.15) provide a numerical iteration approach to approximate the optimal value function $J(\mathbf{x})$, it provides little insight into the explicit form of the optimal ordering and production policy. It has been well recognized that the structural characteristics of the optimal policy are very useful to managers in order to implement optimal or near-optimal policies (Sethi and Zhang 1994; Gershwin 1994).

2.4 Structural Properties of the Value Function

To establish the explicit characteristics of the optimal integrated policy, we will first investigate the structural properties of the optimal value function. Throughout this book, increasing and decreasing are used in non-strict sense, that is, "increasing" means nondecreasing and "decreasing" means nonincreasing. To further simplify the narrative, we assume $\beta + v = 1$ without any loss of generality.

To make use of general results in the literature (Weber and Stidham 1987; Veatch and Wein 1992; Koole 1998), we introduce the definition of submodularity.

Definition 2.2. The function $f: X \to R$ is *submodular* w.r.t. D on X if $f(\mathbf{x} + d_i) + f(\mathbf{x} + d_j) - f(\mathbf{x}) - f(\mathbf{x} + d_i + d_j) \geq 0$, for all $d_i \neq d_j \in D$ and $\mathbf{x}$, such that all four points are in X.

Lemma 2.1. *If $J(x)$ is submodular w.r.t. D, then:*

(i) $J(\mathbf{x} + e_1) - J(\mathbf{x})$ is increasing in x_2.
(ii) $J(\mathbf{x} + e_2) - J(\mathbf{x})$ is increasing in x_1.
(iii) $J(\mathbf{x} - e_1 + e_2) - J(\mathbf{x})$ is decreasing in x_1.
(iv) $J(\mathbf{x} - e_1 + e_2) - J(\mathbf{x})$ is increasing in x_2.
(v) $J(\mathbf{x} + e_1) - J(\mathbf{x})$ is increasing in x_1.
(vi) $J(\mathbf{x} + e_2) - J(\mathbf{x})$ is increasing in x_2.

Proof. Assertions (i) and (ii) are the same as the submodularity of $J(\mathbf{x})$ on d_1 and d_3. Assertions (iii) and (iv) are directly from the submodularity of $J(\mathbf{x})$ on d_1 and d_2 and on d_2 and d_3, respectively. From the submodularity of $J(\mathbf{x})$ and the definitions of the transition directions, we have

$$J(\mathbf{x} + e_1) - J(\mathbf{x}) = J(\mathbf{x} + d_1) - J(\mathbf{x}) \geq J(\mathbf{x} + d_1 + d_2) - J(\mathbf{x} + d_2)$$

$$\geq J(\mathbf{x} + d_1 + d_2 + d_3) - J(\mathbf{x} + d_2 + d_3) = J(\mathbf{x}) - J(\mathbf{x} - e_1)$$

$$J(\mathbf{x} - e_2) - J(\mathbf{x}) = J(\mathbf{x} + d_3) - J(\mathbf{x}) \geq J(\mathbf{x} + d_3 + d_1) - J(\mathbf{x} + d_1)$$

$$\geq J(\mathbf{x} + d_3 + d_1 + d_2) - J(\mathbf{x} + d_1 + d_2) = J(\mathbf{x}) - J(\mathbf{x} + e_2).$$

The above two inequalities yield that the assertions (v) and (vi) are true. This completes the proof.   □

It should be pointed out that $g(\mathbf{x})$ is submodular w.r.t. $D(\mathbf{x})$ if it takes the form of (2.2). Physically, Lemma 2.1(i) and (v) indicates that the incremental cost of holding one more unit of the raw material is increasing as the inventory-on-hand at RM warehouse or FG warehouse increases. This can be explained by the fact that more inventory-on-hand results in longer time for the additional raw material staying in the system. Lemma 2.1(ii) and (vi) can be similarly interpreted. Lemma 2.1(iii) and (iv) states that the incremental cost of converting a raw material into a finished product is decreasing as the inventory-on-hand of the raw material increases and increasing as the inventory-on-hand of the finished goods increases.

Proposition 2.1. *If $g(\mathbf{x})$ is submodular w.r.t. $D(\mathbf{x})$, then $J(\mathbf{x})$ is submodular w.r.t. $D(\mathbf{x})$.*

Proof. Part of the following proof makes use of Lemma 1 and Theorem 1 in Veatch and Wein (1992). The proof can be shown by induction on k using iteration equations (2.14) and (2.15). Since $J_0(\mathbf{x}) = 0$ for any $\mathbf{x} \in X$, so the assertions are true for $k = 0$. Now suppose $J_k(\mathbf{x})$ is submodular w.r.t. $D(\mathbf{x})$, then we want to show that the assertion is also true for $k + 1$.

The first term in RHS of (2.14) is submodular by the submodularity of $g(\mathbf{x})$. The second term in RHS of (2.14) is submodular w.r.t. $D(\mathbf{x})$ because $D(\mathbf{x}) \subseteq D(\mathbf{x} + d_3)$ and the inductive hypothesis. The third term in RHS of (2.14) is submodular from the inductive hypothesis and the boundary condition. The fourth term in RHS of (2.14) is submodular w.r.t. $D(\mathbf{x})$ from Lemma 1 in Veatch and Wein (1992).

Therefore, it suffices to show that the fifth term in RHS of (2.14) is submodular w.r.t. $D(\mathbf{x})$. Theorem 1 in Veatch and Wein (1992) does not apply to the fifth term because the replenishment chooses between multiple transitions, for example, e_1, $2e_1, \ldots, (M - x_1)e_1$. Dropping the constant λ, we want to show the following three inequalities (note that $D = \{d_1, d_2, d_3\}$):

(I) $\min \{ J_k(\mathbf{x} + q \cdot d_1 + d_1) \,|\, q \in [0, M - x_1 - 1] \}$

$\quad + \min \{ J_k(\mathbf{x} + q \cdot d_1 + d_2) \,|\, q \in [0, M + 1 - x_1] \}$

$\quad - \min \{ J_k(\mathbf{x} + q \cdot d_1) \,|\, q \in [0, M - x_1] \}$

$\quad - \min \{ J_k(\mathbf{x} + q \cdot d_1 + d_1 + d_2) \,|\, q \in [0, M - x_1] \} \geq 0;$ (2.16)

(II) $\min \{ J_k(\mathbf{x} + q \cdot d_1 + d_1) \,|\, q \in [0, M - x_1 - 1] \}$

$\quad + \min \{ J_k(\mathbf{x} + q \cdot d_1 + d_3) \,|\, q \in [0, M - x_1] \}$

$\quad - \min \{ J_k(\mathbf{x} + q \cdot d_1) \,|\, q \in [0, M - x_1] \}$

$\quad - \min \{ J_k(\mathbf{x} + q \cdot d_1 + d_1 + d_3) \,|\, q \in [0, M - x_1 - 1] \} \geq 0;$ (2.17)

(III) $\min\{J_k(\mathbf{x} + q \cdot d_1 + d_2) \,|\, q \in [0, M + 1 - x_1]\}$

$\quad + \min\{J_k(\mathbf{x} + q \cdot d_1 + d_3) \,|\, q \in [0, M - x_1]\}$

$\quad - \min\{J_k(\mathbf{x} + q \cdot d_1) \,|\, q \in [0, M - x_1]\}$

$\quad - \min\{J_k(\mathbf{x} + q \cdot d_1 + d_2 + d_3) \,|\, q \in [0, M + 1 - x_1]\} \geq 0. \qquad (2.18)$

To prove (I). This is equivalent to show

$\min\{J_k(\mathbf{x} + qe_1) \,|\, q \in [1, M - x_1]\} + \min\{J_k(\mathbf{x} + qe_1 + e_2) \,|\, q \in [-1, M - x_1]\}$

$\quad - \min\{J_k(\mathbf{x} + qe_1) \,|\, q \in [0, M - x_1]\} - \min\{J_k(\mathbf{x} + qe_1 + e_2) \,|\, q \in [0, M - x_1]\} \geq 0.$

$$(2.19)$$

If $J_k(\mathbf{x} - e_1 + e_2) - J_k(\mathbf{x} + e_2) \leq 0$, then we can obtain the following three inequalities from Lemma 2.1(v), Lemma 2.1(iii), and the second inequality and Lemma 2.1(v), respectively:

$J_k(\mathbf{x} + qe_1 + e_2) - J_k(\mathbf{x} + qe_1 + e_1 + e_2) \leq 0, \quad \text{for } q \in [-1, M - 1 - x_1],$

$J_k(\mathbf{x}) - J_k(\mathbf{x} + e_1) \leq 0,$

$J_k(\mathbf{x} + qe_1) - J_k(\mathbf{x} + qe_1 + e_1) \leq 0, \quad \text{for } q \in [0, M - 1 - x_1].$

The above three inequalities simplify (2.19) to be

$$J_k(\mathbf{x} + e_1) + J_k(\mathbf{x} - e_1 + e_2) - J_k(\mathbf{x}) - J_k(\mathbf{x} + e_2) \geq 0. \qquad (2.20)$$

This is true by the inductive hypothesis and Lemma 2.1(iii).

If $J_k(\mathbf{x} - e_1 + e_2) - J_k(\mathbf{x} + e_2) > 0$, then (2.19) is simplified as

$\min\{J_k(\mathbf{x} + qe_1) \,|\, q \in [1, M - x_1]\} - \min\{J_k(\mathbf{x} + qe_1) \,|\, q \in [0, M - x_1]\} \geq 0.$

$$(2.21)$$

This is obviously true. (2.20) and (2.21) yield (2.19).

To prove (II). This is equivalent to show

$\min\{J_k(\mathbf{x} + qe_1) \,|\, q \in [1, M - x_1]\} + \min\{J_k(\mathbf{x} + qe_1 - e_2) \,|\, q \in [0, M - x_1]\}$

$\quad - \min\{J_k(\mathbf{x} + qe_1) \,|\, q \in [0, M - x_1]\} - \min\{J_k(\mathbf{x} + qe_1 - e_2) \,|\, q \in [1, M - x_1]\} \geq 0.$

$$(2.22)$$

If $J_k(\mathbf{x} - e_2) - J_k(\mathbf{x} + e_1 - e_2) \leq 0$, then we have the following three inequalities from Lemma 2.1(i), the first inequality and Lemma 2.1(v), and Lemma 2.1(v), respectively:

$J_k(\mathbf{x}) - J_k(\mathbf{x} + e_1) \le 0,$

$J_k(\mathbf{x} + qe_1) - J_k(\mathbf{x} + qe_1 + e_1) \le 0, \quad \text{for } q \in [0, M - 1 - x_1],$

$J_k(\mathbf{x} + qe_1 - e_2) - J_k(\mathbf{x} + qe_1 + e_1 - e_2) \le 0, \quad \text{for } q \in [0, M - 1 - x_1].$

The above three inequalities simplify (2.22) to be

$$J_k(\mathbf{x} + e_1) + J_k(\mathbf{x} - e_2) - J_k(\mathbf{x}) - J_k(\mathbf{x} + e_1 - e_2) \ge 0. \qquad (2.23)$$

This is true by the inductive hypothesis and Lemma 2.1(i).

If $J_k(\mathbf{x} - e_2) - J_k(\mathbf{x} + e_1 - e_2) > 0$, then (2.22) is simplified as (2.21), which is obviously true. Therefore, (2.22) holds.

To prove (III). This is equivalent to show

$$\min\{J_k(\mathbf{x} + qe_1 + e_2) \,|\, q \in [-1, M - x_1]\} + \min\{J_k(\mathbf{x} + qe_1 - e_2) \,|\, q \in [0, M - x_1]\}$$

$$- \min\{J_k(\mathbf{x} + qe_1) \,|\, q \in [0, M - x_1]\} - \min\{J_k(\mathbf{x} + qe_1) \,|\, q \in [-1, M - x_1]\} \ge 0.$$
$$(2.24)$$

There must exist $i \in [-1, M - x_1]$ such that $J_k(\mathbf{x} + ie_1) = \min\{J_k(\mathbf{x}), J_k(\mathbf{x} + e_1), \ldots, J_k(\mathbf{x} + Me_1)\}$. This implies that

$$J_k(\mathbf{x} + ie_1) - J_k(\mathbf{x} + ie_1 - e_1) \le 0 \le J_k(\mathbf{x} + ie_1 + e_1) - J_k(\mathbf{x} + ie_1). \quad (2.25)$$

It follows (together with Lemma 2.1(iii), (i), (i), and (iii), respectively)

$$J_k(\mathbf{x} + ie_1 - e_1 + e_2) - J_k(\mathbf{x} + ie_1 - 2e_1 + e_2) \le J_k(\mathbf{x} + ie_1) - J_k(\mathbf{x} + ie_1 - e_1) \le 0$$

$$0 \le J_k(\mathbf{x} + ie_1 + e_1) - J_k(\mathbf{x} + ie_1) \le J_k(\mathbf{x} + ie_1 + e_1 + e_2) - J_k(\mathbf{x} + ie_1 + e_2)$$

$$J_k(\mathbf{x} + ie_1 - e_2) - J_k(\mathbf{x} + ie_1 - e_1 - e_2) \le J_k(\mathbf{x} + ie_1 - e_2) - J_k(\mathbf{x} + ie_1 - e_1 - e_2) \le 0$$

$$0 \le J_k(\mathbf{x} + ie_1 + e_1) - J_k(\mathbf{x} + ie_1) \le J_k(\mathbf{x} + ie_1 + 2e_1 - e_2) - J_k(\mathbf{x} + ie_1 + e_1 - e_2).$$

If $i = -1$, the inequality (2.24) can be simplified as (from the above second and fourth inequalities and Lemma 2.1(iii))

$$J_k(\mathbf{x} - e_1 + e_2) + J_k(\mathbf{x} - e_2) - J_k(\mathbf{x}) - J_k(\mathbf{x} - e_1) \ge 0 \qquad (2.26)$$

This is true by the inductive hypothesis and Lemma 2.1(iv).

If $i \ge 0$, the inequality (2.24) can be simplified as

$$\min\{J_k(\mathbf{x} + ie_1 - e_1 + e_2), J_k(\mathbf{x} + ie_1 + e_2)\} + \min\{J_k(\mathbf{x} + ie_1 - e_2),$$

$$J_k(\mathbf{x} + ie_1 + e_1 - e_2)\} - J_k(\mathbf{x} + ie_1) - J_k(\mathbf{x} + ie_1) \ge 0.$$
$$(2.27)$$

Evaluate all combinations in LHS of (2.27), we have

$$J_k(\mathbf{x} + ie_1 - e_1 + e_2) + J_k(\mathbf{x} + ie_1 - e_2) - J_k(\mathbf{x} + ie_1) - J_k(\mathbf{x} + ie_1)$$
$$= [J_k(\mathbf{x} + ie_1 - e_1 + e_2) - J_k(\mathbf{x} + ie_1) - J_k(\mathbf{x} + ie_1 - e_1) + J_k(\mathbf{x} + ie_1 - e_2)]$$
$$+ [J_k(\mathbf{x} + ie_1 - e_1) - J_k(\mathbf{x} + ie_1)] \geq 0 \qquad (2.28)$$

$$J_k(\mathbf{x} + ie_1 - e_1 + e_2) + J_k(\mathbf{x} + ie_1 + e_1 - e_2) - J_k(\mathbf{x} + ie_1) - J_k(\mathbf{x} + ie_1)$$
$$= [J_k(\mathbf{x} + ie_1 - e_1 + e_2) - J_k(\mathbf{x} + ie_1)] - [J_k(\mathbf{x} + ie_1)$$
$$- J_k(\mathbf{x} + ie_1 + e_1 - e_2)] \geq 0 \qquad (2.29)$$

$$J_k(\mathbf{x} + ie_1 + e_2) + J_k(\mathbf{x} + ie_1 - e_2) - J_k(\mathbf{x} + ie_1) - J_k(\mathbf{x} + ie_1) \geq 0 \quad (2.30)$$

$$J_k(\mathbf{x} + ie_1 + e_2) + J_k(\mathbf{x} + ie_1 + e_1 - e_2) - J_k(\mathbf{x} + ie_1) - J_k(\mathbf{x} + ie_1)$$
$$= [J_k(\mathbf{x} + ie_1 + e_2) - J_k(\mathbf{x} + ie_1 + e_1) - J_k(\mathbf{x} + ie_1) + J_k(\mathbf{x} + ie_1 + e_1 - e_2)]$$
$$+ [J_k(\mathbf{x} + ie_1 + e_1) - J_k(\mathbf{x} + ie_1)] \geq 0. \qquad (2.31)$$

Inequality (2.28) is followed from Lemma 2.1(iv) and (2.25), (2.29) is from Lemma 2.1(iii) and (iv), (2.30) is from Lemma 2.1(vi), and (2.31) is from Lemma 2.1(iv) and (2.25). The above four inequalities yield (2.27), together with (2.26); it follows that the inequality (2.24) holds.

Therefore, the fifth term in RHS of (2.14) is submodular. From the above induction proofs and (2.15), it follows that $J(x)$ is submodular w.r.t. $D(x)$. This completes the proof. $\qquad\square$

Proposition 2.2. *If $g(\mathbf{x})$ satisfies that $g(\mathbf{x} + d_2) - g(\mathbf{x}) \leq 0$ for $x_1 > 0$ and $x_2 < 0$, then $J(\mathbf{x} + d_2) - J(\mathbf{x}) \leq 0$ for any $x_1 > 0$ and $x_2 < 0$.*

Proof. This can be proved by the induction approach similar to Proposition 2.1. Suppose the assertion is true for k and we want to show it also holds for $k + 1$. Consider

$$J_{k+1}(\mathbf{x} + d_2) - J_{k+1}(\mathbf{x}) = [g(\mathbf{x} + d_2) - g(\mathbf{x})] + \mu \cdot [J_k(\mathbf{x} + d_2 + d_3) - J_k(\mathbf{x} + d_3)]$$
$$+ r \cdot [J_k(\mathbf{x} + d_2) \cdot I\{x_1 = 1 \text{ or } x_2 + 1 = N\}$$
$$+ \min\{J_k(\mathbf{x} + 2d_2), J_k(\mathbf{x} + d_2)\} \cdot I\{x_1 > 1 \text{ and } x_2 + 1 < N\}$$
$$- J_k(\mathbf{x}) \cdot I\{x_1 = 0 \text{ or } x_2 = N\}$$
$$- \min\{J_k(\mathbf{x} + d_2), J_k(\mathbf{x})\} \cdot I\{x_1 > 0 \text{ and } x_2 < N\}]$$
$$+ \lambda \cdot [\min\{J_k(\mathbf{x} + d_2 + qd_1) | q \in [-1, M - x_1]\}$$
$$- \min\{J_k(\mathbf{x} + qd_1) | q \in [0, M - x_1]\}]. \qquad (2.32)$$

On the RHS of (2.32), the first term in square bracket [] is negative for $x_1 > 0$ and $x_2 < 0$ due to the condition on $g(\mathbf{x})$. The second term is nonpositive due to the induction hypothesis. If $x_1 = 1$, the third term is equal to 0 due to $J_k(\mathbf{x} + d_2) - J_k(\mathbf{x}) \le 0$. If $x_1 > 1$, the third term is simplified as $\min\{J_k(\mathbf{x} + 2d_2), J_k(\mathbf{x} + d_2)\} - J_k(\mathbf{x} + d_2)$ by the induction hypothesis. This is obviously nonpositive. The last term is nonpositive because $J_k(\mathbf{x} + d_2 + i d_1) - J_k(\mathbf{x} + i d_1) \le 0$ for $i = 0, 1, \ldots, M$ by the induction hypothesis. Since every term on the RHS of (2.32) is nonnegative when $x_1 > 0$ and $x_2 < 0$, this completes the induction proof. $\qquad\square$

Clearly, if $g(\mathbf{x})$ takes the form of (2.2), it satisfies the condition of Proposition 2.2. The physical meaning of Proposition 2.2 is that the manufacturer should always produce with the maximum capacity if there are backordered demands and raw materials are available in the RM warehouse. This is obviously in agreement with the intuition.

Proposition 2.3. *Suppose $g(x)$ is submodular w.r.t. $D(x)$. For fixed x_2, if $J(j, x_2) = \min\{J(x_1, x_2) \,|\, x_1 \in [0, M]\}$, then :*

(i) $\min\{J(j, x_2 - 1), J(j + 1, x_2 - 1)\} = \min\{J(x_1, x_2 - 1) \,|\, x_1 \in [0, M]\}$.
(ii) $\min\{J(j - 1, x_2 + 1), J(j, x_2 + 1)\} = \min\{J(x_1, x_2 + 1) \,|\, x_1 \in [0, M]\}$.

Proof. The condition yields: $J(j, x_2) - J(j - 1, x_2) \le 0 \le J(j + 1, x_2) - J(j, x_2)$. From this inequality, together with Proposition 2.1 and Lemma 2.1(i), (v), (iv), and (i), we have

$$J(j, x_2 - 1) - J(j - 1, x_2 - 1) \le J(j, x_2) - J(j - 1, x_2) \le 0$$

$$0 \le J(j + 1 x_2) - J(j, x_2) \le J(j + 2, x_2 - 1) - J(i + 1, x_2 - 1)$$

$$J(j - 1, x_2 + 1) - J(j - 2, x_2 + 1) \le J(j, x_2) - J(j - 1, x_2) \le 0$$

$$0 \le J(j + 1 x_2) - J(j, x_2) \le J(j + 1, x_2 + 1) - J(j, x_2 + 1).$$

Using Lemma 2.1(v), the first two inequalities yield $J(j, x_2 - 1) \le J(j - 1, x_2 - 1) \le \cdots \le J(0, x_2 - 1)$ and $J(j + 1, x_2 - 1) \le J(j + 2, x_2 - 1) \le \cdots \le J(M, x_2 - 1)$. This implies that the assertion (i) is true. Similarly, the last two inequalities yield the assertion (ii). This completes the proof. $\qquad\square$

2.5 Characterization of Optimal Policy

To characterize the optimal integrated policy, define two switching curves as follows:

Definition 2.3.

- $S_p(x_1) := \max\{x_2 < N \,|\, J(\mathbf{x} + d_2) - J(\mathbf{x}) \le 0\}$ for $x_1 \in [1, M]$;
  $S_p(x_1) := 0$ for $x_1 < 1$; and $S_p(x_1) := S_p(x_1 - 1)$ for $x_1 > M$.

- $S_o(x_2) := \max\{1 \leq x_1 \leq M \mid J(\mathbf{x}) - J(\mathbf{x} - d_1) \leq 0\}$ for $x_2 \leq N$ and
  $S_o(x_2) := 0$ otherwise.

From Proposition 2.2 and the definition of $S_p(x_1)$, we have $-1 \leq S_p(x_1) < N$. From the definition of the switching curve $S_o(x_2)$, we have $0 \leq S_o(x_2) \leq M$. Together with Lemma 2.1(v), we know that for fixed x_2, $J(x_1, x_2)$ reaches its minimum at $x_1 = S_o(x_2)$.

Proposition 2.4. *(Switching control structure). If g(x) is submodular w.r.t. D(x), then the optimal integrated ordering and production policy can be characterized by switching curves $S_p(x_1)$ and $S_o(x_2)$:*

$$u^*(\mathbf{x}) = \begin{cases} r & \text{if } x_2 \leq S_p(x_1), x_1 > 0 \\ 0 & \text{otherwise} \end{cases}, \tag{2.33}$$

$$q^*(\mathbf{x}) = \begin{cases} S_o(x_2) - x_1 & \text{if } x_1 \leq S_o(x_2) \\ 0 & \text{otherwise} \end{cases}. \tag{2.34}$$

Proof. From the definition of $S_p(x_1)$ and Lemma 2.1(iv), we have

$$J(\mathbf{x} + d_2) - J(\mathbf{x}) \leq 0 \quad \text{for any } x_2 \leq S_p(x_1)$$

$$J(\mathbf{x} + d_2) - J(\mathbf{x}) > 0 \quad \text{for any } x_2 > S_p(x_1).$$

Recall the optimal production policy in (2.12); the above two inequalities yield (2.33). Similarly, from the definition of $S_o(x_2)$ and Lemma 2.1(v), we have

$$\min\{J(\mathbf{x} + q \cdot d_1) \mid q \in [0, M - x_1]\} = \begin{cases} J(S_o(x_2), x_2) & \text{if } x_1 \leq S_o(x_2) \\ J(x_1, x_2) & \text{if } x_1 > S_o(x_2) \end{cases}.$$

Recall the optimal ordering policy given in (2.13); it follows that $q^*(\mathbf{x}) = S_o(x_2) - x_1$ if $x_1 \leq S_o(x_2)$ and $q^*(\mathbf{x}) = 0$ if $x_1 > S_o(x_2)$. This completes the proof.   □

Proposition 2.4 provides easy-to-operate managerial insights, that is, the manufacturer should produce with its maximum capacity whenever its finished goods inventory level is below the switching curve $S_p(x_1)$, and produce nothing otherwise; the manufacturer should order up to $S_o(x_2)$ whenever its raw material inventory level falls below the switching curve $S_o(x_2)$, and order nothing otherwise. In other words, the optimal production decision follows a set of base-stock policies, while the optimal ordering decision follows a set of order-up-to-point policies.

Proposition 2.5. *(Transition monotone). If g(x) is submodular w.r.t. D(x), then the switching curves $S_p(x_1)$ and $S_o(x_2)$ have the following monotonic properties:*

(i) $S_p(x_1)$ is monotonic increasing in x_1.
(ii) $S_o(x_2)$ is skip-free monotonic decreasing in x_2. Here skip-free means that $S_o(x_2)$ will change at most by one unit as x_2 increases one unit.

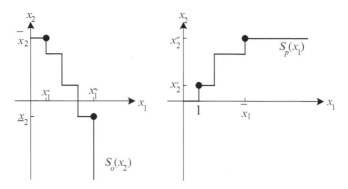

Fig. 2.3 Illustration of the characteristics of the switching curves

Proof. From the definition of $S_p(x_1)$ and Lemma 2.1(iii), we have

$$J(\mathbf{x} + e_1 + d_2) - J(\mathbf{x} + e_1) \le J(\mathbf{x} + d_2) - J(\mathbf{x}) \le 0 \quad \text{for any } x_2 \le S_p(x_1).$$

It follows that $S_p(x_1) \le S_p(x_1 + 1)$. The assertion (i) holds. Moreover, from the definition of $S_o(x_2)$ and Lemma 2.1(i), we have

$$J(\mathbf{x} + e_2) - J(\mathbf{x} + e_2 - d_1) \ge J(\mathbf{x}) - J(\mathbf{x} - d_1) > 0 \quad \text{for any } x_1 > S_o(x_2).$$

This leads to $S_o(x_2 + 1) \le S_o(x_2)$. In addition, Proposition 2.3 yields that either $S_o(x_2 + 1) = S_o(x_2)$ or $S_o(x_2 + 1) = S_o(x_2) - 1$, which implies that $S_o(x_2)$ is a skip-free monotonic decreasing function. This completes the proof. □

Since $S_o(x_2)$ takes values between 0 and M, according to the monotone convergence principle, it must converge to a finite integer as x_2 tends to negative infinity. In addition, $S_p(x_1)$ takes values in $[-1, N)$. Therefore, there exist finite integers, $\bar{x}_2, x'_1, \underline{x}_2$, and x''_1, such that $S_o(x_2) = 0$ if $x_2 > \bar{x}_2$, $S_o(\bar{x}_2) = x'_1$, and $S_o(x_2) = x''_1$ if $x_2 \le \underline{x}_2$. There exist $x'_2, \bar{x}_1$, and x''_2 such that $S_p(1) = x'_2$ and $S_p(x_1) = x''_2$ if $x_1 \ge \bar{x}_1$. The optimal switching curves $S_o(x_2)$ and $S_p(x_1)$ can then be more specifically illustrated in Fig. 2.3. It should be pointed out that $x''_1 - x'_2 \le \bar{x}_2 - \underline{x}_2$ due to the fact that $S_o(x_1)$ is a skip-free monotonic decreasing function. If two switching curves are put into the same graph, they will divide the state space into four decision zones.

To further explore the structural characteristics of the optimal policy, it is interesting to study the asymptotic behaviors of the optimal value function. To simplify the narrative, let $k := x''_1$, that is, $J(k, x_2) = \min\{J(x_1, x_2) \,|\, x_1 \in [0, M]\}$ for any $x_2 \le \underline{x}_2$. Due to the monotonic properties given in Proposition 2.1 and Lemma 2.1, it is proper to define the following limits:

- $a_i := \lim_{x_2 \to -\infty} (J(i, x_2 + 1) - J(i, x_2))$ for $i \in [0, M]$.

- $b_i := \lim\limits_{x_2 \to -\infty} (J(i-1, x_2+1) - J(i, x_2))$ for $i \in [1, M]$.
- $f_i := \lim\limits_{x_2 \to -\infty} (J(i+1, x_2) - J(i, x_2))$ for $i \in [0, M-1]$.

Proposition 2.6. *If $g(x)$ takes the form (2.2), then the optimal value function has the following asymptotic behavior:*

(i) $a_0 = a_1 = \cdots = a_M = -c_2^- / \beta$.
(ii) $b_1 = (-c_1 - c_2^- + \lambda a_0) / (\beta + \lambda + r)$.
(iii) $b_i = (-c_1 - c_2^- + r b_{i-1} + \lambda a_0) / (\beta + \lambda + r)$ for $i = 2, 3, \ldots, k$.
(iv) $b_i = (-c_1 - c_2^- + r b_{i-1}) / (\beta + r)$ for $i = k+1, k+2, \ldots, M$.
(v) $f_0 = (c_1 + r a_0) / (\beta + \lambda + r)$.
(vi) $f_i = (c_1 + r f_{i-1}) / (\beta + \lambda + r)$ for $i = 1, 2, \ldots, k-1$.
(vii) $f_i = (c_1 + r f_{i-1}) / (\beta + r)$ for $i = k, k+1, \ldots, M-1$.

Proof. For sufficiently large negative number x_2, Proposition 2.2 and the definition of k yield

$$J(x_1 - 1, x_2 + 1) - J(x_1, x_2) \le 0 \quad \text{for } x_1 = 1, 2, \ldots, M,$$

$$J(k, x_2) = \min\{J(i, x_2), J(i+1, x_2), \ldots, J(M, x_2)\} \text{ for } i = 0, 1, \ldots, k,$$

$$J(i, x_2) = \min\{J(i, x_2), J(i+1, x_2), \ldots, J(M, x_2)\} \text{ for } i = k+1, k+2, \ldots, M.$$

From the above conditions, together with the definition of a_i and the optimality Eq. (2.11), we have

$$(\beta + v) a_0 = -c_2^- + \mu a_0 + r a_0 + \lambda a_k,$$

$$(\beta + v) a_i = -c_2^- + \mu a_i + r a_{i-1} + \lambda a_k \quad \text{for } i = 1, 2, \ldots, k,$$

$$(\beta + v) a_i = -c_2^- + \mu a_i + r a_{i-1} + \lambda a_i \quad \text{for } i = k+1, k+2, \ldots, M.$$

Solving the above equations, it follows $a_0 = a_1 = \cdots = a_m = -c_2^- / \beta$. From the definition of b_i and the optimality equation (2.11), we have

$$(\beta + v) b_1 = -c_1 - c_2^- + \mu b_1 + \lambda a_0,$$

$$(\beta + v) b_i = -c_1 - c_2^- + \mu b_i + r b_{i-1} + \lambda a_0 \quad \text{for } i = 2, 3, \ldots, k,$$

$$(\beta + v) b_i = -c_1 - c_2^- + \mu b_i + r b_{i-1} + \lambda b_i \quad \text{for } i = k+1, k+2, \ldots, M.$$

This leads to the assertions (ii)–(iv). Similarly, from the definition of f_i and the optimality equation (2.11), we have

$$(\beta + v) f_0 = c_1 + \mu f_0 + r a_0,$$

$$(\beta + v) f_i = c_1 + \mu f_i + r f_{i-1} \quad \text{for } i = 1, 2, \ldots, k-1,$$

$$(\beta + v) f_i = c_1 + \mu f_i + r f_{i-1} + \lambda f_i \quad \text{for } i = k, k+1, \ldots, M-1.$$

This leads to assertions (v)–(vii). The proof is completed.   □

Proposition 2.6 can be interpreted as follows: Suppose there are a very large number of backordered demands. Having one more unit of backordered demand incurs an additional cost c_2^- over the time until the excessive demand has been satisfied. In the limit, the net present value of this additional cost is given by $c_2^- \int_0^\infty e^{-\beta t} dt = c_2^- / \beta = -a_0$. The intuition of b_j and f_j in Proposition 2.6 can be similarly interpreted.

From Proposition 2.6, it is easy to show that $0 > b_1 > b_2 > \cdots > b_M$ and $f_0 > f_1 > \cdots > f_{M-1}$. From the definition of $S_o(x_2)$ and x''_1 (i.e., k), we have:

Proposition 2.7. *If $g(x)$ takes the form (2.2), then $x''_1 := max\{S_o(x_2) \mid x_2 \leq N\}$ is the minimum integer i such that $f_i > 0$, where $f_0 = (c_1 + ra_0)/(\beta + \lambda + r)$ and $f_i = (c_1 + rf_{i-1})/(\beta + \lambda + r)$ for $0 < i < M$. If $f_i \leq 0$ for $0 \leq i < M$, then $x''_1 = M$.*

Note that the definition of f_i for $i \geq x''_1$ is slightly different from that in Proposition 2.6. However, the sign of the value is exactly the same, and therefore, the found x''_1 is valid. Since f_i is strictly decreasing in i and has a closed form in Proposition 2.6, Proposition 2.7 provides a simple analytical method to determine x''_1. From the managerial point of view, the value of x''_1 can be used to set up the raw material warehouse capacity since the inventory level of raw materials will never exceed the point x''_1 under the optimal integrated ordering and production policy. This information is apparently useful when the manufacturer is using a third party's warehouse or designing own RM warehouse.

The structural characteristics of the optimal integrated policy are useful in designing near-optimal but simple control policies, which will be addressed in Chap. 8. Numerical examples to verify the structural results and the comparison with simple near-optimal policies will also be provided in Chap. 8.

2.6 Discussions

There are some practical issues that a supply chain manager has to deal with in real-life systems. This section first provides a brief discussion on the interpretation and extension of the basic supply chain, then discusses several important practical issues related to supply chain integration including information sharing, channel coordination, and cost and benefit sharing.

2.6.1 Interpretation and Extension

With a slight modification of the parameter settings, the basic integrated supply chain in Fig. 2.1 can be interpreted as a manufacturer–retailer supply chain as shown in Fig. 2.4, in which $1/\lambda_1$ and $1/\lambda_2$ represent the average lead times that the manufacturer and the retailer replenish their inventories, respectively, while $q_1(t)$ and $q_2(t)$ represent their corresponding order quantities.

Fig. 2.4 A manufacturer–retailer supply chain

This manufacturer–retailer supply chain can be formulated in the same way as that in Sect. 2.2. The optimal integrated ordering policy for the manufacturer and for the retailer can then be obtained. It is also possible to establish the structural properties of the optimal policy. For example, Song and Dinwoodie (2008) examine the control structure of the optimal integrated inventory management (IIM) policy for a supply chain that is similar to Fig. 2.4.

The first two Assumptions 2.1 and 2.2 will be relaxed in Chap. 3 to generalize the model. For example, the supplier may not allow the manufacturer to modify the quantity of an in-processing production order and an order in transportation. In addition, the supplier may allow the manufacturer to place multiple parallel orders in order to achieve quicker responses and improve customer satisfaction. The exponentially distributed delivery lead time is relaxed by more centralized distributions in Chap. 3.

Another direction of extension is the supply chain infrastructure, for example, including more entities in the supply chain systems. The extension to multistage serial supply chain will be addressed in Chap. 3.

Fully integrated supply chain may be difficult to achieve in practice, whereas partial integration of supply chain systems is more achievable. Depending on the degree of integration, partial supply chain integration may be classified into two types: information sharing and channel coordination.

2.6.2 Information Sharing

For the basic integrated supply chain in Fig. 2.1, it is clear that in order to obtain the optimal integrated control policy for the supply chain, all the items of information in the systems are required and should be appropriately utilized, for example, the inventory levels of raw materials and finished goods; the rates and distributions of raw material lead times, product processing times, and customer demands; the warehouse capacities; and the unit costs of inventories and backlog.

A number of studies have been conducted to explain the value of information sharing theoretically or empirically. The most commonly shared information includes demand and inventory. Lee et al. (2000) demonstrated the value of sharing demand information. Croson and Donohue (2006) observed that inventory information helps upstream chain members better manage inventory and reduce the bullwhip effect. Moinzadeh (2002) reported that the supplier can make use of the retailers' demand and inventory to design a better replenishment policy. Huang and Iravani (2005) considered a supply chain consisting of one manufacturer and two retailers and examined how system factors affect the benefit (cost saving) of information sharing and the relative values of information from each retailer.

In general, information sharing results in inventory reductions and cost savings (Yu et al. 2001). The value of information sharing is increasing as inventory-holding costs, demand variability, and replenishment lead time increase (Bourland et al. 1996; Chen et al. 2000; Moinzadeh 2002).

Under the information sharing mechanism, the entities in the supply chain can still make their decisions independently, which may result in the lack of channel coordination. Nevertheless, information sharing is generally regarded as the fundamental requirement to achieve supply chain integration.

2.6.3 Channel Coordination

A further step toward supply chain integration is the channel coordination. Channel coordination refers to the collaborative decision making in supply chains for better management of channel relationship. Depending on the partnership agreement between channel members, the degree and scope of coordination vary. The highest partnership is termed as vertical integration, for example, the supply chain members have the same ownership (Harrison and Hoek 2005), which implies that the management of the supply chain can be fully coordinated in principle. In practice, the partnership is often formed using contracts, which specify the parameters such as quantity, price, time, quality, return, obligation, and responsibility. We introduce two commonly used channel coordination practices below: vendor-managed inventory and collaborative planning, forecasting, and replenishment.

2.6.3.1 Vendor-Managed Inventory (VMI)

Vendor-managed inventory (VMI) is one of the most well-known types of channel coordination mechanisms, in which the supplier (vendor) takes the responsibility to manage the inventory level and the replenishment for the customer based on the demand and inventory information provided by the customer (Song and Dinwoodie 2008). There are many different forms of VMI ranging from basic replenishment systems to more complex computerized approaches. Their common feature is that the upstream entities essentially take the responsibility of inventory management on behalf of downstream entities.

Lee and Chu (2005) reviewed several common forms of VMI including VMI with full refund, consignment VMI, and the vendor hub. The VMI with full refund permits downstream entities to return excess goods to the upstream entities, which effectively shifts the inventory carrying responsibility to the upstream entities. In return, the upstream has more power to control the supply chain and takes more responsibility and risk to carry inventories in the supply chain. This is similar to the buyback contract, in which the retailer is allowed to return the unsold inventory to some fixed amount at agreed prices; the manufacturer accepts the returns from the retailer when the production costs are sufficiently low and demand uncertainty is not too great (Padmanabhan and Png 1997). With consignment VMI, the supplier maintains and manages a stock of its products at the customer's site as vendor consignment stock and carries out replenishment planning on behalf of the customer. For example, Wal-Mart, the VMI pioneer and one of the largest retailers in the world, moved to a consignment VMI arrangement. By doing so, Wal-Mart only owns the goods for a brief moment in time as the goods are passing through the check-out point. On the other hand, the supplier has more visible information of inventory, customer demands, requirements, and market scale so that it can rapidly adjust its production plan and product development. The vendor hub is a popular business process implemented between many electronics/computer manufacturers and their suppliers. The manufacturer requests a supplier to keep a minimum stock in supplier's warehouse or hub and only pays the supplier when goods are pulled and consumed. Therefore, suppliers have to manage the agreed-upon level of inventory at the hub.

VMI offers benefits for both downstream and upstream entities. For retailers, VMI provides higher product availability and service level with lower inventory monitoring and ordering costs due to shifting the inventory management role to upstream entities (Waller et al. 1999). For suppliers, VMI offers opportunities to better synchronize the supply chain, better align the production plan with the inventory replenishment, and reduce the bullwhip effect (Waller et al. 1999; Disney and Towill 2003). Some studies have quantified the benefits of using VMI in comparison with traditional retailer-managed inventory (RMI) policies (e.g., Mishra and Raghunathan 2004; Song and Dinwoodie 2008).

In VMI programs, retailer acts more likely as a passive entity in the supply chain, since it does not involve in the demand forecasting and ordering decision making. Its role is to share sales and inventory data with suppliers. According to marketing channel theory, retailers are the last link to marketplace, and they should have better knowledge about customer behaviors and product preference. Such unique knowledge is not utilized in most VMI programs (Sari 2008).

2.6.3.2   Collaborative Planning, Forecasting, and Replenishment (CPFR)

Another popular type of channel coordination mechanisms is the collaborative planning, forecasting, and replenishment (CPFR), which is a business practice that combines the intelligence of multiple trading partners in the planning and fulfillment of customer demand (VICS 2004). Essentially, it requires all channel members to jointly develop demand forecasts, production and procurement plans, and inventory replenishments (Aviv 2002; Larsen et al. 2003). In that sense, CPFR can be regarded as a further development of VMI.

The benefits of collaborative forecasting and CPFR have been well evidenced in a number of studies (e.g., Aviv 2001, 2002, 2007; Aghazadeh 2003; Fliedner 2003). Sari (2008) conducted a comparative study to quantify the benefits of CPFR and VMI using simulation method.

The main objectives of channel coordination and supply chain contracts are the following: to increase the total supply chain profit, to reduce overstock and out-of-stock costs, and to share the risks among the supply chain partners (Tsay 1999). Such practices may result in the reduction of the costs of individual SC members and the total supply chain costs, compared to non-coordination mechanisms (Arshinder et al. 2008).

2.6.4 Cost and Benefit Sharing

Supply chain partnership is the base for information sharing or channel coordination. The main driver for supply chain partnership is that the potential benefits that channel members can gain should be able to offset the cost incurred in forming the partnership. However, information sharing or channel coordination does not necessarily lead to Pareto improvement, that is, one party is better off and the other is no worse off. This raises an important issue: how channel members share the costs and the benefits in the process toward supply chain integration.

Supply chain contracts are a useful tool to coordinate the behaviors of channel members, in which a revenue sharing mechanism can be embedded to deal with costs and benefits sharing issues. For example, the supplier offers the buyer a low wholesale price when the retailer shares a fraction of its revenue with the supplier (Giannoccaro and Pontrandolfo 2004; Cachon and Lariviere 2005). Chen and Chen (2005) presented a saving-sharing mechanism, through a quantity discount scheme, in which Pareto improvements can be achieved among channel members. In addition, there are some implicit benefits to channel members as well, for example, the vendor has more control of the supply chain whereas the retailer decreases the risk of holding inventory.

Cooperative game approach has been used to examine the appropriate revenue sharing scheme, for example, Leng and Parlar (2009) constructed a three-person cooperative game to investigate the allocation of cost savings from sharing demand information in a three-level supply chain; Li et al. (2009) developed a cooperative game model to implement profit sharing between a manufacturer and a retailer, in which the manufacturer chooses the delivery quantity and the retail price and the retailer sets the revenue shares. Ding et al. (2011) developed a cooperative game model for a three-echelon supply chain system to explore the possible cooperative solutions to profit allotment among partners, in which the profit is gained from information sharing.

Theoretically, information sharing, channel coordination, and supply chain integration will lead to cost reduction for individual entities and the entire supply chain. In practice, apart from the implementation cost and benefit allotment, there are other intangible issues to consider such as data confidentiality and relationship management. Nevertheless, supply chain integration has demonstrated to be a strong trend of the business process development to gain competitive advantage.

2.7 Notes

This chapter is mainly based on Song (2009). In the literature, two groups of research are closely related to the basic integrated supply chain examined in this chapter. The first is on the optimal production/service control policies, and the second is on the optimal replenishment/ordering policies.

The first group focuses on the optimal production control with uncertainties in production and demands. Under the continuous-review mode, a base-stock policy was shown to be optimal for a single stage system in Gavish and Graves (1980) and Li (1992). The policy states that there is a base-stock level below which it is optimal to produce at full speed and stop otherwise. Veatch and Wein (1992) established the monotone optimal control of production rates in a serial make-to-stock (MTS) system based on the concept *submodularity*. Veatch and Wein (1994) demonstrated that the optimal policy has the switching structure in a two-station tandem system. Various suboptimal policies were presented and compared numerically. With an additional dimension of uncertainty, that is, machine failure and repair, the optimality of a hedging point policy (also termed threshold control policy sometimes) in a single machine manufacturing system was established in Song and Sun (1999), Feng and Yan (2000), and Feng and Xiao (2002). The steady-state probability and the optimal hedging point were obtained in Feng and Xiao (2002) and Ching et al. (1997) for the systems with completely or finitely backordered unmet demands. Song and Sun (1998) established the monotonic and asymptotic characteristics of the optimal feedback control policy in a stochastic serial production line with failure-prone machines. In some of the above work (Veatch and Wein 1994; Song and Sun 1998), internal buffers between machines are considered. However, the buffer sizes are assumed to be infinite, part flow is in size of one through all workstations, and the material supply is not explicitly modeled.

The second group focuses on the optimal inventory replenishment policies for stochastic production systems. Under the continuous-review mode, He et al. (2002a) considered a special two-echelon model in which raw material inventories exist, production is make-to-order (MTO) (i.e., no finished goods inventory exists), and raw material replenishment lead time is zero. They examined the structure of the optimal replenishment policy and constructed simpler heuristics policies. He and Jewkes (2000) further conducted the performance analysis for a set of replenishment polices in a make-to-order inventory system by using matrix analytic methods. He et al. (2002b) demonstrated the value of unfilled demand information to the optimal replenishment policy in a two-level make-to-order inventory-production system. Berman and Kim (2001) characterized an optimal ordering policy as a monotonic threshold structure in a supply chain consisting of single class of customers, company, and supplier. Berman and Kim (2004) extended the above model to a profit maximization problem with stochastic service, demand, and lead time. Kim (2005) characterized the optimal replenishment policy as a monotonic threshold function of reorder point under the discounted cost criterion in a lost-sale service system with finite waiting room capacity. Haji et al. (2011) applied base-stock (one-for-one ordering) policy in a two-level lost-sales inventory system and derived the steady-state distributions of the system state. In the above work, He et al. (2002a, b) allowed different order quantities, and Berman and Kim (2001) and Kim (2005) made decisions on either ordering nothing or ordering a fixed pre-specified quantity. This group concentrates on the optimal material ordering/replenishment policy, and the production or service rates were uncontrollable.

For the stochastic inventory systems on the periodic-review basis, there has been a rich literature on the optimal inventory policies and their structural properties (e.g., Clark and Scarf (1960), Zipkin (2000), and Porteus (2002)). Optimal production control for continuous material flows with uncertainties such as machine break-downs can be referred to Gershwin (1994), Sethi and Zhang (1994), Hu (1995), and Martinelli and Valigi (2004).

References

Aghazadeh, S.M.: Going toward a better production by CPFR. J. Acad. Bus. Econ. **2**(1), 123–131 (2003)

Arshinder, K., Kanda, A., Deshmukh, S.G.: Supply chain coordination: perspectives, empirical studies and research directions. Int. J. Prod. Econ. **115**(2), 316–335 (2008)

Aviv, Y.: The effect of collaborative forecasting on supply chain performance. Manage. Sci. **47**(10), 1326–1343 (2001)

Aviv, Y.: Gaining benefits of joint forecasting and replenishment process: the case of auto-correlated demand. Manuf. Serv. Oper. Manage. **4**(1), 55–74 (2002)

Aviv, Y.: On the benefits of collaborative forecasting partnerships between retailers and manufacturers. Manage. Sci. **53**(5), 777–794 (2007)

Berman, O., Kim, E.: Dynamic order replenishment policy in internet-based supply chains. Math. Method Oper. Res. **53**, 371–390 (2001)

Berman, O., Kim, E.: Dynamic inventory strategies for profit maximization in a service facility with stochastic service, demand and lead time. Math. Method Oper. Res. **60**, 497–521 (2004)

Bertsekas, D.P.: Dynamic Programming: Deterministic and Stochastic Models. Prentice-Hall, Englewood Cliffs (1987)

Bourland, K.E., Powell, S.G., Pyke, D.F.: Exploiting timely demand information to reduce inventories. Eur. J. Oper. Res. **92**(2), 239–253 (1996)

Buzacott, J.A., Shanthikumar, J.G.: Stochastic Models of Manufacturing Systems. Prentice Hall, Englewood Cliffs (1993)

Cachon, G.P., Lariviere, M.A.: Supply chain coordination with revenue sharing contracts: strengths and limitations. Manage. Sci. **51**(1), 30–44 (2005)

Chen, T.H., Chen, J.M.: Optimizing supply chain collaboration based on joint replenishment and channel coordination. Transp. Res. Part E **41**(4), 261–285 (2005)

Chen, F., Drezner, Z., Ryan, J.K., Levi, D.S.: Quantifying the bullwhip effect in a simple supply chain: the impact of forecasting, lead times, and information. Manage. Sci. **46**(3), 436–443 (2000)

Ching, W.K., Chan, R.H., Zhou, X.Y.: Circulant preconditioners for Markov-modulated Poisson processes and their applications to manufacturing systems. SIAM J. Matrix Anal. Appl. **18**(2), 464–481 (1997)

Clark, A.J., Scarf, H.E.: Optimal policies for a multi-echelon inventory problem. Manage. Sci. **6**, 475–490 (1960)

Croson, R., Donohue, K.: Behavioral causes of the bullwhip effect and the observed value of inventory information. Manage. Sci. **52**(3), 323–336 (2006)

Ding, H., Guo, B., Liu, Z.: Information sharing and profit allotment based on supply chain cooperation. Int. J. Prod. Econ. **133**(1), 70–79 (2011)

Disney, S.M., Towill, D.R.: The effect of vendor management inventory dynamics on the bullwhip effect in supply chains. Int. J. Prod. Econ. **85**(2), 199–215 (2003)

Feng, Y.Y., Xiao, B.C.: Optimal threshold control in discrete failure-prone manufacturing systems. IEEE Trans. Autom. Control **47**(7), 1167–1174 (2002)

Feng, Y.Y., Yan, H.M.: Optimal production control in a discrete manufacturing system with unreliable machines and random demands. IEEE Trans. Autom. Control **45**(12), 2280–2296 (2000)

Fliedner, G.: CPFR: an emerging supply chain tool. Ind. Manage. Data Syst. **103**(1), 14–21 (2003)

Gavish, B., Graves, S.: A one-product production/inventory problem under continuous review policy. Oper. Res. **28**, 1228–1236 (1980)

Gershwin, S.B.: Manufacturing Systems Engineering. Prentice-Hall, Englewood Cliffs (1994)

Giannoccaro, I., Pontrandolfo, P.: Supply chain coordination by revenue sharing contracts. Int. J. Prod. Econ. **89**(2), 131–139 (2004)

Haji, R., Haji, A., Saffari, M.: Queueing inventory system in a two-level supply chain with one-for-one ordering policy. J. Ind. Syst. Eng. **5**(1), 337–347 (2011)

Harrison, A., Hoek, R.I.: Logistics Management and Strategy. Prentice Hall, London (2005)

He, Q.M., Jewkes, E.M.: Performance measures of a make-to-order inventory-production system. IIE Trans. **32**, 409–419 (2000)

He, Q.M., Jewkes, E.M., Buzacott, J.: Optimal and near-optimal inventory control policies for a make-to-order inventory-production system. Eur. J. Oper. Res. **141**, 113–132 (2002a)

He, Q.M., Jewkes, E.M., Buzacott, J.: The value of information used in inventory control of a make-to-order inventory-production system. IIE Trans. **34**, 999–1013 (2002b)

Hu, J.Q.: Production rate control for failure-prone production systems with no-backlog permitted. IEEE Trans. Autom. Control **40**(2), 291–295 (1995)

Huang, B., Iravani, S.M.R.: Production control policies in supply chains with selective-information sharing. Oper. Res. **53**, 662–674 (2005)

Kim, E.: Optimal inventory replenishment policy for a queueing system with finite waiting room capacity. Eur. J. Oper. Res. **161**, 256–274 (2005)

Koole, G.M.: Structural results for the control of queueing systems using event-based dynamic programming. Queueing Syst. **30**, 323–339 (1998)

Larsen, T.S., Thernoe, C., Anderson, C.: Supply chain collaboration theoretical perspective and empirical evidence. Int. J. Phys. Distrib. Logist. **33**(6), 531–549 (2003)

Lee, C.C., Chu, W.H.J.: Who should control inventory in a supply chain? Eur. J. Oper. Res. **164**, 158–172 (2005)

Lee, H.L., So, K.C., Tang, C.S.: The value of information sharing in a two-level supply chain. Manage. Sci. **46**(5), 626–643 (2000)

Leng, M., Parlar, M.: Allocation of cost savings in a three-level supply chain with demand information sharing: a cooperative-game approach. Oper. Res. **57**, 200–213 (2009)

Li, L.: The role of inventory in delivery-time competition. Manage. Sci. **38**, 182–197 (1992)

Li, S., Zhu, Z., Huang, L.: Supply chain coordination and decision making under consignment contract with revenue sharing. Int. J. Prod. Econ. **120**(1), 88–99 (2009)

Martinelli, F., Valigi, P.: Hedging point policies remain optimal under limited backlog and inventory space. IEEE Trans. Autom. Control **49**(10), 1863–1869 (2004)

Mishra, B.K., Raghunathan, S.: Retailer vs. vendor managed inventory and brand competition. Manage. Sci. **50**, 445–457 (2004)

Moinzadeh, K.: A multi-echelon inventory system with information exchange. Manage. Sci. **48**(3), 414–426 (2002)

Padmanabhan, V., Png, I.P.L.: Manufacturer's returns policies and retailer's competition. Mark. Sci. **16**(1), 81–94 (1997)

Porteus, E.L.: Foundations of Stochastic Inventory Theory. Stanford University Press, Stanford (2002)

Sari, K.: On the benefits of CPFR and VMI: a comparative simulation study. Int. J. Prod. Econ. **113**, 575–586 (2008)

Sethi, S., Zhang, Q.: Hierarchical Decision Making in Stochastic Manufacturing Systems. Birkhauser, Boston (1994)

Song, D.P.: Optimal integrated ordering and production policy in a supply chain with stochastic lead-time, processing-time and demand. IEEE Trans. Autom. Control **54**(9), 2027–2041 (2009)

Song, D.P., Dinwoodie, J.: Quantifying the effectiveness of VMI and integrated inventory management in a supply chain with uncertain lead-times and uncertain demands. Prod. Plan. Control **19**(6), 590–600 (2008)

Song, D.P., Sun, Y.X.: Optimal service control of a serial production line with unreliable workstations and random demand. Automatica **34**(9), 1047–1060 (1998)

Song, D.P., Sun, Y.X.: Optimal control structure of an unreliable manufacturing system with random demands. IEEE Trans. Autom. Control **44**(3), 619–622 (1999)

Tsay, A.: The quantity flexibility contract and supplier–customer incentives. Manage. Sci. **45**(10), 1339–1358 (1999)

Veatch, M., Wein, L.: Monotone control of queueing networks. Queueing Syst. **12**, 391–408 (1992)

Veatch, M., Wein, L.: Optimal-control of a 2-station tandem production inventory system. Oper. Res. **42**(2), 337–350 (1994)

VICS.: Collaborative Planning, Forecasting and Replenishment (CPFR): an overview. Voluntary Interindustry Commerce Standards (VICS), available at http://www.vics.org/docs/committees/cpfr/CPFR_Overview_US-A4.pdf (2004). Accessed Mar 2012

Waller, M.A., Johnson, M.E., Davis, T.: Vendor-managed inventory in the retail supply chain. J. Bus. Logist. **20**(1), 183–203 (1999)

Weber, R., Stidham, S.: Optimal control of service rates in networks of queues. Adv. Appl. Probab. **19**(1), 202–218 (1987)

Yu, Z., Yan, H., Cheng, T.C.E.: Benefits of information sharing with supply chain partnerships. Ind. Manage. Data Syst. **101**(3), 114–119 (2001)

Zipkin, P.: Foundations of Inventory Management. McGraw-Hill, Boston (2000)

Chapter 3
Optimal Control of Supply Chains in More General Situations

3.1 Introduction

For the basic supply chain system described in Chap. 2, it consists of three entities (i.e., supplier, manufacturer, and customer) and two finite capacitated warehouses (raw material warehouse and finished goods warehouse). One key assumption in Chap. 2 is that there is at most one outstanding order to the supplier at any time and its size can be changed without incurring any cost. Such assumption may be appropriate in a fully integrated supply chain, for example, a single ownership, or under an agreement of appropriate distribution of final revenue along all the entities in the supply chain. However, a more realistic situation is that placed orders are not allowed to be modified freely since some work may have been planned and effort may have been committed. The alteration of the order size can cause disruption to the supply chain and damage the relationship between the entities. On the other hand, suppliers may allow customers to have parallel orders in order to achieve quick responses.

Moreover, in reality, order replenishment lead time is less likely to be an exponential distribution due to the physical distance between the manufacturer and the supplier. The uncertainty in the lead times may be caused by various waiting times during the production processes and/or during the physical transportation from the supplier to the manufacturer. A more centralized distribution such as gamma or normal distributions may be more appropriate.

Further, the basic supply chain in Chap. 2 considers two inventory stocking points, that is, raw material warehouse and finished goods warehouse. A more general supply chain infrastructure would include multiple stages with multiple stocking points.

In this chapter, we consider four types of relaxations. The first relaxation is to consider the case of at most one outstanding order but with its order size not changeable once issued. The second relaxation is to allow two parallel outstanding orders with their order sizes not changeable once issued. The third relaxation is that the replenishment lead time follows a multistage Erlang distribution (which is

D.-P. Song, *Optimal Control and Optimization of Stochastic Supply Chain Systems*, Advances in Industrial Control, DOI 10.1007/978-1-4471-4724-4_3, © Springer-Verlag London 2013

a special class of gamma distribution). The forth relaxation is to extend the basic supply chain infrastructure into a multistage serial supply chain in which multiple entities or workstations can be explicitly modeled and managed in an integrated way.

3.2 One Outstanding Order with Its Size Not Changeable Once Issued

In this section, we relax Assumption 2.1 to be the case that there is at most one outstanding order but the order size cannot be modified once it was issued. In order to model this system, we have to introduce another state variable, y, to represent the order status. More specifically, $y = 0$ represents that there is no outstanding order (i.e., ready to place a new order); $y = q$ (>0) represents there is an outstanding order with size q, namely, an order has been issued but not yet reached the manufacturer. Clearly, the outstanding order size q must satisfy $q \in [0, M - x_1]$ due to the capacity constraint of the raw material warehouse.

The system state space becomes $X = \{(\mathbf{x}, y) \,|\, \mathbf{x} = (x_1, x_2), x_1 \in [0, M], x_2 \in (-\infty, N], \text{ and } y \in [0, M - x_1]\}$. The manufacturer needs to make two types of decisions: the production rate $u \in \{0, r\}$ at any state (x_1, x_2, y) and the raw material order quantity $q \in [0, M - x_1]$ at any state $(x_1, x_2, 0)$.

The state transition map at a state (x_1, x_2, y) under the control u is depicted in Fig. 3.1a, and the state transition at a state $(x_1, x_2, 0)$ under the control (u, q) is depicted in Fig. 3.1b. The uniform transition rate is $v = \mu + r + \lambda$. It should be pointed out that after uniformization, the self-transition rate at a state (x_1, x_2, y) is $r - r \cdot I\{u > 0\}$, whereas there is no self-transition at the state $(x_1, x_2, 0)$ if $q > 0$. Instead, there is a transition from the state $(x_1, x_2, 0)$ into (x_1, x_2, q) with a rate $r + \lambda - r \cdot I\{u > 0\} - \lambda \cdot I\{x_1 + q \le M\}$. Such transitions are partly shown in Fig. 3.1.

Following the uniformization technique and the stochastic dynamic programming theory, the Bellman optimality equation is given as follows (note that there are two sequential *min* operators in (3.2) that reflect the coupled decisions of ordering and production):

$$J(x_1, x_2, y) = \frac{1}{\beta + v} \left[g(x_1, x_2) + \mu \cdot J(x_1, x_2 - 1, y) + \lambda \cdot J(x_1 + y, x_2, 0) \right.$$

$$\left. + r \cdot \min \{J(x_1, x_2, y), J(x_1 - 1, x_2 + 1, y)\} \right], \quad \text{for } y > 0;$$

$$(3.1)$$

$$J(x_1, x_2, 0)$$

$$= \frac{1}{\beta + v} \left[g(x_1, x_2) + \min_{0 \le q \le M - x_1} \{\mu \cdot J(x_1, x_2 - 1, q) + \lambda \cdot J(x_1, q, x_2, 0) \right.$$

$$\left. + r \cdot \min \{J(x_1, x_2, q), J(x_1 - 1, x_2 + 1, q)\} \,|\, q \in [0, M - x_1]\} \right]$$

$$(3.2)$$

Fig. 3.1 System state
transition map in the basic
supply chain with a
non-changeable outstanding
order. (**a**) At state (x_1, x_2, y)
with control u. (**b**) At state
$(x_1, x_2, 0)$ with control (u, q)

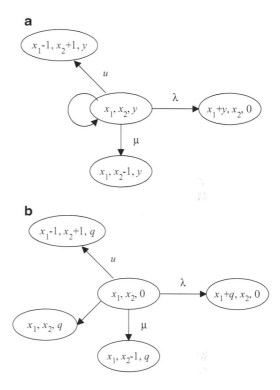

To simplify the narrative, we omit the indicator function in the above equations (also in some of other equations in the rest of this chapter). Therefore, state $(x_1 - 1, x_2 + 1, y)$ should be understood as (x_1, x_2, y) if it falls out of the state space X. The same applies to state $(x_1 - 1, x_2 + 1, q)$.

From Eqs. (3.1) and (3.2), the optimal integrated ordering and production policy can be expressed in terms of the optimal value functions. Since the optimal discounted-cost function can be approximated using the value iteration algorithm, the optimal integrated ordering and production policy can be obtained at least numerically.

Intuitively, as the placed orders cannot be adjusted at any time, the system becomes less flexible and less robust to uncertainties. The raw material ordering decisions are made less frequently. Therefore, it is believed that the placed orders may have larger sizes compared to the situation with changeable order size.

Although we cannot rigorously establish the control structure of the optimal policy, a range of experimental examples (e.g., Fig. 3.7) reveal that the integrated ordering and production policy under this relaxation has the similar switching

Fig. 3.2 Illustration of the characteristics of the optimal policy in the basic supply chain with a non-changeable outstanding order. (**a**) Ordering policy at $(x_1, x_2, 0)$. (**b**) Production policy at (x_1, x_2, y)

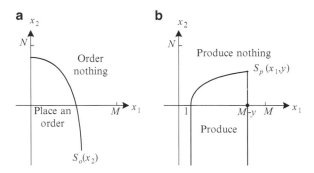

structure to that in Chap. 2. More specifically, there exists a set of switching curves $S_o(x_2)$ and $S_p(x_1, y)$ for $y = 1, 2, \ldots, M - x_1$. The optimal policy states that:

1. We should place an order if the system state $(x_1, x_2, 0)$ locates below the curve $S_o(x_2)$ in the $x_1 - x_2$ plane, otherwise, place no order.
2. For any given outstanding order with size y (issued but not arrived), we should produce products at full speed if the system state (x_1, x_2, y) locates below the curve $S_p(x_1, y)$ in the $x_1 - x_2$ plane, otherwise, produce nothing.
3. The switching curve $S_o(x_2)$ is monotonic decreasing in x_2, and $S_p(x_1, y)$ is monotonic increasing in x_1.

On the other hand, there are a few differences observed compared to the case under Assumption 2.1. Firstly, the placed order size does not exactly follow the order-up-to-point rule given in (2.34). In other words, the optimal order size $q^*(x_1, x_2, 0)$ may remain the same as x_1 changes. Nevertheless, it is observed that $q^*(x_1, x_2, 0)$ is not increasing in x_1. Secondly, the optimal order size $q^*(x_1, x_2, 0)$ may not take very small values when x_2 is adequately negative. This is in agreement with the intuition that placing a very small order will reduce the system flexibility because new orders cannot be placed before its arrival. Thirdly, the production decisions are now characterized by a set of switching curves $\{S_p(x_1, y) \mid y = 1, 2, \ldots, M - x_1\}$ rather than a single curve. This obviously makes the implementation more complicated. However, it is observed that $S_p(x_1, y)$ is almost the same for different y, which implies that a single switching curve $S_p(x_1, 0)$ is probably good enough to determine the production decision.

The optimal ordering and production policy in the basic supply chain with a non-changeable outstanding order can be illustrated in Fig. 3.2. It should be pointed out that the production control region is bounded by $x_1 = 1$ and $x_1 = M - y$ because the outstanding order is constrained by the spare capacity of the raw material warehouse.

3.3 Two Outstanding Orders with Their Sizes Not Changeable Once Issued

Now we further relax Assumption 2.1 to be that there are at most two outstanding orders but with the order sizes that cannot be modified once they were issued (the case of allowing more than two parallel orders could be similarly formulated). The system state space can now be described by $X = \{(x_1, x_2, y_1, y_2) | x_1 \in [0, M], x_2 \in (-\infty, N], y_1 \geq 0, y_2 \geq 0, \text{ and } y_1 + y_2 \leq M - x_1\}$, where y_1 and y_2 represent the status of two orders, that is, whether issued and in what quantity.

The manufacturer needs to make the following decisions: the production rate $u \in \{0, r\}$ at any state (x_1, x_2, y_1, y_2), the raw material order quantity $q_1 \in [0, M - x_1 - y_2]$ at any state $(x_1, x_2, 0, y_2)$, and the raw material order quantity $q_2 \in [0, M - x_1 - y_1]$ at any state $(x_1, x_2, y_1, 0)$. We assume that at a state $(x_1, x_2, 0, 0)$, we only place one new order if necessary with its size $q_1 \in [0, M - x_1]$. This is reasonable since placing a single large order is often preferable to placing two separate orders simultaneously.

The state transition map at a state (x_1, x_2, y_1, y_2) under the control u is depicted in Fig. 3.3a, the state transition at a state $(x_1, x_2, 0, y_2)$ under the control (u, q_1) is depicted in Fig. 3.3b, and the state transition at a state $(x_1, x_2, y_1, 0)$ under the control (u, q_2) is depicted in Fig. 3.3c.

Since we allow two outstanding orders, there are two independent events of order arrivals. Assume these two orders have the same arrival rate λ. The uniform transition rate has to be redefined as $\nu = \mu + r + 2\lambda$. After applying the uniformization technique, the self-transition rate at a state (x_1, x_2, y_1, y_2) is $r - r \cdot I\{u > 0\}$, whereas there is no self-transition at the state $(x_1, x_2, y_1, 0)$ if $q_2 > 0$, and no self-transition at the state $(x_1, x_2, 0, y_2)$ if $q_1 > 0$. Instead, there is a transition from the state $(x_1, x_2, y_1, 0)$ into (x_1, x_2, y_1, q_2) with a rate $r + \lambda - r \cdot I\{u > 0\} - \lambda \cdot I\{x_1 + y_1 + q_2 \leq M\}$ and a transition from the state $(x_1, x_2, 0, y_2)$ into (x_1, x_2, q_1, y_2) with a rate $r + \lambda - r \cdot I\{u > 0\} - \lambda \cdot I\{x_1 + y_2 + q_1 \leq M\}$.

From the transition map and the uniformization technique, the Bellman optimality equations are given by

$$J(x_1, x_2, y_1, y_2) = \frac{1}{\beta + \nu} [g(x_1, x_1) + \mu \cdot J(x_1, x_2 - 1, y_1, y_2)$$

$$+ \lambda \cdot J(x_1 + y_1, x_2, 0, y_2) + \lambda \cdot J(x_1 + y_2, x_2, y_1, 0)$$

$$+ r \cdot \min\{J(x_1, x_2, y_1, y_2), J(x_1 - 1 + x_2 + 1, y_1, y_2)\}],$$

$$\text{for } y_1 > 0, y_2 > 0; \tag{3.3}$$

Fig. 3.3 System state
transition map in the basic
supply chain with two
non-changeable outstanding
orders. (**a**) At state (x_1, x_2, y_1, y_2) with control u. (**b**) At state
$(x_1, x_2, y_1, 0)$ with control (u, q_2). (**c**) At state $(x_1, x_2, 0, y_2)$ with control (u, q_1)

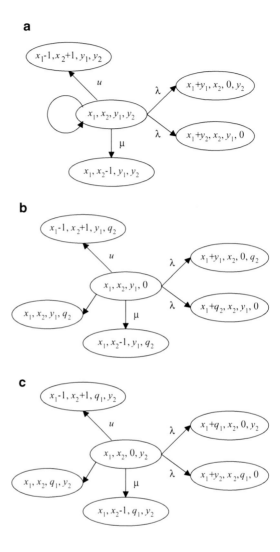

$$J(x_1, x_2, y_1, 0)$$

$$= \frac{1}{\beta + \nu}\Big[g(x_1, x_2) + \min_{0 \leq q_1 \leq M - x_1 - y_1}\{\mu \cdot J(x_1, x_2 - 1, y_1, q_2)$$

$$+ \lambda \cdot J(x_1 + y_1, x_2, 0, q_2) + \lambda \cdot J(x_1 + q_2, x_2, y_1, 0)$$

$$+ r \cdot \min\{J(x_1, x_2, y_1, q_2), J(x_1 - 1, x_2 + 1, y_1, q_2)\}\}\Big], \quad \text{for } y_1 > 0;$$

$$(3.4)$$

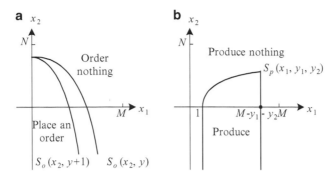

Fig. 3.4 Illustration of the characteristics of the optimal policy in the basic supply chain with two non-changeable outstanding orders. (**a**) Ordering policy at $(x_1, x_2, 0, y)$ or $(x_1, x_2, y, 0)$. (**b**) Production policy at (x_1, x_2, y_1, y_2)

$$J(x_1, x_2, 0, y_2)$$

$$= \frac{1}{\beta + v} \Big[g(x_1, x_2) + \min_{0 \leq q_1 \leq M - x_1 - y_2} \{ \mu \cdot J(x_1, x_2 - 1, q_1, y_2)$$

$$+ \lambda \cdot J(x_1 + y_2, x_2, q_1, 0) + \lambda \cdot J(x_1 + q_1, x_2, 0, y_2)$$

$$+ r \cdot \min \{ J(x_1, x_2, q_1, y_2), J(x_1 - 1, x_2 + 1, q_1, y_2) \} \} \Big], \quad \text{for } y_2 \geq 0.$$

$$(3.5)$$

Through numerical examples (cf. Fig. 3.8), it is observed that there exist similar structural properties to those of the case in Sect. 3.2. The differences are:

1. The ordering policy is now characterized by a set of switching curves $\{S_o(x_2, y) \mid y \in [0, M - x_1]\}$, where y represents the size of the outstanding order. The switching curve $S_o(x_2, y)$ is moving toward left as y increases. This corresponds to the intuition that if a larger order has been placed, the new order tends to be smaller.
2. The ordering policy at the state $(x_1, x_2, y_1, 0)$ is the same as that at the state $(x_1, x_2, 0, y_2)$.
3. The production policy is characterized by a set of switching curves $\{S_p(x_1, y_1, y_2) \mid y_1 \geq 0, y_2 \geq 0, \text{ and } y_1 + y_2 \leq M - x_1 \}$. It is also noticed that the production switching curve $S_p(x_1, y_1, y_2)$ only changes very slightly for different y_1 and y_2.

More intuitively, the optimal ordering and production policy in the basic supply chain with two non-changeable outstanding orders can be illustrated in Fig. 3.4. The production control region is bounded by $x_1 = 1$ and $x_1 = M - y_1 - y_2$ because the sum of two outstanding orders is constrained by the spare capacity of the raw material warehouse.

3.4   One Outstanding Order with Its Size Not Changeable and Its Lead Time Following an Erlang Distribution

In this section, we relax both Assumptions 2.1 and 2.2 and consider the case of at most one outstanding order with its size not changeable and an Erlang-distributed lead time with L stages and a rate γ. As we know, when the number of stages L goes to infinity while its mean remains a constant (i.e., $L/\gamma =$ constant), the lead time converges (weakly) to a deterministic variable. Therefore, it is believed that Erlang-distributed lead times with reasonably large L are more centralized and realistic since the mean lead time of the Erlang distribution is given by L/γ. We take $\gamma = L\lambda$ to keep the mean lead time consistent with the cases in Sects. 3.2 and 3.3.

Note that the Erlang distribution is the sum of L independent exponential random variables with rate $L\lambda$. Therefore, the system state space can be described by $X = \{(x_1, x_2, y, l) \,|\, x_1 \in [0, M], x_2 \in (-\infty, N), y \in [0, M - x_1], \text{ and } l \in [0, L - 1]\}$, where y represents the size of the outstanding order, and $l + 1$ represents the stage that the outstanding order enters (only when $y > 0$). The actual order arrival occurs when the outstanding order leaves stage L.

Clearly, the raw material ordering decision is only made at a state $(x_1, x_2, 0, 0)$ and the production decision is made at any state (x_1, x_2, y, l). The state transition map at a state (x_1, x_2, y, l) under the control u is depicted in Fig. 3.5a, the state transition at a state $(x_1, x_2, y, L - 1)$ under the control u is depicted in Fig. 3.5b, and the state transition at a state $(x_1, x_2, 0, 0)$ under the control (u, q) is depicted in Fig. 3.5c.

Redefine the uniform transition rate as $v = \mu + r + L\lambda$. Under the uniformization, the self-transition rate at a state (x_1, x_2, y, l) for $l = 0, 1, \ldots, L - 1$ and $y > 0$ is $r - r \cdot I\{u > 0\}$, whereas there is no self-transition at the state $(x_1, x_2, 0, 0)$ if $q > 0$. Instead, there is a transition from the state $(x_1, x_2, 0, 0)$ into $(x_1, x_2, q, 0)$ with a rate $r + L\lambda - r \cdot I\{u > 0\} - L\lambda \cdot I\{x_1 + q \leq M\}$.

From the transition map and the uniformization technique, the Bellman optimality equations are given by

$$J(x_1, x_2, y, l)$$

$$= \frac{1}{\beta + v} [g(x_1, x_2) + \mu \cdot J(x_1, x_2 - 1, y, l) + L\lambda \cdot J(x_1, x_2, y, l + 1)$$

$$+ r \cdot \min\{J(x_1, x_2, y, l), J(x_1 - 1, x_2 + 1, y, l)\}],$$

$$\text{when } y > 0 \text{ and } 0 \leq l < L - 1; \tag{3.6}$$

$$J(x_1, x_2, y, L - 1)$$

$$= \frac{1}{\beta + v} [g(x_1, x_2) + \mu \cdot J(x_1, x_2 - 1, y, L - 1) + L\lambda \cdot J(x_1 + y, x_2, 0, 0)$$

$$+ r \cdot \min\{J(x_1, x_2, y, L - 1), J(x_1 - 1, x_2 + 1, y, L - 1)\}], \quad \text{when } y > 0; \tag{3.7}$$

Fig. 3.5 System state transition map in the basic supply chain with a non-changeable outstanding order and Erlang lead times. (**a**) At state (x_1, x_2, y, l) with control u. (**b**) At state $(x_1, x_2, y, L-1)$ with control u. (**c**) At state $(x_1, x_2, 0, 0)$ with control (u, q)

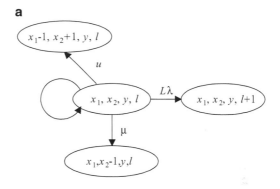

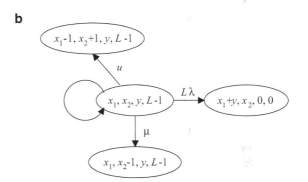

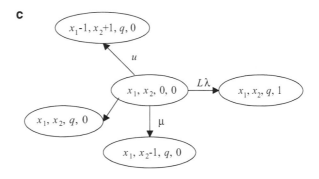

$$J(x_1, x_2, 0, 0) = \frac{1}{\beta + v}\left[g(x_1, x_2) + \min_{0 \le q \le M - x_1} \{\mu \cdot J(x_1, x_2 - 1, q, 0)\right.$$

$$+ L\lambda \cdot J(x_1, x_2, q, 1) \cdot I\{q > 0\} + L\lambda \cdot J(x_1, x_2, q, 0) \cdot I\{q = 0\}$$

$$\left. + r \cdot \min\{J(x_1, x_2, q, 0), J(x_1 - 1, x_2 + 1, q, 0)\}\}\right]. \qquad (3.8)$$

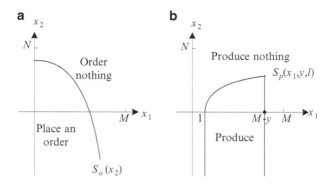

Fig. 3.6 Illustration of the characteristics of the optimal policy in the basic supply chain with a non-changeable outstanding order and Erlang lead times. (**a**) Ordering policy at $(x_1, x_2, 0, 0)$. (**b**) Production policy at (x_1, x_2, y, l)

From the numerical examples (cf. Fig. 3.9), it is observed that the control structure of the integrated ordering and production policy in this case is very similar to the case in Sect. 3.2. However, the switching curves are not exactly the same, and the production policy is characterized by more switching curves $\{S_p(x_1, y, l) \mid y \in [0, M - x_1], \text{ and } l \in [0, L - 1]\}$.

The optimal ordering and production policy in the basic supply chain with one non-changeable outstanding order and Erlang lead times can be illustrated in Fig. 3.6. The production control region is bounded by $x_1 = 1$ and $x_1 = M - y$ because the outstanding order is constrained by the spare capacity of the raw material warehouse.

3.5 Numerical Examples

In this section numerical examples are given to illustrate the validity of the control structure in three generalized cases discussed in this chapter. We use the value iteration algorithm to evaluate the performance (Bertsekas 1987).

Consider a base case with the system parameters as follows: the lead-time rate $\lambda = 1.0$, the maximum production rate $r = 1.0$, demand rate $\mu = 0.7$, the raw material inventory unit cost $c_1 = 1$, the finished goods inventory unit cost $c_2^+ = 2$, and the backordering cost $c_2^- = 10$. The warehouse capacities are set as $M = N = 6$, and the discount factor $\beta = 0.5$. The system state space is limited into a finite area with $x_2 \in [N - 50, N]$. The iterative procedure will be terminated when the value difference is less than 10^{-3} or the number of iterations exceeds 5,000.

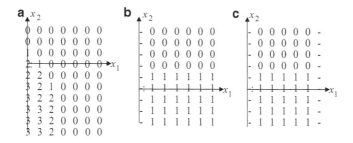

Fig. 3.7 Control structure of the optimal policy for the case of at most one outstanding order with its size not changeable once issued. (**a**) Ordering policy. (**b**) Production policy when $y = 0$. (**c**) Production policy when $y = 1$.

Example 3.1. The above base case is extended to the case of at most one outstanding order with its size not changeable once issued. The control structure of the optimal policy is partially illustrated in Fig. 3.7.

Example 3.2. The base case is extended to the case of at most two outstanding orders with its size not changeable once issued. The control structure of the optimal policy is partially illustrated in Fig. 3.8.

Example 3.3. The base case is extended to the case of at most one outstanding order with its size not changeable once issued and Erlang-distributed lead times. The Erlang distribution is assumed to have five stages, that is, $L = 5$. The control structure of the optimal policy is partially illustrated in Fig. 3.9.

The numbers in Figs. 3.7a, 3.8a–c, and 3.9a represent the optimal order sizes at the corresponding states; the numbers in Figs. 3.7b–c, 3.8d–f, and 3.9b–f indicate the production actions (e.g., 0 represents "not produce," and 1 represents "produce" with the maximum production rate) at a given state; the dash symbol represents that production cannot be issued due to the constraints such as availability of raw materials or finished warehouse capacity.

It can be observed from Figs. 3.7, 3.8, and 3.9 that the main structural properties of the optimal integrated ordering and production policy established in Chap. 2 are preserved, for example, the monotonic switching structure (i.e., two regions with positive number and with zero are separated by switching curves), the asymptotic behaviors, and the monotonic properties of the order quantity with respect to the raw material inventory level or the finished goods inventory level (e.g., the order size is decreasing in x_1 and x_2 in Figs. 3.7a, 3.8a–c, and 3.9a). However, the order-up-to-point rule is not preserved after the relaxations.

Another important difference is that the optimal ordering policy tends not to place very small orders, particularly when the system has a relatively large number of backordered demands, for example, Figs. 3.7a, 3.8b, and 3.9a. An interesting point is that in all three generalized cases, the production policy remains the same and does not change for different outstanding order sizes or different stages of the

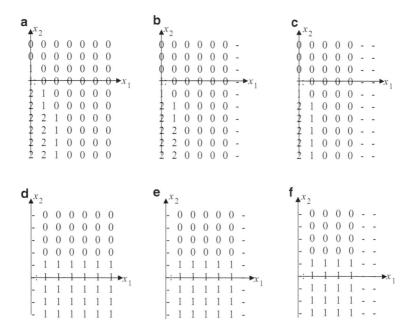

Fig. 3.8 Control structure of the optimal policy for the case of at most two outstanding orders with its size not changeable once issued. (**a**) Ordering policy with no outstanding order. (**b**) Ordering policy with an outstanding order size $= 1$. (**c**) Ordering policy with an outstanding order size $= 2$. (**d**) Production policy when $y_1 = 0$, $y_2 = 0$. (**e**) Production policy when $y_1 + y_2 = 1$. (**f**) Production policy when $y_1 + y_2 = 2$

Erlang distribution, for example, Figs. 3.7b–c, 3.8d–f, and 3.9b–f. This implies that we could construct much simpler near-optimal policies for those generalized cases, which will be addressed in Chap. 8.

3.6 Multistage Serial Supply Chain Systems

The basic supply chain in Fig. 2.1 can be extended to multistage serial supply chain systems. Assume there are n entities between supplier and customer. Figure 3.10 shows a multistage serial supply chain structure, in which $1/\lambda_i$ represents the average replenishment/production lead times and $q_i(t)$ represents the ordering/production quantity. The entities between supplier and customer can be regarded as workstations in a manufacturing firm in which the lead times can be interpreted as machine processing times and the quantities as the production batches.

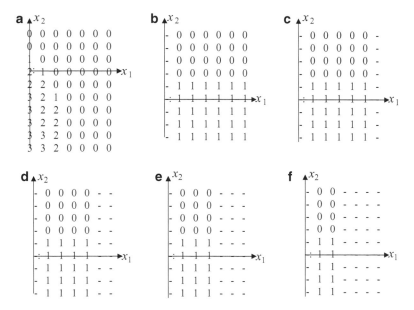

Fig. 3.9 Control structure of the optimal policy for the case of at most one outstanding order with its size not changeable once issued and Erlang-distributed lead times with $L = 5$. (**a**) Ordering policy. (**b**) Production policy when $y = 0$, $l = 0$. (**c**) Production policy when $y = 1$, $l = 0$, $\ldots$, $L - 1$. (**d**) Production policy when $y = 2$, $l = 0$, $\ldots$, $L - 1$. (**e**) Production policy when $y = 3$, $l = 0$, $\ldots$, $L - 1$. (**f**) Production policy when $y = 4$, $l = 0$, $\ldots$, $L - 1$

Fig. 3.10 A multistage serial supply chain

Let Q_i be the maximum ordering/production quantity for entity i. Suppose the supply chain systems are subject to Assumptions 2.1–2.4, but the warehouse capacity for each entity is assumed to be sufficiently large. Then the system state space can be described by $X = \{\mathbf{x} := (x_1, x_2, \ldots, x_n) \,|\, x_i \in Z^+ \text{ for } i = 1, 2, \ldots, n - 1, \text{ and } x_n \in Z\}$. The admissible control set $\Omega = \{\mathbf{u}(t) := (q_1(t), q_2(t), \ldots, q_N(t)) \,|\, 0 \le q_1(t) \le Q_1 \text{ and } 0 \le q_i(t) \le \min\{x_{i-1}(t), Q_i\} \text{ for } i > 1, t \ge 0\}$. Here we assume that the supplier has infinite materials available.

The optimal control problem of the multistage serial supply chain is to find the optimal integrated policy $\mathbf{u} \in \Omega$ to minimize the infinite-horizon expected discounted cost as follows:

$$J(\mathbf{x}_0) = \min_{\mathbf{u}} E\left[\int_0^\infty e^{-\beta t} g(\mathbf{x}(t), \mathbf{u}(t)) \, dt \,\Big|\, \mathbf{x}(0) = \mathbf{x}_0\right] \qquad (3.9)$$

where $g(\mathbf{x}, \mathbf{u}) = c_1 x_1 + \cdots + c_{n-1} x_{n-1} + c_n^+ x_n^+ + c_n^- x_n^-$, representing the cost function penalizing holding inventories and backlogs. Define the uniform transition rate as $v = \mu + \lambda_1 + \lambda_2 + \cdots + \lambda_n$. The optimality equation is given by

$$J(\mathbf{x}) = \frac{1}{\beta + v} \left[g(\mathbf{x}) + \mu \cdot J(\mathbf{x} + d_{n+1}) + \sum_i \lambda_i \cdot \min\{J(\mathbf{x} + q_i \cdot d_i) \right.$$

$$\left. |0 \le q_i \le Q_i \text{ if } i = 1; 0 \le q_i \le \min\{x_{i-1}, Q_i\} \text{ if } i > 1\} \right] \qquad (3.10)$$

where $d_1 = e_1$, $d_i = e_i - e_{i-1}$ for $i = 2, 3, \ldots, n$, and $d_{N+1} = -e_n$. From (3.10), the optimal feedback control policy $\mathbf{u}^* = \{q_i^*(\mathbf{x})\}$ is given by

$$q_1^*(\mathbf{x}) = \text{argmin}_j \{J(\mathbf{x} + j \cdot d_1) | 0 \le q_1 \le Q_1\}, \qquad (3.11)$$

$$q_i^*(\mathbf{x}) = \text{argmin}_j \{J(\mathbf{x} + j \cdot d_i) | 0 \le q_i \le \min\{x_{i-1}, Q_i\}\}, \quad \text{for } i > 1. \qquad (3.12)$$

In the remainder of this section, we consider a special case with $Q_i = 1$ and present some theoretical results that have been established in the literature. Following Definition 2.1, let $D := \{d_1, d_2, \ldots, d_n, d_{n+1}\}$ be the transition set and $D(\mathbf{x}) := \{d \in D \mid \mathbf{x} + d \in X\}$ be the feasible transition directions from state $\mathbf{x}$. The following Proposition 3.1 is the direct result from Veatch and Wein (1992) and Song and Sun (1998), and Proposition 3.2 can be obtained from Proposition 3.1.

Proposition 3.1. *Assume $Q_i = 1$ for any i. $J(\mathbf{x})$ is submodular w.r.t. $D(\mathbf{x})$. That is, $J(\mathbf{x} + d_l) + J(\mathbf{x} + d_j) \ge J(\mathbf{x}) - J(\mathbf{x} + d_l + d_j)$ for any $l, j \in \{1, 2, \ldots, n+1\}$, and $l \ne j$.*

Proposition 3.2. *Assume $Q_i = 1$ for any i. $J(\mathbf{x} + d_i) - J(\mathbf{x})$ is increasing in x_j for $j = i, i+1, \ldots, n$ and is decreasing in x_j for $j = 1, 2, \ldots, i-1$, where $i = 1, 2, \ldots, n$.*

Physically, together with (3.11) and (3.12), Proposition 3.2 states that if the optimal decision for entity i is to order/produce a unit of item at the state $\mathbf{x} = (x_1, x_2, \ldots, x_n)$, then so is it at the state $(x_1, \ldots, x_{j-1}, x_j + 1, x_{j+1}, \ldots, x_n)$, where $1 \le j < i$; if the optimal decision entity i is to stop ordering/producing a unit of item at the state $(x_1, x_2, \ldots, x_n)$, then so is it at the state $(x_1, \ldots, x_{j-1}, x_j + 1, x_{j+1}, \ldots, x_n)$, where $i \le j < n$. Intuitively, upstream entities tend to order/produce nothing when inventories of downstream entities increase.

Definition 3.1. Define the switching manifolds as follows:

- $S_1(\mathbf{x} \backslash x_1) := S_1(x_2, x_3, \ldots, x_n) := \min\{x_1 \mid J(\mathbf{x} + d_1) - J(\mathbf{x}) > 0\}$; if $J(\mathbf{x} + d_1) - J(\mathbf{x}) \le 0$ for any $x_1 \in Z^+$, then we define $S_1(\mathbf{x} \backslash x_1) = +\infty$.
- $S_i(\mathbf{x} \backslash x_i) := S_i(x_1, \ldots, x_{i-1}, x_{i+1}, \ldots, x_n) := \min\{x_i \mid J(\mathbf{x} + d_i) - J(\mathbf{x}) > 0, x_{i-1} > 0\}$ for $i = 2, 3, \ldots, n$.

From Song and Sun (1998), the switching manifolds defined above have the following monotonic properties in Proposition 3.3, which yields the explicit structure of the optimal control policy in Proposition 3.4.

Proposition 3.3. *The switching manifold $S_i(x \backslash x_i)$ is increasing in x_j ($j = 1, 2, \ldots$, $i - 1$) and decreasing in x_j ($j = i + 1, 2, \ldots, n$), where $i = 1, 2, \ldots, n$.*

Proposition 3.4. *The optimal control policy for the multistage serial supply chain with maximum order size one, $\{q_i^*(x), \text{ for } i = 1, 2, \ldots, n\}$, has the switching structure and can be determined by a set of control regions:*

$$q_i^*(\mathbf{x}) = \begin{cases} 1 & x \in R_i \\ 0 & x \notin R_i \end{cases}, \quad \text{for } i = 1, 2, \ldots, n; \tag{3.13}$$

where control region $R_i := \{x = (x_1, x_2, \ldots, x_{n+1}) \in X \mid x_i < S_i(x \backslash x_i)\}$ for $i = 1, 2, \ldots, n$.

It should be pointed out that the results in Propositions 3.1–3.4 also hold for the multistage serial production line with failure-prone machines at each stage (Song and Sun 1998).

3.7 Discussions and Notes

This chapter has extended the basic supply chain in Chap. 2 in several directions. Sections 3.2, 3.3, 3.4, and 3.5 are based on Song (2009) and Sect. 3.6 is an extension from Song and Sun (1998). In this section, we discuss a few relevant issues and note the literature related to optimal control of stochastic production/inventory systems in various contexts, including deterministic lead times and random demands, stochastic lead times and outstanding order, multiple replenishment channels and order information, and ordering capacity and storage capacity.

3.7.1 Deterministic Lead Times and Random Demands

The issues of optimal ordering and inventory policies in multi-echelon (or multi-stage) inventory systems with random demands and deterministic lead times have been studied for many years. The majority studies took the periodic-review scheme.

Clark and Scarf (1960) considered a serial multi-echelon system with deterministic lead times and random customer demands. In a finite-horizon setting, Clark and Scarf (1960) showed that the echelon base-stock policies are optimal for the uncapacitated situation (i.e., there is no upper limit on the order quantity in a single period). The echelon stock at a stage is defined as the inventory position of the subsystem consisting of the current stage and all its downstream stages. Here the inventory position refers to the sum of inventory-on-hand (nonnegative), inventory on order (nonnegative), and backorders (negative). Federgruen and Zipkin (1984) extended the multi-echelon results to a stationary infinite-time horizon setting. Chen and Zheng (1994) showed that the echelon stock policy (r, nQ) is also optimal in continuous-review systems with compound Poisson demand and in periodic-review systems with independent identically distributed (i.i.d.) demands. The (r, nQ) policy states that whenever the echelon stock drops below the reorder point r, an order of nQ units is placed or produced, where n is the minimum integer number required to bring the echelon stock to above r and Q is the base quantity. Chen (2000) generalized the classical multi-echelon model by introducing the batch-ordering constraint into each stage and established the optimality of the echelon stock (r, nQ) policy. Van Houtum et al. (2007) considered a multi-echelon serial system with fixed replenishment interval constraints. They proved that the echelon base-stock policy is optimal, and derived the newsvendor-type characteristics for the optimal base-stock levels. Rosling (1989) showed that a general assembly system could be converted into an equivalent serial system. Chao and Zhou (2009) generalized the work of Chen (2000) and Van Houtum et al. (2007) to derive the optimal ordering policy for the multi-echelon system with both batch-ordering and fixed replenishment intervals. All of the above papers assume i.i.d. demands and deterministic lead times. Muharremoglu and Tsitsiklis (2008) used an alternative approach based on item-customer decomposition to confirm the optimality of base-stock policies, in which order replenishment lead times are stochastic and non-crossing; both demands and lead times are modulated by an exogenous Markov process. A common assumption in the above studies is uncapacitated ordering decision, that is, there is no upper limit on the order quantity in a single period.

The research for optimal control of inventory systems with limited ordering capacity under periodic review is mostly constrained to one-echelon systems (Parker and Kapuscinski 2004). Federgruen and Zipkin (1986a, b) showed the optimality of modified base-stock policies for single-stage capacitated system with deterministic lead times and stationary demand for infinite-horizon average-cost and discounted-cost criteria, respectively. This work was extended to the case of periodic demand processes and Markov-modulated demand processes by Aviv and Federgruen (1997) and Kapuscinski and Tayur (1998), respectively. Tayur (1993) provided an algorithm to compute the base-stock level (critical number) and the associated cost for the average-cost criteria in the infinite-horizon inventory model with finite production capacity and stochastic demands.

Relatively few results have been obtained about the structure of the optimal policy for a serial system with finite capacities. Parker and Kapuscinski (2004) demonstrated that a modified echelon base-stock policy is optimal in a two-stage system when there is a smaller capacity at the downstream facility, and the optimal structure holds for both stationary and nonstationary stochastic customer demands. Finite-horizon and infinite-horizon results are included under discounted-cost and average-cost criteria. The results confirmed the effectiveness of the heuristics (Kanban policies and order-up-to-point policies) proposed by Veatch and Wein (1994). Janakiraman and Muckstadt (2009) studied a class of two-echelon serial systems with identical ordering/production capacities for both echelons. For the case where the lead time to the upstream echelon is one period, the optimality of state-dependent modified echelon base-stock policies is proved using a decomposition approach. For the case where the upstream lead time is two periods, they introduced a new class of policies called "two-tier base-stock policies" and proved their optimality. Janakiraman and Muckstadt (2009) argued that a generalization of two-tier base-stock policies, termed as the multi-tier base-stock policy, is optimal for multi-echelon serial systems with identical capacities at all stages and arbitrary lead times. Huh et al. (2010) studied a periodically reviewed multi-echelon serial inventory system, which is identical to the one in Clark and Scarf (1960) except that there is a capacity limit on the order quantity by each stage. Under the echelon base-stock policy, they established the useful relationship between the multi-echelon capacitated system and several suitably defined single-stage capacitated systems.

However, the optimal policy is not known for capacitated supply chains with general (e.g., tree) network structures (Simchi-Levi and Zhao 2005). In fact, it is well known that a base-stock policy is not optimal for supply chains with a tree structure (Zipkin 2000).

3.7.2 Stochastic Lead Times and Outstanding Orders

The study of inventory control with stochastic lead times and random demands dated back to early 1960s. The uncertainty in lead times may cause order-crossing issues. Namely, an order placed in a later period may arrive before the one that was ordered in an earlier period. The relevant literature on stochastic lead times and random demands may be classified into two groups. The first group mainly focuses on periodic decision making for inventory replenishment without assuming independent identical distribution (i.i.d.) of lead times. The second group assumes a specific probability distribution of lead times.

The majority of the first group literature assumes that the arrivals of outstanding orders do not cross in time. Hadley and Whitin (1963) considered a single-stage problem and identified the optimal inventory control policies for some special cases with restrictive assumptions, for example, orders do not cross each other and they are independent. Kaplan (1970) provided a simple model of stochastic lead times that prevents order crossing, while keeping the probability that an outstanding order arrives in the next time period independent of the current status of other outstanding orders. He showed that the deterministic lead-time results carry over to the stochastic lead-time cases for a periodic-review system, for example, the sequential multidimensional minimization problem can be reduced to a sequence of one-dimensional minimizations, in which the minimization is done over a scalar state variable representing the sum of inventory-on-hand plus all outstanding orders. Ehrhardt (1984) extended Kaplan's results and provided conditions for the optimality of myopic base-stock policies and for the optimality of (s, S) policies for both finite and infinite planning horizons. Zipkin (1986) investigated stochastic lead times in continuous-time single-stage inventory models. Song and Zipkin (1996) studied a single-stage system with Markov-modulated lead times. They characterized an optimal inventory control policy that does not depend on order progress information based on the assumption of non-order crossing.

Svoronos and Zipkin (1991) evaluated one-for-one replenishment policies in the serial system setting. They focused on computing the steady-state behaviors of the system. Simchi-Levi and Zhao (2005) considered the safety stock positioning problem in multistage supply chains with tree network structure with stochastic demands and lead times, in which a continuous-time base-stock policy is used in each stage to control its inventory. Muharremoglu and Yang (2010) applied the base-stock policy to single and multistage inventory systems with stochastic lead times, which include Kaplan's lead times with no order crossing and i.i.d. lead times with order crossing, and provided a method to determine base-stock levels and to compute the cost of a given base-stock policy.

Muharremoglu and Tsitsiklis (2008) showed the optimality of state-dependent echelon base-stock policies in uncapacitated serial inventory systems with Markov-modulated demand and lead times in the absence of order crossing. They stated that there has been little optimality result for serial systems under any type of stochastic lead times. In addition, the optimal inventory control policy when order crossing is allowed has not been well researched (Gaukler et al. 2008).

In the second group, a common assumption of lead-time distributions is exponential or Erlang distribution. Berman and Kim (2001) considered the optimal dynamic ordering problem in a supply chain with Erlang-distributed lead times, exponential service times, and Poisson customer arrival process. With the assumption of at most one outstanding order, they showed that the optimal ordering policy has a monotonic threshold structure. Berman and Kim (2004) extended the above model to including revenue generated upon the service considering both exponential and Erlang lead times. Kim (2005) further extended the above model to a queueing system with finite waiting room for customers and characterized the optimal replenishment policy as a monotonic threshold function of reorder point under the discounted-cost criterion. Berman and Kim (2004) and Kim (2005) also discussed the multiple outstanding orders case with some conjectured results. He et al. (2002a) considered a two-echelon make-to-order system with Poisson demand, exponential processing times, and zero lead times for ordering raw materials. They explored the structure of the optimal replenishment policy. He et al. (2002b) assumed that the raw material replenishment process allows multiple outstanding orders of any sizes with Erlang lead times and demonstrated that the classical (r, Q) policy may not perform well if information about the number of demands is partially or fully available. In this group, the decision making is on continuous-time basis and depends on the system state. The optimality of monotonic control policies often relies on the submodularity of the value function (e.g., Glasserman and Yao 1994). However, when there is a capacity constraint in the system, the analysis may be complex and few results have been reported (Berman and Kim 2001). Multiple outstanding orders with order crossover can be modeled, but the structure of the optimal policies may be difficult to establish. Yang (2005) studied a periodic-review production control problem where both the raw material supply and product demand are exogenous and random, and the raw material can be stored, purchased, and sold to an outside market. He was able to establish the partial characterizations of the optimal policies under both strict convex and linear raw material purchasing/selling costs.

There have been some studies on the order crossover issue. For example, Robinson et al. (2001) argued that order crossover is becoming more prevalent and analyzed the dangers of ignoring it. Bradley and Robinson (2005) pointed out when order crossovers are allowed, the optimal reorder levels should be set with regard to the inventory shortfall distribution rather than the lead-time demand distribution.

3.7.3 Multiple Replenishment Channels and Order Information

In practice, it is common that an inventory system may have more than one replenishment channel or supply modes. For example, a company may have a normal supply channel but also a supplementary channel for placing emergency orders. Having a separate, faster, and presumably more costly replenishment channel can offer opportunity to buffer against the uncertainties in customer demands or in the normal supply channel; meanwhile it also provides more flexibility to take an action in response to order progress information. However, multiple replenishment channels add another dimension to control decisions such as when the supplementary channel should be used, what the order size should be, and how to make use of the outstanding order's progress information.

Moinzadeh and Nahmias (1988) proposed an approximate model of an inventory control system in which there exist two replenishment modes with different but deterministic lead times. They stated that an optimal replenishment policy for this setting is extremely complex and proposed a reasonable extension to the traditional (r, Q) policy. Moinzadeh and Schmidt (1991) extended the above work and considered a one-for-one ordering policy for placing emergency orders that uses information about the age of outstanding orders. The results are further extended to a multi-echelon inventory system where all stages have the option to replenish their inventory through either a normal or a more expensive emergency supply channel (Moinzadeh and Aggarwal 1997). Chiang (2001) considered the optimal policies for a periodic-review inventory system in which emergency orders can be placed at the start of each period, while regular orders are placed at the start of an order cycle, which is equal to a number of periods. Bylka (2005) studied the structure of optimal ordering policy for a periodic-review capacitated inventory system with limited backlogging, in which emergency orders can be placed to satisfy the demands as soon as a shortage occurs.

The benefits and cost savings of two replenishment channels compared to a system without emergency replenishment option have been discussed in Alfredsson and Verrijdt (1999) and Tagaras and Vlachos (2001).

Veeraraghavan and Scheller-Wolf (2008) introduced a dual-index policy in a capacitated, periodically reviewed, single-stage inventory system with two replenishment channels and demonstrated its closeness to the complex globally optimal state-dependent policy. Sheopuri et al. (2010) generalized the above dual-index policy by taking the form of order-up-to structure for the emergency mode or the regular mode. Song and Zipkin (2009) considered an inventory system with multiple supply sources, Poisson demand, and stochastic lead times. They focused on performance evaluation for a family of state-dependent ordering policies. Gaukler et al. (2008) investigated a retailer facing a stochastic lead time that allows for releasing emergency orders in response to the order progress information and showed that the structure of the optimal ordering policy is given by a sequence of threshold values depending on the order progress information.

3.7.4 Ordering Capacity and Storage Capacity

Ordering capacity refers to the upper limit imposed on the order quantity when making replenishment decisions. Another closely related concept is the storage or warehouse capacity for storing materials. When the order sizes are constrained by bounds, the problem is called capacitated. When the inventories are constrained by bounds, it is called bounded/limited inventory problem (Gutierrez et al. 2003).

Clearly, storage capacity often implies the ordering capacity since the ordering size should not exceed the maximum storage capacity minus the current inventory level. On the other hand, ordering capacity constraint does not guarantee the satisfaction of storage capacity constraints. The reason is that decision makers may ignore the limit of own warehouse, which results in the cumulative inventories of materials exceeding the limit and incurs expensive cost to rent spaces. In Sect. 3.7.1, we have reviewed the relevant literature on the optimal policy for serial supply chains with random demands and finite ordering capacities. The following literature addresses the optimal production/ordering policies for inventory systems with storage/warehouse capacity.

In deterministic situations, Love (1973) studied the optimal schedule for a multi-period single-facility inventory model with constant storage capacity allowing backlogging. Gutierrez et al. (2003) considered the similar problem without allowing stock-out. They established new characterization of the optimal policies and provided faster computational algorithms. The above results were further extended to the backlogging case in Gutierrez et al. (2007). Liu and Tu (2008) considered the production planning in a situation where the production quantity is limited by inventory capacity rather than production capacity. Some properties of the optimal solution are established and utilized to develop a polynomial algorithm for the problem. Lee and Wang (2008) formulated an integrated inventory control model for a single-manufacturer and single-buyer supply chain, in which the joint lot size decisions of production batch and replenishment lot subject to buyer's warehouse capacity constraint should be made.

In stochastic situations, Mohebbi (2006, 2008) considered the production control for a facility with limited storage capacity and compound Poisson demand. Under the parameterized control policies, the limiting distribution of the inventory level is derived, which facilitates the evaluation of a variety of performance measures. Minner and Silver (2007) addressed the replenishment policies for multiple products with compound Poisson demand where the aggregated level of inventory at any time is limited by the warehouse capacity. Heuristics were presented to set reorder points and reorder quantities.

Due to a high level of complexity, analytical treatment of production/inventory planning and control operations under the influence of a multistate random environment such as stochastic demand, production, and supply remains largely unexplored (Mohebbi 2008).

References

Alfredsson, P., Verrijdt, J.: Modeling emergency supply flexibility in a two-echelon inventory system. Manage. Sci. **45**(10), 1416–1431 (1999)

Aviv, Y., Federgruen, A.: Stochastic inventory models with limited production capacity and periodically varying parameters. Probab. Eng. Inf. Sci. **11**(1), 107–135 (1997)

Berman, O., Kim, E.: Dynamic order replenishment policy in internet-based supply chains. Math. Method Oper. Res. **53**, 371–390 (2001)

Berman, O., Kim, E.: Dynamic inventory strategies for profit maximization in a service facility with stochastic service, demand and lead time. Math. Method Oper. Res. **60**, 497–521 (2004)

Bertsekas, D.P.: Dynamic Programming: Deterministic and Stochastic Models. Prentice-Hall, Englewood Cliffs (1987)

Bradley, J.R., Robinson, L.W.: Improved base-stock approximations for independent stochastic lead times with order crossover. Manuf. Serv. Oper. Manage. **7**(4), 319–329 (2005)

Bylka, S.: Turnpike policies for periodic review inventory model with emergency orders. Int. J. Prod. Econ. **93–94**(8), 357–373 (2005)

Chao, X., Zhou, S.X.: Optimal policy for a multiechelon inventory system with batch ordering and fixed replenishment intervals. Oper. Res. **57**(2), 377–390 (2009)

Chen, F.: Optimal policies for multi-echelon inventory problems with batch ordering. Oper. Res. **48**, 376–389 (2000)

Chen, F., Zheng, Y.S.: Evaluating echelon stock (R, nQ) policies in serial production/inventory systems with stochastic demand. Manage. Sci. **40**, 1262–1275 (1994)

Chiang, C.: A note on optimal policies for a periodic inventory system with emergency orders. Comput. Oper. Res. **28**, 93–103 (2001)

Clark, A.J., Scarf, H.E.: Optimal policies for a multi-echelon inventory problem. Manage. Sci. **6**, 475–490 (1960)

Ehrhardt, R.: (s, S) policies for a dynamic inventory model with stochastic lead times. Oper. Res. **32**(1), 121–132 (1984)

Federgruen, A., Zipkin, P.: Computational issues in an infinite-horizon, multiechelon inventory model. Oper. Res. **32**(4), 818–836 (1984)

Federgruen, A., Zipkin, P.: An inventory model with limited production capacity and uncertain demands I. The average-cost criterion. Math. Oper. Res. **11**(2), 193–207 (1986a)

Federgruen, A., Zipkin, P.: An inventory model with limited production capacity and uncertain demands II. The discounted-cost criterion. Math. Oper. Res. **11**(2), 208–215 (1986b)

Gaukler, G., Ozer, O., Hausman, W.H.: Order progress information: improved dynamic emergency ordering policies. Prod. Oper. Manage. **17**(6), 599–613 (2008)

Glasserman, P., Yao, D.: Monotone Structure in Discrete-Event Systems. Wiley, New York (1994)

Gutierrez, J., Sedeno-Noda, A., Colebrook, M., Sicilia, J.: A new characterization for the dynamic lot size problem with bounded inventory. Comput. Oper. Res. **30**(3), 383–395 (2003)

Gutierrez, J., Sedeno-Noda, A., Colebrook, M., Sicilia, J.: A polynomial algorithm for the production/ordering planning problem with limited storage. Comput. Oper. Res. **34**(4), 934–937 (2007)

Hadley, G., Whitin, T.M.: Analysis of Inventory Systems. Prentice-Hall, Englewood Cliffs (1963)

He, Q.M., Jewkes, E.M., Buzacott, J.: Optimal and near-optimal inventory control policies for a make-to-order inventory-production system. Eur. J. Oper. Res. **141**, 113–132 (2002a)

He, Q.M., Jewkes, E.M., Buzacott, J.: The value of information used in inventory control of a make-to-order inventory-production system. IIE Trans. **34**, 999–1013 (2002b)

Huh, W.T., Janakiraman, G., Nagarajan, M.: Capacitated serial inventory systems: sample path and stability properties under base-stock policies. Oper. Res. **58**(4), 1017–1022 (2010)

Janakiraman, G., Muckstadt, J.A.: A decomposition approach for a class of capacitated serial systems. Oper. Res. **57**(6), 1384–1393 (2009)

Kaplan, R.S.: A dynamic inventory model with stochastic lead times. Manage. Sci. **16**(7), 491–507 (1970)

Kapuscinski, R., Tayur, S.: A capacitated production-inventory model with periodic demand. Oper. Res. **46**(6), 899–911 (1998)

Kim, E.: Optimal inventory replenishment policy for a queueing system with finite waiting room capacity. Eur. J. Oper. Res. **161**, 256–274 (2005)

Lee, W., Wang, S.P.: Managing level of consigned inventory with buyer's warehouse capacity constraint. Prod. Plan. Control **19**(7), 677–685 (2008)

Liu, X., Tu, Y.: Production planning with limited inventory capacity and allowed stockout. Int. J. Prod. Econ. **111**(1), 180–191 (2008)

Love, S.F.: Bounded production and inventory models with piecewise concave costs. Manage. Sci. **20**(3), 313–318 (1973)

Minner, S., Silver, E.A.: Replenishment policies for multiple products with compound-Poisson demand that share a common warehouse. Int. J. Prod. Econ. **108**(1–2), 388–398 (2007)

Mohebbi, E.: A production-inventory model with randomly changing environmental conditions. Eur. J. Oper. Res. **174**, 539–552 (2006)

Mohebbi, E.: A note on a production control model for a facility with limited storage capacity in a random environment. Eur. J. Oper. Res. **190**(2), 562–570 (2008)

Moinzadeh, K., Aggarwal, P.: An information based multiechelon inventory system with emergency orders. Oper. Res. **45**(5), 694–701 (1997)

Moinzadeh, K., Nahmias, S.: A continuous review model for an inventory model with two supply modes. Manage. Sci. **34**(6), 761–773 (1988)

Moinzadeh, K., Schmidt, C.: An (S − 1, S) inventory system with emergency orders. Oper. Res. **39**(2), 308–321 (1991)

Muharremoglu, A., Tsitsiklis, J.N.: A single-unit decomposition approach to multiechelon inventory systems. Oper. Res. **56**, 1089–1103 (2008)

Muharremoglu, A., Yang, N.: Inventory management with an exogenous supply process. Oper. Res. **58**, 111–129 (2010)

Parker, R.P., Kapuscinski, R.: Optimal policies for a capacitated two-echelon inventory system. Oper. Res. **52**(5), 739–755 (2004)

Robinson, L.W., Bradley, J.R., Thomas, L.J.: Consequences of order crossover under order-up-to inventory policies. Manuf. Serv. Oper. Manage. **3**(3), 175–188 (2001)

Rosling, K.: Optimal inventory policies for assembly systems under random demands. Oper. Res. **37**, 565–579 (1989)

Sheopuri, A., Janakiraman, G., Seshadri, S.: New policies for the stochastic inventory control problem with two supply sources. Oper. Res. **58**(3), 734–745 (2010)

Simchi-Levi, D., Zhao, Y.: Safety stock positioning in supply chains with stochastic lead times. Manuf. Serv. Oper. Manage. **7**(4), 295–318 (2005)

Song, D.P.: Optimal integrated ordering and production policy in a supply chain with stochastic lead-time, processing-time and demand. IEEE Trans. Autom. Control **54**(9), 2027–2041 (2009)

Song, D.P., Sun, Y.X.: Optimal service control of a serial production line with unreliable workstations and random demand. Automatica **34**(9), 1047–1060 (1998)

Song, J.S., Zipkin, P.: Inventory control with information about supply conditions. Manage. Sci. **42**(10), 1409–1419 (1996)

Song, J.S., Zipkin, P.: Inventories with multiple supply sources and networks of queues with overflow bypasses. Manage. Sci. **55**(3), 362–372 (2009)

Svoronos, A., Zipkin, P.: Evaluation of one-for-one replenishment policies for multiechelon inventory systems. Manage. Sci. **37**(1), 68–83 (1991)

Tagaras, G., Vlachos, D.: A periodic review inventory system with emergency replenishments. Manage. Sci. **47**(3), 415–429 (2001)

Tayur, S.: Computing the optimal policy for capacitated inventory models. Commun. Stat. Stoch. Models **9**(4), 585–598 (1993)

Van Houtum, G.J., Scheller-Wolf, A., Yi, J.: Optimal control of serial inventory systems with fixed replenishment intervals. Oper. Res. **55**, 674–687 (2007)

Veatch, M., Wein, L.: Monotone control of queueing networks. Queueing Syst. **12**, 391–408 (1992)

Veatch, M., Wein, L.: Optimal-control of a 2-station tandem production inventory system. Oper. Res. **42**(2), 337–350 (1994)

Veeraraghavan, S., Scheller-Wolf, A.: Now or later: a simple policy for effective dual sourcing in capacitated systems. Oper. Res. **56**(4), 850–864 (2008)

Yang, J.: Production control in the face of storable raw material, random supply, and an outside market. Oper. Res. **52**(2), 293–311 (2005)

Zipkin, P.: Stochastic lead times in continuous-time inventory models. Nav. Res. Logist. **33**, 763–774 (1986)

Zipkin, P.: Foundations of Inventory Management. McGraw-Hill, Boston (2000)

Chapter 4
Optimal Control of Supply Chain Systems with Backordering Decisions

4.1 Introduction

In the literature, there are two common ways to handle unmet customer demands, that is, fully backlogged and lost sales. A few of them assumed limited backlog, which is regarded as an exogenously given constraint. There is a lack of research of treating the backordering as a manufacturer's decision that could be made together with the other decisions such as raw material ordering and production rate control in stochastic supply chain systems, particularly in failure-prone manufacturing supply chains (Song 2006).

In reality, it is reasonable for a supplier or manufacturer to make an integrated decision of ordering, production, and backordering to cope with the difficulties caused by limited resource capacity and various uncertainties such as random demand arrivals, unreliable machines, stochastic processing times, and uncertainty in material supply. It becomes even more important when the demand arrival rate is higher than the average production rate, in which the system could turn out to be unstable if the unmet demands are completely backlogged.

This chapter tackles two problems. The first is the joint optimal control problem for ordering, production, and backordering decisions in the basic supply chain system in Chap. 2. We will present the mathematical formulation and explore its optimal control structure. This is an extension of the problem in Chap. 2 to the situation with additional backordering decisions.

The second is the joint optimal control problem for production and backordering in a failure-prone manufacturing supply chain. Apart from the system being subject to random customer demand and stochastic processing times, it is also subject to the machine breakdown, which may force the manufacturer to stop production. Such disruption will obviously affect the supply chain's production and backordering policy. Failure-prone manufacturing systems have attracted much attention in the last three decades due to its severe impact on the entire supply chains. However, little literature has considered the backordering as a manufacturer's decision integrated with the production rate control. We will focus a simpler supply chain with a

D.-P. Song, *Optimal Control and Optimization of Stochastic Supply Chain Systems*, Advances in Industrial Control, DOI 10.1007/978-1-4471-4724-4_4, © Springer-Verlag London 2013

manufacturer and its customers so that we are able to establish the explicit structure of the optimal control structure. The decisions include determining how much the production rate should take when the machine is up and whether a demand should be backordered or rejected in order to minimize the costs associated with inventory, backlog, and rejecting customer demands.

4.2 Optimal Control in a Supply Chain with Backordering Decisions

Consider a supply chain that consists of three entities (supplier, manufacturer, and customer) and two warehouses (a raw material (RM) warehouse and a finished goods (FG) warehouse) as shown in Fig. 4.1. A customer demand is satisfied immediately if there are finished goods stored as inventory. The system is subject to Assumptions 2.1–2.5. The only difference is that customer demands can be backordered or rejected. The rejected demands are lost and incur a lost-sale cost. There are three types of decisions to make: the ordering decision $q(t)$ to the supplier, the production rate $u(t)$, and the backordering decision. The backordering decision is denoted by $a(t)$, which takes 1 if the arriving demand is accepted and takes 0 otherwise.

Define an admissible control set $\Omega = \{(u(t), q(t), a(t)) \mid u(t) \in \{0, r\}, \text{ if } x_1(t) > 0$ and $x_2(t) < N$, $u(t) = 0$ if $x_1(t) \leq 0$ or $x_2(t) \geq N$; $q(t) \in [0, M - x_1(t)]$; $a(t) \in \{0, 1\}$; $t \geq 0\}$. The system state space can be described by $X = \{\mathbf{x}:= (x_1, x_2) \mid x_1 \in [0, M]$ and $x_2 \in (-\infty, N)\}$, where M and N are the warehouse capacity for raw materials and finished goods, respectively.

The problem is to find the optimal integrated policy $(u(t), q(t), a(t)) \in \Omega$ from 0 to infinity to minimize the infinite horizon expected discounted cost (depending on the initial state)

$$J(\mathbf{x}_0) = \min_{u,q,a} E \left[\int_0^\infty e^{-\beta t} h(\mathbf{x}(t), u(t), q(t), a(t)) \, dt \mid \mathbf{x}(0) = \mathbf{x}_0 \right], \qquad (4.1)$$

where $0 < \beta < 1$ is a discounted factor and $h(.)$ is a cost function to penalize inventory, backlog, and rejecting demands. For example, a commonly used form of $h(.)$ is

$$h(\mathbf{x}, u, q, a) = g(\mathbf{x}) + c_r(1 - a), \qquad (4.2)$$

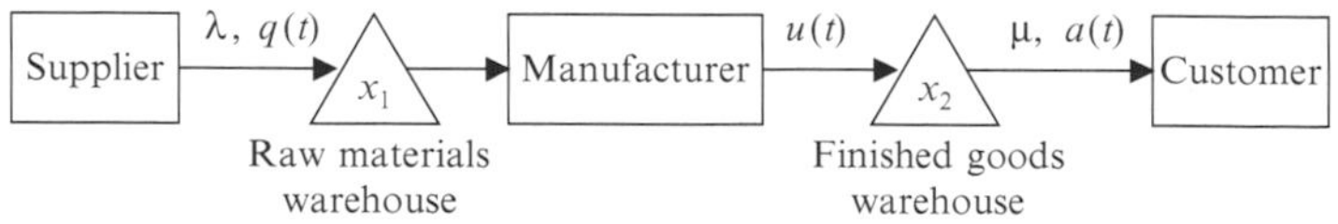

Fig. 4.1 A supply chain with backordering decisions

Fig. 4.2 System state
transition in a supply chain
with backordering decisions

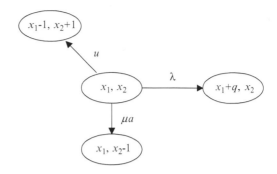

$$g(\mathbf{x}) = c_1 x_1 + c_2{}^+ x_2{}^+ + c_2{}^- x_2{}^-, \tag{4.3}$$

where c_1, $c_2{}^+$, and $c_2{}^-$ are the unit costs of raw material inventory, finished goods
inventory, and backordered demands, respectively; c_r represents the penalty of
rejecting a customer demand; and $x_2{}^+ := \max(x_2, 0)$ and $x_2{}^- := \max(-x_2, 0)$. The
state transition map at a state (x_1, x_2) under the control (u, q, a) is depicted in Fig. 4.2.

Let $v = \mu + r + \lambda$ be the uniform transition rate. Under a stationary control
policy $(u, q, a) \in \Omega$, the one-step transition probability functions from (x_1, x_2) are
given by

$$\text{Prob}\{(x_1 + q, x_2) | (x_1, x_2), (u, q, a)\} = \lambda \cdot I\{q > 0\}/v; \tag{4.4}$$

$$\text{Prob}\{(x_1 - 1, x_2 + 1) | (x_1, x_2), (u, q, a)\} = r \cdot I\{u > 0\}/v; \tag{4.5}$$

$$\text{Prob}\{(x_1, x_2 - 1) | (x_1, x_2), (u, q, a)\} = \mu \cdot I\{a > 0\}/v; \tag{4.6}$$

$$\text{Prob}\{(x_1, x_2) | (x_1, x_2), (u, q, a)\} = \left[v - \lambda \cdot I\{q > 0\} - r \cdot I\{u > 0\}\right.$$
$$\left. - \mu \cdot I\{a > 0\}\right]/v. \tag{4.7}$$

From the uniformization technique and the stochastic dynamic programming
approach, the Bellman optimality equation is given as follows:

$$J(x_1, x_2) = \frac{1}{\beta + v} \left[g(x_1, x_2) + r \cdot J(x_1, x_2) \cdot I\{x_1 = 0 \text{ or } x_2 = N\}\right.$$
$$+ r \cdot \min\{J(x_1 - 1, x_2 + 1), J(x_1, x_2)\} \cdot I\{x_1 > 0 \text{ and } x_2 < N\}$$
$$+ \lambda \cdot \min\{J(x_1 + q, x_2) | q \in [0, M - x_1]\}$$
$$\left. + \min\{\mu J(x_1, x_2 - 1), c_r + \mu J(x_1, x_2)\}\right]. \tag{4.8}$$

From Eq. (4.8), we have the following result.

Proposition 4.1. *The optimal integrated ordering, production, and backordering policy $(u^*(x_1, x_2), q^*(x_1, x_2), a^*(x_1, x_2))$ for the supply chain system in Fig. 4.1 is given by*

$$u^* (x_1, x_2) = \begin{cases} r & J(x_1 - 1, x_2 + 1) \leq J(x_1, x_2), & \text{if } x_1 > 0, \ x_2 < N \\ 0 & & \text{otherwise} \end{cases}, \quad (4.9)$$

$$q^* (x_1, x_2) = j, \ \text{if } J(x_1 + j, x_2) = \min \{J(x_1, x_2), J(x_1 + 1, x_2), \ldots, J(M, x_2)\}, \quad (4.10)$$

$$a^* (x_1, x_2) = \begin{cases} 1 & \mu J(x_1, x_2 - 1) \leq c_r + \mu J(x_1, x_2) \\ 0 & \text{otherwise} \end{cases}. \quad (4.11)$$

The optimal decisions for ordering and production in (4.9) and (4.10) have the same format as (2.12) and (2.13). The difference is the backordering decision in (4.11).

The optimal discounted cost function can be approximated numerically using the value iteration algorithm. The convergence of the k-stage policy and cost function to the infinite-horizon optimal policy and cost follows from the fact that only finitely many controls are considered at each state.

The following result provides the condition on when the completely backordering decision, that is, $a^*(x_1, x_2) \equiv 1$, is optimal.

Proposition 4.2. *The completely backordering policy is optimal if and only if $c_r \geq \mu c_2^- / \beta$.*

Proof. The necessity of the condition: If the completely backordering policy is adopted, the system becomes the same as the one in Chap. 2. From Proposition 2.6 and Lemma 2.1, we know that $J(i, x_2 + 1) - J(i, x_2)$ is decreasing and converging to $-c_2^- / \beta$, as x_2 tends to $-\infty$ for any $i \in [0, M]$. The optimality of the completely backordering policy implies that $J(i, x_2 + 1) - J(i, x_2) \geq -c_r / \mu$. It follows $-c_2^- / \beta \geq -c_r / \mu$. That is, $c_r \geq \mu c_2^- / \beta$.

The sufficiency of the condition: This can be proved by the induction method on the k-stage cost function in the value iteration procedure. Note that the optimality of the completely backordering policy is equivalent to $J(x_1, x_2 - 1) - J(x_1, x_2) \leq c_r / \mu$ for any (x_2, x_2) by (4.11). The assertion is obviously true when $k = 0$ and $J_0(x_1, x_2) \equiv 0$. Suppose $J_k(x_1, x_2 - 1) - J_k(x_1, x_2) \leq c_r / \mu$. We want to show $J_{k+1}(x_1, x_2 - 1) - J_{k+1}(x_1, x_2) \leq c_r / \mu$. From the value iteration algorithm, we have

$$(\beta + v) (J_{k+1}(x_1, x_2 - 1) - J_{k+1}(x_1, x_2)) = [g_k(x_1, x_2 - 1) - g_k(x_1, x_2)]$$
$$+ r \cdot [J_k(x_1, x_2 - 1) \cdot I\{x_1 = 0 \text{ or } x_2 - 1 = N\}$$
$$+ \min \{J_k(x_1 - 1, x_2), J_k(x_1, x_2 - 1)\} \cdot I\{x_1 > 0 \text{ and } x_2 - 1 < N\}$$
$$- J_k(x_1, x_2) \cdot I\{x_1 = 0 \text{ or } x_2 = N\}$$

$$- \min \{ J_k (x_1 - 1, x_2 + 1) , J_k (x_1, x_2) \} \cdot I\{x_1 > 0 \text{ and } x_2 < N\}]$$
$$+ \lambda \cdot \left[\min \{ J_k (x_1 + q, x_2 - 1) \, | q \in [0, M - x_1] \} \right.$$
$$- \min \{ J_k (x_1 + q, x_2) \, | q \in [0, M - x_1] \} \right]$$
$$+ [\min \{ \mu J_k (x_1, x_2 - 2) , c_r + \mu J_k (x_1, x_2 - 1) \}$$
$$- \min \{ \mu J_k (x_1, x_2 - 1) , c_r + \mu J_k (x_1, x_2) \}] . \tag{4.12}$$

From (4.3), we know that the first term on the RHS of (4.12) is not greater than c_2^-, that is,

$$[g_k (x_1, x_2 - 1) - g_k (x_1, x_2)] \le c_2^- . \tag{4.13}$$

From the induction hypothesis $J_k (x_1, x_2 - 1) - J_k (x_1, x_2) \le c_r / \mu$, we have:

- The second term on the RHS of (4.12) is not greater than $r \cdot c_r / \mu$.
- The third term on the RHS of (4.12) is not greater than $\lambda \cdot c_r / \mu$.
- The last term on the RHS of (4.12) is not greater than c_r.

From (4.12), (4.13), and the above three bulletin points, it follows

$$(\beta + v) (J_{k+1} (x_1, x_2 - 1) - J_{k+1} (x_1, x_2)) \le c_2^- + r \cdot c_r / \mu + \lambda \cdot c_r / \mu + c_r$$
$$\le \beta c_r / \mu + r \cdot c_r / \mu + \lambda \cdot c_r / \mu + c_r = (\beta + v) c_r / \mu . \tag{4.14}$$

Therefore, the assertion holds for $k + 1$. This completes the proof.

Physically, Proposition 4.2 can be interpreted as follows. Suppose there are a very large number of backlogged demands. Backordering one more demand incurs an additional cost c_2^- over the time until it is satisfied. In the limit, the expected additional cost is $c_2^- \int_0^\infty e^{-\beta t} dt = c_2^- / \beta$. As long as this value is not greater than the one-off cost incurred by rejecting a customer demand (i.e., c_r / μ), the completely backordering policy is optimal.

Proposition 4.3. $J(x_1 - 1, x_2 + 1) - J(x_1, x_2) \le 0$ for any $x_1 > 0$ and $x_2 < 0$.

Proof. This can be proved by the induction method on the k-stage cost function in the value iteration procedure. Suppose the assertion is true for k and we want to show it also holds for $k + 1$. Consider (by using the notation defined in Sect. 2.3)

$$J_{k+1} (\mathbf{x} + d_2) - J_{k+1} (\mathbf{x}) = [g (\mathbf{x} + d_2) - g (\mathbf{x})]$$
$$+ r \cdot [J_k (\mathbf{x} + d_2) \cdot I\{x_1 = 1 \text{ or } x_2 + 1 = N\}$$
$$+ \min \{ J_k (\mathbf{x} + 2d_2) , J_k (\mathbf{x} + d_2) \} \cdot I\{x_1 > 1 \text{ and } x_2 + 1 < N\}$$
$$- J_k (\mathbf{x}) \cdot I\{x_1 = 0 \text{ or } x_2 = N\}$$
$$- \min \{ J_k (\mathbf{x} + d_2) , J_k (\mathbf{x}) \} \cdot I\{x_1 > 0 \text{ and } x_2 < N\}]$$

$$+ \lambda \cdot [\min \{J_k (\mathbf{x} + d_2 + q d_1) \, | q \in [-1, M - x_1]\}$$
$$- \min \{J_k (\mathbf{x} + q d_1) \, | q \in [0, M - x_1]\}]$$
$$+ [\min \{\mu J_k (\mathbf{x} + d_2 + d_3), c_r + \mu J_k (\mathbf{x} + d_2)\}$$
$$- \min \{\mu J_k (\mathbf{x} + d_3), c_r + \mu J_k (\mathbf{x})\}] . \tag{4.15}$$

On the RHS of (4.15), the first term in square bracket [] is negative for $x_1 > 0$ and $x_2 < 0$ due to the format of $g(\mathbf{x})$. If $x_1 = 1$, the second term is equal to 0 due to $J_k(\mathbf{x} + d_2) - J_k(\mathbf{x}) \leq 0$. If $x_1 > 1$, the second term is simplified as $\min\{J_k(\mathbf{x} + 2d_2),$ $J_k(\mathbf{x} + d_2)\} - J_k(\mathbf{x} + d_2)$ by the induction hypothesis. This is obviously nonpositive. The third term is nonpositive because $J_k(\mathbf{x} + d_2 + id_1) - J_k(\mathbf{x} + id_1) \leq 0$ for $i = 0, 1, \ldots, M$ by the induction hypothesis. The last term is nonpositive due to $J_k(\mathbf{x} + d_2 + d_3) - J_k(\mathbf{x} + d_3) \leq 0$ and $J_k(\mathbf{x} + d_2) - J_k(\mathbf{x}) \leq 0$ from the induction hypothesis. Since every term on the RHS of (4.15) is nonnegative when $x_1 > 0$ and $x_2 < 0$, this completes the induction proof.

With the similar arguments to that of Proposition 4.3 by the induction method, we have the following result.

Proposition 4.4. $\mu J_k(x_1, x_2 - 1) \leq c_r + \mu J_k(x_1, x_2)$ for any $x_2 > 0$.

Physically, Proposition 4.3 implies that we should always produce finished goods if the raw materials are available and there are backlogged demands. Proposition 4.4 means that customer demands should always be accepted if there are inventories of finished goods available. These two results are intuitively true.

From numerical experiments, it appears that the optimal integrated ordering, production, and backordering policy can be characterized by a set of switching curves that are similar to those in Chap. 2. More specifically, there exist three switching curves $S_o(x_2)$, $S_p(x_1)$, and $S_b(x_1)$. The optimal policy states that:

1. We should place an order if the system state (x_1, x_2) locates below the curve $S_o(x_2)$ in the $x_1 - x_2$ plane (otherwise, place no order) and $S_o(x_2)$ is monotonic increasing in x_2. The placed order size follows the order-up-point rule.
2. We should produce products at full speed if the system state (x_1, x_2) locates below the curve $S_p(x_1)$ in the $x_1 - x_2$ plane (otherwise, produce nothing) and $S_p(x_1)$ is monotonic increasing in x_1.
3. We should accept customer orders if the system state (x_1, x_2) locates below the curve $S_b(x_1)$ in the $x_1 - x_2$ plane (otherwise, rejecting the customer orders) and the switching curve $S_b(x_1)$ is monotonic decreasing in x_1.

It appears that the main difference between the above results and those in Chap. 2 is that we now have an additional switching curve to control whether a customer order should be accepted or rejected. However, it should be pointed that although the switching curves for ordering and production have the same structure as that in Chap. 2, they are not necessarily the same due to the impact of additional backordering decisions. The optimal integrated ordering, production, and backordering policy can be illustrated in Fig. 4.3.

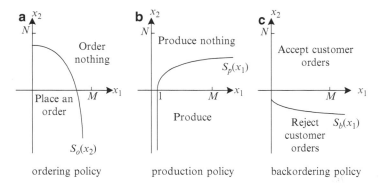

Fig. 4.3 Illustration of the characteristics of the optimal policy in a serial supply chain with backordering decisions

Example 4.1. Consider the base case in Chap. 3, that is, the lead-time rate $\lambda = 1.0$, the maximum production rate $r = 1.0$, demand rate $\mu = 0.7$, the raw material inventory unit cost $c_1 = 1$, the finished goods inventory unit cost $c_2^+ = 2$, and the backordering cost $c_2^- = 10$. The warehouse capacities are set as $M = N = 6$ and the discount factor $\beta = 0.5$. The system state space is limited into a finite area with $x_2 \in [N - 50, N]$. The iterative procedure will be terminated when the value difference is less than 10^{-3} or the number of iterations exceeds 5,000. Let the order rejecting cost c_r take four different values: 5, 10, 12, and 15. Figure 4.4 illustrates the structure of the optimal policies.

From Fig. 4.4, it can be observed that the switching curve for the backordering decisions is moving downward as the unit lost-sale cost is increasing. Note that $\mu c_2^- / \beta = 0.7 * 10 / 0.5 = 14$. The completely backordering policy is optimal when $c_r = 15$ as shown in Fig. 4.4d, whereas the completely backordering policy is not optimal when $c_r < 14$ from Proposition 4.2, supported by Fig. 4.4a–c.

4.3   Optimal Control in a Failure-Prone Manufacturing Supply Chain with Backordering Decisions

This section considers the optimal production and backordering policy in a failure-prone manufacturing supply chain by taking into account the impact of unreliable machine on the system management. To simplify the analysis and make the problem analytically tractable, we neglect the raw material supply part in this section.

4.3.1   Problem Formulation

Consider a failure-prone manufacturing system producing one part-type with backordering decisions as shown in Fig. 4.5. The part-processing time is exponentially

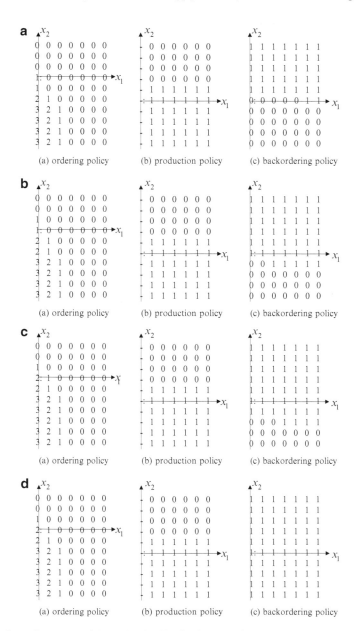

Fig. 4.4 Control structure of the optimal policy for the case of at most one outstanding order with its size not changeable once issued. (**a**) The case with $c_2^- = 10$ and $c_r = 5$. (**b**) The case with $c_2^- = 10$ and $c_r = 10$. (**c**) The case with $c_2^- = 10$ and $c_r = 12$. (**d**) The case with $c_2^- = 10$ and $c_r = 15$

Fig. 4.5 A failure-prone
manufacturing supply chain
with backordering decisions

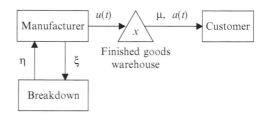

distributed with a rate u when the machine is up, where u is a controllable variable
that takes a value in $[0, r]$. The machine uptime and downtime are exponentially
distributed with failure rate ξ and repair rate η, respectively. The failure process is
assumed to be time dependent. The customer demand process is a homogeneous
Poisson flow with arrival rate μ. A demand is satisfied immediately if there are
finished goods stored as inventory. Unmet customer demands can be backordered
(if $a(t) = 1$) or rejected (if $a(t) = 0$). The rejected demands are lost that incurs a
penalty cost.

Let $x(t)$ denote the difference of the cumulative number of finished products and
the cumulative number of accepted demands. Let $\alpha(t)$ denote the machine state, that
is, $\alpha(t) = 1$ if the machine is up and $\alpha(t) = 0$ if the machine is down. The system
state space can be described by $X = \{(\alpha, x) \mid \alpha = 0, 1 \text{ and } x \in Z\}$.

There are two types of decisions to make: the production rate $u(t)$ and the
backordering decision $a(t)$. Define an admissible control set $\Omega = \{(u(t), a(t)) \mid$
$u(t) \in [0, r] \text{ if } \alpha(t) = 1 \text{ and } u(t) = 0 \text{ if } \alpha(t) = 0; a(t) \in \{0, 1\}, t \in (0, +\infty)\}$. The
problem is to find the optimal control $u(t) \in \Omega$ from time 0 to infinity to minimize
the infinite-horizon expected discounted cost (depending on the initial state)

$$J(\alpha, x) = \min_{u,a} E\left[\int_0^\infty e^{-\beta t} h(x(t), u(t))\, dt \mid \alpha(0) = \alpha, x(0) = x\right], \quad (4.16)$$

where $0 < \beta < 1$ is a discounted factor and $h(.)$ is a cost function to penalize
inventory, backlog, and rejecting demands. For example, $h(x(t), a(t)) = g(x(t)) + c_r$
$(1 - a(t))$, and

$$g(x(t)) = c^+ \max(x(t), 0) + c^- \max(-x(t), 0), \quad (4.17)$$

where c^+, c^- are the costs per unit of product over per unit of time for inventory
and backlog, respectively, and c_r is the cost of rejecting a demand.

The evolution of the system state is driven by four types of events, that is, demand
arrivals, production completion, machine failure, and machine repaired. Due to the
memoryless properties of Poisson process and exponential distribution, unfinished
production interrupted by an event is statistically equivalent to that of restarting. The
system state changes if and only if one of the above events occurs. In reality, it is
often the stationary control policies (i.e., state-feedback control) that are of interest.
A stationary control policy makes decision only based on the current system state.
We aim to seek the optimal stationary policy. The state transition map at a state
(α, x) under the control (u, a) is depicted in Fig. 4.6.

Fig. 4.6 State transition in a
failure-prone supply chain
with backordering decisions

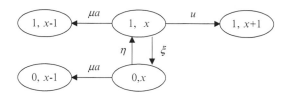

4.3.2 Optimal Control Policy

By the uniformization technique, the problem can be formulated into a Markov decision process. Let $v = \mu + r + \eta + \xi$ be the uniform transition rate. Under a stationary control policy $u = (u, a)$, the one-step transition probability functions from $(x, 1)$ and $(x, 0)$ are given by

$$\text{Prob}\left((x, 0) \,|\, (x, 1)\right) = \xi / v,$$

$$\text{Prob}\left((x + 1, 0) \,|\, (x, 1)\right) = u / v,$$

$$\text{Prob}\left((x - 1, 1) \,|\, (x, 1)\right) = \mu \cdot a / v,$$

$$\text{Prob}\left((x, 1) \,|\, (x, 1)\right) = (\eta + r + \mu - u - \mu \cdot a) / v,$$

$$\text{Prob}\left((x, 1) \,|\, (x, 0)\right) = \eta / v,$$

$$\text{Prob}\left((x - 1, 1) \,|\, (x, 0)\right) = \mu \cdot a / v,$$

$$\text{Prob}\left((x, 0) \,|\, (x, 0)\right) = (\xi + r + \mu - \mu \cdot a) / v.$$

Following the stochastic dynamic programming approach, the Bellman optimality equation is given as follows:

$$J(1, x) = \frac{1}{\beta + v} \left[c^+ x^+ + c^- x^- + \xi J(0, x) + \eta J(1, x) \right.$$

$$\left. + r \cdot \min\left(J(1, x+1), J(1, x)\right) + \min\left(\mu J(1, x-1), c_r + \mu J(1, x)\right) \right], \tag{4.18}$$

$$J(0, x) = \frac{1}{\beta + v} \left[c^+ x^+ + c^- x^- + (\xi + r) J(0, x) + \eta J(1, x) \right.$$

$$\left. + \min\left(\mu J(0, x - 1), c_r + \mu J(0, x)\right) \right]. \tag{4.19}$$

From the above equations, it is clear that the optimal production and backordering policy $(u^*(\alpha, x), a^*(\alpha, x))$ is given by

$$u^*(\alpha, x) = \begin{cases} r & J(\alpha, x + 1) \le J(\alpha, x), \quad \alpha = 1 \\ 0 & \text{otherwise} \end{cases}, \tag{4.20}$$

$$a^* (\alpha, x) = \begin{cases} 0 & c_r + \mu J (\alpha, x) \leq \mu J (\alpha, x - 1) \\ 1 & \text{otherwise} \end{cases}. \qquad (4.21)$$

The optimal cost function can be approximated numerically by the value iteration algorithm. The convergence of the k-stage policy and cost function to the infinite-horizon optimal policy and cost follows from the fact that only finitely many controls are considered at each state. However, the form of the optimal policy in (4.20) and (4.21) is implicit. It would be more interesting to explore the explicit format of the optimal policy, which will be addressed in the next section.

4.3.3   Characterization of the Optimal Policy

We first establish the properties of the optimal value function and then investigate the characteristics of the optimal policies.

Proposition 4.5. *(i) $J(\alpha, x + 1) - J(\alpha, x)$ is increasing in x; (ii) $J(\alpha, x + 1) - J(\alpha, x)$ ≤ 0 for $x \leq -1$.*

Proof. The proof can be shown by induction on k using the value iteration procedure. Since $J_0(\alpha, x) = 0$ for any $(\alpha, x) \in X$, so the assertions are true for $k = 0$. Now suppose they hold for k, then we want to show that the assertions are also true for $k + 1$. First, consider $\alpha = 1$:

$$J_{k+1} (1, x + 1) - J_{k+1} (1, x) = \frac{1}{\beta + v} \{[-c^- \cdot I\{x < 0\} + c^+ \cdot I\{x \geq 0\}]$$

$$+ \xi (J_k (0, x + 1) - J_k (0, x)) + \eta (J_k (1, x + 1) - J_k (1, x))$$

$$+ r \cdot [\min (J_k (1, x + 2), J_k (1, x + 1)) - \min (J_k (1, x + 1), J_k (1, x))]$$

$$+ [\min (\mu J_k (1, x), c_r + \mu J_k (1, x+1)) - \min (\mu J_k (1, x-1), c_r + \mu J_k (1, x))]\}. \qquad (4.22)$$

Clearly, on the right-hand side (RHS) of the above equation, the first term is increasing in x; the second and third terms are also increasing in x due to the induction hypothesis. The fourth term of RHS in (4.22) can be rewritten as (the constant r is dropped)

$$\min (J_k (1, x + 2) - J_k (1, x + 1), 0) + \max (0, J_k (1, x + 1) - J_k (1, x)). \qquad (4.23)$$

Note that the functions $\min(.)$ and $\max(.)$ preserve the monotonicity; (4.23) is increasing in x from the induction hypothesis. The last term of RHS in (4.22) can be rewritten as

$$\min\left(0,\ c_r + \mu J_k(1, x{+}1) - \mu J_k(1, x)\right) + \max\left(\mu J_k(1,) - \mu J_k(1, x-1),\ -c_r\right).$$
(4.24)

Again, (4.24) is increasing in x from the induction hypothesis and the fact that $\min(.)$ and $\max(.)$ preserve the monotonicity. Therefore, $J_{k+1}(1, x+1) - J_{k+1}(1, x)$ is increasing in x.

In addition, when $x \leq -1$, the first three terms of RHS in (4.22) are not greater than zero by the induction hypothesis. It is also clear that both (4.23) and (4.24) are not greater than zero when $x \leq -1$ by the induction hypothesis. It follows that $J_{k+1}(1, x+1) - J_{k+1}(1, x) \leq 0$ for $x \leq -1$.

With the similar arguments, we can show that $J_{k+1}(0, x+1) - J_{k+1}(0, x)$ is increasing in x and $J_{k+1}(0, x+1) - J_{k+1}(0, x) \leq 0$ for $x \leq -1$. Hence, the assertions are true for $k+1$. This completes the induction proof.

Proposition 4.6. *(i) $J(1, x) - J(0, x)$ is increasing in x; (ii) $J(1, x) - J(0, x) \leq 0$ for any x.*

Proof. Similar to the proof of Proposition 4.5, we show the assertions by induction on k using the value iteration procedure. The assertions are true for $k = 0$. Now suppose they hold for k, then it suffices to show that the assertions are also true for $k+1$:

$$J_{k+1}(1, x) - J_{k+1}(0, x) = \frac{1}{\beta + v}\{r \cdot (J_k(1, x) - J_k(0, x))$$

$$+ r \cdot \min\left(J_k(1, x+1) - J_k(1, x),\ 0\right)$$

$$+ [\min\left(\mu J_k(1, x{-}1),\, c_r + \mu J_k(1, x)\right) - \min\left(\mu J_k(0, x{-}1),\, c_r + \mu J_k(0, x)\right)]\}.$$
(4.25)

The first two terms of RHS in (4.25) are increasing in x according to the induction hypothesis and Proposition 4.5. We want to show that the last term of RHS in (4.25) is also increasing in x, that is,

$$\min\left(\mu J_k(1, x-1),\, c_r + \mu J_k(1, x)\right) - \min\left(\mu J_k(0, -1),\, c_r + \mu J_k(0, x)\right)$$

$$\leq \min\left(\mu J_k(1, x),\, c_r + \mu J_k(1, x{+}1)\right) - \min\left(\mu J_k(0, x),\, c_r + \mu J_k(0, x+1)\right).$$
(4.26)

From the induction hypothesis and Proposition 4.5, we have

$$J_k(0, x) - J_k(0, x-1) \leq J_k(1, x) - J_k(1, x-1) \leq J_k(1, x+1) - J_k(1, x),$$
(4.27)

$$J_k(0, x) - J_k(0, x-1) \leq J_k(0, x+1) - J_k(0, x) \leq J_k(1, x+1) - J_k(1, x).$$
(4.28)

If $J_k(0, x) - J_k(0, x - 1) \geq -c_r/\mu$ or $J_k(1, x + 1) - J_k(1, x) \leq -c_r/\mu$, then the inequality (4.26) is obviously true from (4.27) and (4.28). On the other hand, if $J_k(0, x) - J_k(0, x - 1) \leq -c_r/\mu \leq J_k(1, x + 1) - J_k(1, x)$, then (4.26) can be rewritten as

$$\min\left(\mu J_k(1, x - 1), c_r + \mu J_k(1, x)\right) - c_r - \mu J_k(0, x)$$
$$\leq \mu J_k(1, x) - \min\left(\mu J_k(0, x), c_r + \mu J_k(0, x + 1)\right). \qquad (4.29)$$

By simply rearranging the terms and comparing both sides of (4.29), the inequality apparently holds. Therefore, $J_{k+1}(1, x) - J_{k+1}(0, x)$ is increasing in x. Similarly, we can show that $J_{k+1}(1, x) - J_{k+1}(0, x) \leq 0$ for any x. The induction proof is completed.

Proposition 4.7. *The optimal production and backordering policy $(u^*(\alpha, x),$ $a^*(\alpha, x))$ is a threshold-type control, characterized by three threshold parameter l^*, m^*, and n^*, that is,*

$$u^*(1, x) = \begin{cases} r & x < n^* \\ 0 & x \geq n^* \end{cases}, \quad u^*(0, x) \equiv 0, \qquad (4.30)$$

$$a^*(1, x) = \begin{cases} 0 & x \leq l^* \\ 1 & x > l^* \end{cases}, \quad a^*(0, x) = \begin{cases} 0 & x \leq m^* \\ 1 & x > m^* \end{cases}, \qquad (4.31)$$

where $n^* = \max\{x \mid J(1, x) - J(1, x-1) \leq 0\}$; $l^* = \max\{x \mid J(1, x) - J(1, x - 1) \leq -c_r/\mu\}$; and $m^* = \max\{x \mid J(0, x) - J(0, x - 1) \leq -c_r/\mu\}$. In addition, we have $n^* \geq 0$ and $l^* \leq m^* \leq 0$.

Proof. The threshold control structure of the optimal policy is the direct result of Eq. (4.20) and Proposition 4.5. Moreover, Proposition 4.6(ii) yields that $n^* \geq 0$ and Proposition 4.6(i) yields $l^* \leq m^*$. Note that the backordering policy is available only if $x \leq 0$; this implies that $m^* \leq 0$. This completes the proof.

Definition 4.1 We define a general threshold policy by replacing l^*, m^*, and n^* with l, m, and n in (4.30) and (4.31). This policy is denoted by $u_{l,m,n}$.

Proposition 4.8. *(i) $\lim\limits_{x \to +\infty} J(\alpha, x + 1) - J(\alpha, x) = c^+/\beta$; (ii) $\lim\limits_{x \to -\infty} J(\alpha, x + 1) - J(\alpha, x) = -c^-/\beta$.*

Proof. From Proposition 4.5, $J(\alpha, x + 1) - J(\alpha, x)$ is monotone increasing in x. Using the optimality equations (4.18) and (4.19), as x tends to positive or negative infinity, the limit of $J(\alpha, x + 1) - J(\alpha, x)$ can be computed to be c^+/β or $-c^-/\beta$, respectively. This completes the proof.

Proposition 4.9. *The optimal threshold value n^* satisfies $0 \leq n^* < +\infty$. The optimal threshold values l^* and m^* satisfy $l^* = m^* = -\infty$ if $c_r \geq \mu c^-/\beta$ and $l^* \geq m^* > -\infty$ if $c_r < \mu c^-/\beta$. Namely, the completely backordering policy is optimal if and only if $c_r < \mu c^-/\beta$.*

Physically, Proposition 4.7 provides an explicit form of the optimal production and backordering policy for the failure-prone manufacturing supply chain, whereas Propositions 4.8 and 4.9 provide further insights into the boundary and relationship of the threshold parameters that determine the optimal policy.

4.4 Discussions and Notes

For the first problem in Sect. 4.2, that is, the joint optimal control of ordering, production, and backordering, it can be further extended to multistage systems or more general supply chains by relaxing the assumptions as that in Chap. 3.

For the second problem in Sect. 4.3, that is, the joint optimal control of production and backordering in failure-prone manufacturing supply chains, it is also interesting to consider the finite backlogging situation. Suppose the manufacturer has the maximum backlog length L. In other words, when $x > -L$, there are two actions: backlogging the demand and rejecting the demand, but when $x \leq -L$, there is only one action, that is, rejecting the demand. This problem can be similarly formulated, and the optimality equations are the same as (4.18) and (4.19) except at the state $(\alpha, -L)$, in which the last term $\min(.,.)$ on the RHS of (4.18) and (4.19) should be replaced by $c_r + \mu J^\beta(\alpha, -L)$. With the similar arguments, Propositions 4.5 and 4.6 can be proved and lead to Proposition 4.7. Thus, the optimal production and backordering policy in the finite backlogging situation is also of a threshold control.

Section 4.3 is mainly based on Song (2006). In the remainder of this section, we review some relevant literature on backordering decisions and failure-prone manufacturing supply chains.

4.4.1 Backordering Decisions

When the production rate is less than the demand rate in long term, completely backordering demands cannot be profitable since the number of unmet demands would grow without bound in the infinite horizon. One way to deal with such situation is to accept or reject demands in a random manner during the out-of-stock situation independent of the current backlog (Stidham 1985; Moinzadeh 1989). This may reflect the customers' behavior. Another more commonly used practice is to reject customer demands when backlog reaches a certain limit and accept them otherwise. This limit is referred as the base backlog. However, the optimality of such practice is not obvious.

Rabinowitz et al. (1995) considered the partial backordering policy combined with the (r, Q) policy in a single-stage inventory system with Poisson demand and constant lead time. It is demonstrated that the partial backordering policy is better than the completely backordering or lost-sales policies, although the percentage of cost savings depends on the fill rate. Kouikoglou and Phillis (2002) considered a single machine producing one product with Poisson demand and exponential processing time. They emphasized the importance of the coordination between the production control and the quality control departments and demonstrated that the partially backordering policy outperforms the completely backlogging and the lost-sale practices. Ioannidisa and Kouikogloub (2008) studied a single-stage production system producing one product to meet random demands and addressed the questions of when to produce and when to accept an incoming customer demand to maximize net profit. Under the CONWIP (constant work-in-process) policy, they focused on determining the best CONWIP and base backlog levels. Thangam and Uthayakumar (2008) considered a two-level supply chain consisting of a single supplier and multiple identical retailers with constant lead times and Poisson demands. Under the continuous-review fixed reorder policy (r, Q) for all entities and base backlog policy for all retailers, they developed the approximate cost function to find optimal reorder points for given batch sizes for all entities and the optimal value of the base backlog in the identical retailers. Economopoulos et al. (2011) extended the one-stage one-product model to include three types of activities during out-of-stock periods: (1) customer balking, that is, arriving customers may leave immediately; (2) supplier rejecting, that is, the system rejects new customers by following the base backlog policy; and (3) customer reneging, that is, waiting customers may become impatient and withdraw their demands. Combining the base-stock inventory control with the base backlog policy, they concentrated on the calculation of the optimal base-stock and base backlog levels.

There is another group of studies addressing the admission control (which is similar to backordering decisions) when the system is facing two types of customers. Chen and Kulkarni (2007) considered an $M/M/1$ queuing system serving two classes of customers in which class 1 customers have preemptive resume priority over class 2 customers. They investigated the structure of the optimal accepting or rejecting policy in different scenarios. Benjaafar et al. (2010) studied a single-stage system with two classes of customers and showed the threshold control structure of the optimal production and backordering policy. Iravani et al. (2012) considered the optimal production and admission control policies in manufacturing systems producing two types of products assuming that one type must be fully backlogged and the other is produced to order with possible rejection. They are able to characterize the optimal policies with a partial-linear structure.

4.4.2 Failure-Prone Manufacturing Supply Chains

Failure-prone manufacturing systems have been extensively studied in the last three decades. The literature can be classified into two groups (Buzacott and Shathikumar 1993). The first is based on "fluid flow models," where the production is modeled as a continuous production process and the analysis often assumes constant demand rate. It has been shown that the optimal policy is a hedging point control in a single machine system if unmet demands are completely backlogged (Akella and Kumar 1986; Bielecki and Kumar 1988). The results were extended to systems without backlog (Hu 1995) or bounded backlog (Martinelli and Valigi 2004). A comprehensive list of references on this subject can be found in Gershwin (1994), Sethi and Zhang (1994), and Sethi et al. (2002).

The second is "discrete part manufacturing systems," where parts are produced in discrete mode. This group often takes into account the extra randomness in production times and demand arrivals. With the assumption that unmet demands are fully backlogged, the optimality of a hedging point policy (also termed as threshold control policy) in failure-prone manufacturing systems producing one part-type or two part-types was established in (Song and Sun 1999; Feng and Yan 2000; Feng and Xiao 2002). Explicit forms of the discounted cost and the average cost under a threshold control policy were provided in (Feng and Yan 2000; Feng and Xiao 2002).

The majority of the previous research in failure-prone manufacturing systems assumed that unsatisfied demands are fully backordered. A few of them assumed no backlog (lost sales) or limited backlog, which is regarded as an exogenously given constraint. More research is required for the integrated management of complex failure-prone manufacturing supply chains facing backordering decisions.

References

Akella, R., Kumar, P.R.: Optimal control of production rate in a failure prone manufacturing system. IEEE Trans. Autom. Control **31**(1), 116–126 (1986)

Benjaafar, S., ElHafsi, M., Huang, T.: Optimal control of a production-inventory system with both backorders and lost sales. Nav. Res. Logist. **57**(3), 252–265 (2010)

Bielecki, T., Kumar, P.R.: Optimality of zero-inventory policies for unreliable manufacturing systems. Oper. Res. **36**(4), 532–541 (1988)

Buzacott, J.A., Shathikumar, J.G.: Stochastic Models of Manufacturing Systems. Prentice Hall, Englewood Cliffs (1993)

Chen, F., Kulkarni, V.G.: Individual, class-based, and social optimal admission policies in two-priority queues. Stoch. Model. **23**(1), 97–127 (2007)

Economopoulos, A.A., Kouikoglou, V.S., Grigoroudis, E.: The base stock/base backlog control policy for a make-to-stock system with impatient customers. IEEE Trans. Autom. Sci. Eng. **8**(1), 243–249 (2011)

Feng, Y.Y., Xiao, B.C.: Optimal threshold control in discrete failure-prone manufacturing systems. IEEE Trans. Autom. Control **47**(7), 1167–1174 (2002)

Feng, Y.Y., Yan, H.M.: Optimal production control in a discrete manufacturing system with unreliable machines and random demands. IEEE Trans. Autom. Control **45**(12), 2280–2296 (2000)

Gershwin, S.B.: Manufacturing Systems Engineering. Prentice-Hall, Englewood Cliffs (1994)

Hu, J.Q.: Production rate control for failure-prone production systems with no-backlog permitted. IEEE Trans. Autom. Control **40**(2), 291–295 (1995)

Ioannidisa, S., Kouikogloub, V.S.: Revenue management in single-stage CONWIP production systems. Int. J. Prod. Res. **46**(22), 6513–6532 (2008)

Iravani, S.M.R., Liu, T., Simchi-Levi, D.: Optimal production and admission policies in make-to-stock/make-to-order manufacturing systems. Prod. Oper. Manage. **21**(2), 224–235 (2012)

Kouikoglou, V.S., Phillis, Y.A.: Design of product specifications and control policies in a single-stage production system. IIE Trans. **34**(7), 590–600 (2002)

Martinelli, F., Valigi, P.: Hedging point policies remain optimal under limited backlog and inventory space. IEEE Trans. Autom. Control **49**(10), 1863–1869 (2004)

Moinzadeh, K.: Operating characteristics of the (S-1, S) inventory system with partial backorders and constant resupply times. Manage. Sci. **35**(4), 472–477 (1989)

Rabinowitz, G.A., Mehrez, A., Chull, C., Patuwo, B.E.: A partial backorder control for continuous review (r, Q) inventory system with Poisson demand and constant lead time. Comput. Oper. Res. **22**(7), 689–700 (1995)

Sethi, S., Zhang, Q.: Hierarchical Decision Making in Stochastic Manufacturing Systems. Birkhauser, Boston (1994)

Sethi, S.P., Yan, H., Zhang, H., Zhang, Q.: Optimal and hierarchical controls in dynamic stochastic manufacturing systems: a survey. Manuf. Serv. Oper. Manage. **4**(2), 133–170 (2002)

Song, D.P.: Optimal production and backordering policy in failure-prone manufacturing systems. IEEE Trans. Autom. Control **51**(5), 906–911 (2006)

Song, D.P., Sun, Y.X.: Optimal control structure of an unreliable manufacturing system with random demands, *IEEE Trans*. Autom. Control **44**(3), 619–622 (1999)

Stidham, S.J.: Optimal control of admission to a queueing system. IEEE Trans. Autom. Control **30**(8), 705–713 (1985)

Thangam, A., Uthayakumar, R.: A two-level supply chain with partial backordering and approximated Poisson demand. Eur. J. Oper. Res. **187**(1), 228–242 (2008)

Chapter 5
Optimal Control of Supply Chain Systems with Preventive Maintenance Decisions

5.1 Introduction

Supply chain system may be disrupted by many unexpected events, for example, earthquake, accidents, industry actions, and equipment failures. Machine failure is probably one of the most frequently occurred disruptions that could significantly influence on the production process and the supply chain performance. A large number of studies have been conducted to investigate the machine unreliability in the few decades. We have reviewed some relevant literature at the end of the last chapter. This chapter will focus on the optimal control of failure-prone manufacturing supply chains with preventive maintenance decisions. First, we distinguish the machine failure modes.

There are two major types of machine failure modes: time dependent and operation dependent . Both time-dependent failures and operation-dependent failures are common in manufacturing systems (Buzacott and Hanifin 1978; Xie et al. 2004). Time-dependent failures mainly depend on the elapsed time that the machine is turned on, for example, electronic failures. As a result, the machine may fail even when it is not operating on a part. In the literature, time-dependent failure rate is often assumed to be a constant regardless whether the machine is processing a part or how long it has been turned on (e.g., Kimemia and Gershwin 1983; Akella and Kumar 1986; Bielecki and Kumar 1988; Sharifnia 1988; Sethi and Zhang 1994; Gershwin 1994).

Operation-dependent failures depend on the time that the machine spends on working on a part, for example, a tool breakage. This type of failures occurs only if the machine is processing a part, namely, the machine cannot fail if it is idle. Operation-dependent failures can be further divided into the following subtypes: constant failure rate (i.e., the failure rate is a constant as long as the machine is operating on a part), age-dependent failure (i.e., the failure rate depends on how long the machine has been in operation), and production-rate-dependent failure (i.e., the failure rate depends on the instantaneous rate of production). Age-dependent failure can be referred to Boukas and Haurie (1990) and Hu and Xiang (1994).

D.-P. Song, *Optimal Control and Optimization of Stochastic Supply Chain Systems*, Advances in Industrial Control, DOI 10.1007/978-1-4471-4724-4_5, © Springer-Verlag London 2013

Production-rate-dependent failure systems are discussed in Hu et al. (1994) and Liberopoulos and Caramanis (1994). Performance evaluation under operation-dependent failures can be referred to Dallery et al. (1989), Xie (1993), and Fu and Xie (2002).

A common way to tackle the failure-prone machine problem is to improve its reliability such as carrying out preventive maintenance. Although preventive maintenance may affect the actual production capacity, its effect is controllable to a large degree since when to perform preventive maintenance is a decision rather than an unpredictable event.

There are two common types of preventive maintenance. The first is less radical and does not require the stoppage of the production, whereas the second is more fundamental and requires the stoppage of the current operation. In this chapter, we assume the first type of preventive maintenance. Such preventive maintenance includes lubrication, cleaning, and adjustment.

In our context, both production rate and preventive maintenance rate are simultaneously controllable. On the one hand, preventive maintenance can improve the machine's reliability to some extent; on the other hand, it will slightly disrupt the production and therefore reduce the production capacity. Therefore, the decisions on whether and when to perform such preventive maintenance should be carefully balanced.

This chapter is organized as follows. In the next section, we consider the integrated control for raw material ordering, production, and preventive maintenance in the basic stochastic supply chain with the manufacturer having failure-prone machines. Both operation-dependent failure mode and time-dependent failure mode are examined. In Sect. 5.3, we consider a simpler failure-prone manufacturing supply chain without considering the raw material ordering activities and focus on the optimal joint production and maintenance control. We end this chapter with some discussions and notes.

5.2   Optimal Control of Ordering, Production, and Preventive Maintenance in a Supply Chain

Consider the basic supplier–manufacturer–customer supply chain in Chap. 2 with the manufacturer having failure-prone machines as shown in Fig. 5.1. The system is subject to Assumptions 2.1–2.4. The manufacturer can perform

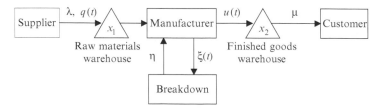

Fig. 5.1 A supply chain with preventive maintenance

preventivemaintenance to reduce the machines' failure rate. The machine uptime and downtime are exponentially distributed with failure rate $\xi(t)$ and repair rate η, respectively, where $\xi(t)$ depends on whether preventive maintenance is performed. The customer demand process is a homogeneous Poisson flow with arrival rate μ. A demand is satisfied immediately if there are finished goods stored as inventory.

In the remainder of this section, we first model the optimal control problem under the operation-dependent failure mode and then under the time-dependent failure mode.

5.2.1 Optimal Control Under Operation-Dependent Failures

In this section, it is assumed that the failure rate depends on whether the machine is operating and whether the preventive maintenance is being performed. This failure process is similar to Boukas and Liu (2001).

More specifically, if the preventive maintenance is performed while the machine is operating, the machine failure rate $\xi(t)$ will become ξ_0, which is smaller than the normal failure rate ξ_1 corresponding to no preventive maintenance situation. However, the maximum production rate will be reduced from a normal rate r_1 to r_0 if the preventive maintenance is being conducted. In other words, when the machine is operating and the preventive maintenance is performed, the production decision $u(t)$ can take a value in $\{0, r_0\}$ with the machine failure rate ξ_0; when the machine is operating without preventive maintenance, the production decision $u(t)$ can take a value in $\{0, r_1\}$ the machine failure rate ξ_1, where $r_0 < r_1$ and $\xi_0 < \xi_1$; and when the machine is producing nothing and idle, it is assumed that the machine will not break down and no preventive maintenance is required.

The raw material ordering decision $q(t)$ does not depend on the machine state. The raw material replenishment lead time is assumed to follow an exponential distribution with the mean $1/\lambda$.

Let $x_1(t)$ and $x_2(t)$ denote the inventory-on-hand of raw materials and finished goods, respectively. Negative $x_2(t)$ represents the backlogged customer demands. Let $\alpha(t)$ denote the machine state, that is, $\alpha(t) = 1$ if the machine is up and $\alpha(t) = 0$ if the machine is down. The system state space can be described by $X = \{(\alpha, x_1, x_2) | \alpha \in \{0, 1\}, x_1 \in [0, M]$ and $x_2 \in (-\infty, N)\}$, where M and N are the warehouse capacity for raw materials and finished goods, respectively. To simplify the narrative, let $\mathbf{x} := (x_1, x_2)$.

There are three types of decisions to make: the preventive maintenance decision $\xi(t)$ when the machine is up, the production rate $u(t)$ when the machine is up, and the ordering decision $q(t)$ to the supplier. The admissible control set $\Omega = \{(\xi(t), u(t), q(t)) | \xi(t) \in \{\xi_0, \xi_1\} \cdot I\{\alpha(t) = 1, u(t) > 0\}, \xi(t) = 0 \cdot I\{u(t) = 0\}; u(t) \in (0, r_1] \cdot I\{x_1(t) > 0, x_2(t) < N\}$ if $\xi(t) = \xi_1, u(t) \in (0, r_0] \cdot I\{x_1(t) > 0, x_2(t) < N\}$ if $\xi(t) = \xi_0, u(t) = 0$ if $x_1(t) \leq 0$ or $x_2(t) \geq N$ or $\alpha(t) = 0; q(t) \in [0, M - x_1(t)]; t \geq 0\}$.

Fig. 5.2 System state
transition in a supply chain
with preventive maintenance

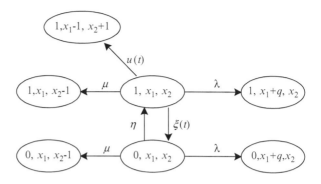

The problem is to find the optimal integrated policy $(\xi(t), u(t), q(t)) \in \Omega$ from time 0 to infinity to minimize the infinite horizon expected discounted cost (depending on the initial state):

$$J(\mathbf{x}_0) = \min_{\xi, u, q} E\left[\int_0^\infty e^{-\beta t} h(\mathbf{x}(t), \xi(t), u(t), q(t)) dt \mid \mathbf{x}(0) = \mathbf{x}_0 \right], \qquad (5.1)$$

where $0 < \beta < 1$ is a discounted factor and $h(.)$ is a cost function to penalize inventory, backlog, and preventive maintenance. For example, a commonly used form of $h(.)$ is

$$h(\mathbf{x}, \xi, u, q) = g(\mathbf{x}) + c_m \cdot (\xi_1 - \xi(t)), \qquad (5.2)$$

$$g(\mathbf{x}) = c_1 x_1 + c_2^+ x_2^+ + c_2^- x_2^-, \qquad (5.3)$$

where c_1, c_2^+, and c_2^- are the unit costs of raw material inventory, finished goods inventory, and backordered demands, respectively; c_m represents the cost of performing preventive maintenance; $x_2^+ := \max(x_2, 0)$; and $x_2^- := \max(-x_2, 0)$. We concentrate on the state-feedback stationary control policies, that is, $(\xi(t), u(t), q(t)) = (\xi(\alpha(t), \mathbf{x}(t)), u(\alpha(t), \mathbf{x}(t)), q(\alpha(t), \mathbf{x}(t)))$, which is simplified as (ξ, u, q). The state transition map at a state (α, x_1, x_2) under the control (ξ, u, q) is depicted in Fig. 5.2.

Let $v = \lambda + \mu + \eta + r_1 + \xi_1$ be the uniform transition rate. The one-step transition probability functions at $(1, \mathbf{x})$ and $(0, \mathbf{x})$ under the control (ξ, u, q) are given by

$$\text{Prob}\{(1, x_1, x_2 - 1) \mid (1, \mathbf{x})\} = \mu / v, \qquad (5.4)$$

$$\text{Prob}\{(0, \mathbf{x}) \mid (1, \mathbf{x})\} = \xi \cdot I\{u > 0\} / v, \qquad (5.5)$$

$$\text{Prob}\{(1, x_1 - 1, x_2 + 1) \mid (1, \mathbf{x})\} = u \cdot I\{u > 0\} / v, \qquad (5.6)$$

$$\text{Prob}\{(1, x_1 + q, x_2) \mid (1, \mathbf{x})\} = \lambda \cdot I\{q > 0\} / v, \tag{5.7}$$

$$\text{Prob}\{(1, \mathbf{x}) \mid (1, \mathbf{x})\} = (\eta + \xi_1 - \xi \cdot I\{u > 0\} + r_1 - u \cdot I\{u > 0\} + \lambda - \lambda \cdot I\{q > 0\}) / v, \tag{5.8}$$

$$\text{Prob}\{(0, x_1, x_2 - 1) \mid (0, \mathbf{x})\} = \mu / v, \tag{5.9}$$

$$\text{Prob}\{(1, \mathbf{x}) \mid (0, \mathbf{x})\} = \eta / v, \tag{5.10}$$

$$\text{Prob}\{(0, x_1 + q, x_2) \mid (0, \mathbf{x})\} = \lambda \cdot I\{q > 0\} / v, \tag{5.11}$$

$$\text{Prob}\{(0, \mathbf{x}) \mid (0, \mathbf{x})\} = (\xi_1 + r_1 + \lambda - \lambda \cdot I\{q > 0\}) / v, \tag{5.12}$$

where $I\{\text{condition}\}$ is an indicator function. It takes 1 if the condition is true and 0 otherwise. The Bellman optimality equation is given as follows:

$$
\begin{aligned}
J(1, \mathbf{x}) = \frac{1}{\beta + v} \min_{\xi, u, q} \Big[& c_1 x_1 + c_2{}^+ x_2{}^+ + c_2{}^- x_2{}^- + c_m \cdot (\xi_1 - \xi) \cdot I\{u > 0\} \\
& + \mu J(1, x_1, x_2 - 1) + \xi J(0, \mathbf{x}) \cdot I\{u > 0\} \\
& + u J(1, x_1 - 1, x_2 + 1) \times I\{u > 0\} + \lambda \cdot J(1, x_1 + q, x_2) \cdot I\{q > 0\} \\
& + (r_1 + \eta + \xi_1 - \xi \cdot I\{u > 0\} - u \cdot I\{u > 0\} + \lambda - \lambda \cdot I\{q > 0\}) J(1, \mathbf{x}) \Big];
\end{aligned}
\tag{5.13}
$$

$$
\begin{aligned}
J(0, \mathbf{x}) = \frac{1}{\beta + v} \min_q \Big[& c_1 x_1 + c_2{}^+ x_2{}^+ + c_2{}^- x_2{}^- + \mu J(0, x_1, x_2 - 1) + \eta J(1, \mathbf{x}) \\
& + \lambda \cdot J(0, x_1 + q, x_2) \times I\{q > 0\} + (r_1 + \xi_1 + \lambda - \lambda \cdot I\{q > 0\}) J(0, \mathbf{x}) \Big].
\end{aligned}
\tag{5.14}
$$

Take into account the control constraints such as $\xi \in \{\xi_0, \xi_1\} \cdot I\{u > 0\}, u \in \{0, r_1\}$ if $\xi = \xi_1$ and $x_1 > 0$ and $x_2 < N$, $u \in \{0, r_0\}$ if $\xi = \xi_0$ and $x_1 > 0$ and $x_2 < N$, and $q(t) \in [0, M - x_1(t)]$; the above equations can be simplified as

$$
\begin{aligned}
J(1, \mathbf{x}) = \frac{1}{\beta + v} \Big[& c_1 x_1 + c_2{}^+ x_2{}^+ + c_2{}^- x_2{}^- + \mu J(1, x_1, x_2 - 1) + \eta J(1, \mathbf{x}) \\
& + (\xi_1 + r_1) \cdot J(1, \mathbf{x}) \cdot I\{x_1 = 0 \text{ or } x_2 = N\} + \min \{(\xi_1 - \xi_0) \cdot c_m \\
& + \xi_0 J(0, \mathbf{x}) + r_0 \cdot J(1, x_1 - 1, x_2 + 1) + (r_1 - r_0 + \xi_1 - \xi_0) J(1, \mathbf{x}), \\
& \xi_1 \cdot J(0, \mathbf{x}) + r_1 \cdot J(1, x_1 - 1, x_2 + 1), (\xi_1 + r_1) \cdot J(1, \mathbf{x})\} \cdot \\
& I\{x_1 > 0 \text{ and } x_2 < N\} + \lambda \cdot \min \{J(1, x_1 + q, x_2) \mid q \in [0, M - x_1]\} \Big],
\end{aligned}
\tag{5.15}
$$

$$J(0, \mathbf{x}) = \frac{1}{\beta + \upsilon} \left[c_1 x_1 + c_2{}^+ x_2{}^+ + c_2{}^- x_2{}^- + \mu J(0, x_1, x_2 - 1) + \eta J(1, \mathbf{x}) \right.$$

$$\left. + (r_1 + \xi_1) J(0, \mathbf{x}) + \lambda \cdot \min \{ J(0, x_1 + q, x_2) \,|\, q \in [0, M - x_1] \} \right].$$

$$(5.16)$$

Proposition 5.1. *The optimal ordering, production, and preventive maintenance policy* $(\xi^*(\alpha, \mathbf{x}), u^*(\alpha, \mathbf{x}), q^*(\alpha, \mathbf{x}))$ *for operation-dependent failure mode is given by*

$$(\xi^*(1, \mathbf{x}), u^*(1, \mathbf{x})) = (\xi_1, r_1), \text{ if } \xi_1 \cdot J(0, \mathbf{x}) + r_1 \cdot J(1, x_1 - 1, x_2 + 1)$$

$$\leq (\xi_1 + r_1) \cdot J(1, \mathbf{x}), (\xi_1 - \xi_0) \cdot J(0, \mathbf{x}) + (r_1 - r_0) \cdot J(1, x_1 - 1, x_2 + 1)$$

$$\leq (\xi_1 - \xi_0) \cdot c_m + (r_1 - r_0 + \xi_1 - \xi_0) J(1, \mathbf{x}), x_1 > 0, \text{ and } x_2 < N;$$

$$(\xi^*(1, \mathbf{x}), u^*(1, \mathbf{x})) = (\xi_0, r_0), \text{ if } (\xi_1 - \xi_0) \cdot c_m + \xi_0 \cdot J(0, \mathbf{x})$$

$$+ r_0 \cdot J(1, x_1 - 1, x_2 + 1) < (\xi_0 + r_0) \cdot J(1, \mathbf{x}), (\xi_1 - \xi_0) \cdot J(0, \mathbf{x})$$

$$+ (r_1 - r_0) \cdot J(1, x_1 - 1, x_2 + 1) > (\xi_1 - \xi_0) \cdot c_m + (r_1 - r_0 + \xi_1 - \xi_0)$$

$$\times J(1, \mathbf{x}), x_1 > 0, \text{ and } x_2 < N;$$

$$\left(\xi^*(1, \mathbf{x}), u^*(1, \mathbf{x}) \right) = (0, 0), \text{ otherwise.} \qquad (5.17)$$

$$q^*(\alpha, \mathbf{x}) = \arg \min_q \{ J(\alpha, x_1 + q, x_2) \,|\, q \in [0, M - x_1] \}. \qquad (5.18)$$

The above proposition states that the optimal raw material ordering decisions depend on the optimal value function and are loosely linked to the production and maintenance decisions. The production and maintenance decisions are coupled into three actions: maximum production with maintenance, maximum production without maintenance, and no production (implied no maintenance). The explicit structure of the optimal integrated policy is difficult to establish analytically. However, it can at least be explored numerically. We present an example below.

Example 5.1. Consider an operation-dependent failure mode supply chain with the following setting: the lead-time rate $\lambda = 0.8$; the warehouse capacities $M = N = 10$; demand rate $\mu = 0.7$; machine repair rate $\eta = 0.9$; the machine failure rate under preventive maintenance $\xi_0 = 0.1$; the maximum production rates $r_0 = 1.0$ (under preventive maintenance) and $r_1 = 1.2$ (without preventive maintenance); the raw material inventory unit cost $c_1 = 1$; the finished goods inventory unit cost $c_2{}^+ = 2$; the backordering cost $c_2{-} = 10$; and the preventive maintenance cost $c_m = 1.0$. The discount factor $\beta = 0.4$. The system state space is limited into a finite area with $x_1 \in [0, 10]$ and $x_2 \in [-30, 10]$. The iterative procedure will be terminated when the number of iterations exceeds 100. Consider three cases with different machine failure rates without preventive maintenance: (a) $\xi_1 = 0.2$, (b) $\xi_1 = 0.3$, and (c) $\xi_1 = 0.4$. Figure 5.3 illustrates the structure of the optimal policies in three cases.

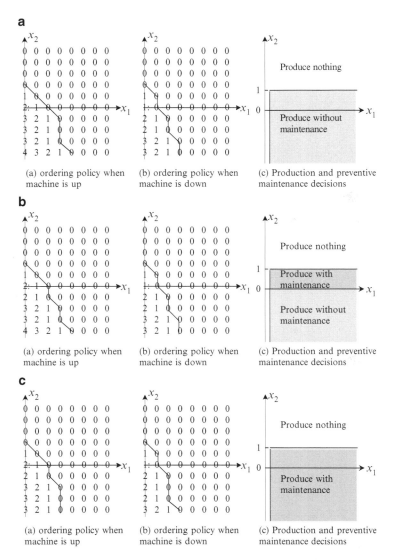

Fig. 5.3 Structure of the optimal ordering, production, and preventive maintenance policy in an operation-dependent failure supply chain. (**a**) Case $\xi_1 = 0.2$. (**b**) Case $\xi_1 = 0.3$. (**c**) Case $\xi_1 = 0.4$

From Fig. 5.3, it can be seen that the optimal policy appears to be of switching structure. The ordering decisions follow the order-up-to-point concept similar to the results in previous chapters. The switching curve of the ordering decisions when the machine is down is located slightly lower than the switching curve when the machine is up. This reflects the fact that when the machine is down, it requires time to repair, and therefore, lower inventory levels of raw materials are possible.

Figure 5.3a indicates that no preventive maintenance is optimal; Fig. 5.3c indicates that performing preventive maintenance whenever possible is optimal, whereas Fig. 5.3b indicates that preventive maintenance should be performed only when the system state belongs to a certain region. More specifically, Fig. 5.3 reveals that the production and preventive maintenance actions are divided into two regions in cases (a) and (c) by the straight line $x_2 = 1$, into three regions in case (b) by two straight lines $x_2 = 1$ and $x_2 = 0$. Note that when $x_1 = 0$, the machine is forced to produce nothing and maintenance is not necessary for the operation-dependent failure mode. The results imply that simple threshold-type control policies could closely approximate the optimal production and maintenance decisions.

As the failure rate without preventive maintenance ξ_1 increases, the switching curves that determine the raw material ordering decisions are moving downward in the state space. This reflects the intuition that lower levels of raw material safety stock are adequate if the machine has a higher failure rate since the machine operates less frequently and offers more time to order raw materials. On the other hand, as ξ_1 increases from case (a) to case (c), the preventive maintenance is being performed more frequently. This reflects the fact that when performing preventive maintenance becomes more beneficial, it will be conducted more frequently, particularly in situations when there are backlogged demands.

5.2.2 Optimal Control Under Time-Dependent Failures

We now assume that the machine failure is time dependent , that is, occurring whether or not the machine is processing a part. However, the failure rate is still controllable by performing preventive maintenance. More specifically, if the preventive maintenance is performed, the machine failure rate $\xi(t)$ will become ξ_0, which is smaller than the normal failure rate ξ_1 corresponding to no preventive maintenance situation.

The Bellman optimality equation for $J(1, \mathbf{x})$ is given as follows (for $J(0, \mathbf{x})$, it is the same as (5.16) because no production and maintenance are involved when the machine is down):

$$J(1, \mathbf{x}) = \frac{1}{\beta + \nu} \left[c_1 x_1 + c_2{}^+ x_2{}^+ + c_2{}^- x_2{}^- + \xi_1 \cdot c_m + \mu J(1, x_1, x_2 - 1) \right.$$

$$+ (r_1 + \eta + \xi_1) \cdot J(1, \mathbf{x}) + \min_{\xi u} \{ \xi \cdot (J(0, \mathbf{x}) - J(1, \mathbf{x}) - c_m)$$

$$+ u \cdot (J(1, x_1 - 1, x_2 + 1) - J(1, \mathbf{x})) \cdot I\{u > 0\} \}$$

$$\left. + \lambda \cdot \min \{ J(1, x_1 + q, x_2) \, | q \in [0, M - x_1] \} \right]. \tag{5.19}$$

Proposition 5.2. *The optimal ordering, production, and preventive maintenance policy* $(\xi^*(\alpha, \mathbf{x}), u^*(\alpha, \mathbf{x}), q^*(\alpha, \mathbf{x}))$ *for the time-dependent failure mode is given by*

$$q^*(\alpha, \mathbf{x}) = \arg \min_q \{ J(\alpha, x_1 + q, x_2) \, | q \in [0, M - x_1] \}. \tag{5.20}$$

If $x_1 = 0$ or $x_2 = N$, $(\xi^* (1, \mathbf{x}), u^* (1, \mathbf{x})) = \begin{cases} (\xi_1, 0) & J(0, \mathbf{x}) \leq J(1, \mathbf{x}) + c_m \\ (\xi_0, 0) & J(0, \mathbf{x}) > J(1, \mathbf{x}) + c_m \end{cases}$.

If $x_1 > 0$ and $x_2 < N$, $(\xi^* (1, \mathbf{x}), u^* (1, \mathbf{x}))$

$$
= \begin{cases}
(\xi_1, r_1) & J(0, \mathbf{x}) \leq J(1, \mathbf{x}) + c_m, \, J(1, x_1 - 1, x_2 + 1) \leq J(1, \mathbf{x}) \\
(\xi_1, 0) & J(0, \mathbf{x}) \leq J(1, \mathbf{x}) + c_m, \, J(1, x_1 - 1, x_2 + 1) > J(1, \mathbf{x}) \\
(\xi_0, 0) & J(0, \mathbf{x}) > J(1, \mathbf{x}) + c_m, \, J(1, x_1 - 1, x_2 + 1) \geq J(1, \mathbf{x}) \\
(\xi_1, r_1) & J(0, \mathbf{x}) > J(1, \mathbf{x}) + c_m, \, J(1, x_1 - 1, x_2 + 1) < J(1, \mathbf{x}), \text{ and } (\xi_1 - \xi_0) \cdot \\
& (J(0, \mathbf{x}) - J(1, \mathbf{x}) - c_m) + (r_1 - r_0) (J(1, x_1 - 1, x_2) - J(1, \mathbf{x})) \leq 0 \\
(\xi_0, r_0) & J(0, \mathbf{x}) > J(1, \mathbf{x}) + c_m, \, J(1, x_1 - 1, x_2 + 1) < J(1, \mathbf{x}), \text{ and } (\xi_1 - \xi_0) \\
& (J(0, \mathbf{x}) - J(1, \mathbf{x}) - c_m) + (r_1 - r_0) (J(1, x_1 - 1, x_2) - J(1, \mathbf{x})) > 0
\end{cases}
$$
$$(5.21)$$

In the time-dependent failure mode, there are four combinations of production and preventive maintenance : (ξ_1, r_1) – no maintenance and maximum production; $(\xi_1, 0)$ – no maintenance and no production; (ξ_0, r_0) – maintenance and maximum production; and $(\xi_0, 0)$ – maintenance and no production. We use a numerical example to explore the structure of the optimal control policy.

Example 5.2. Consider a time-dependent failure mode supply chain with the same parameter setting as Example 5.1. Consider three cases: (a) $\xi_1 = 0.2$, (b) $\xi_1 = 0.3$, and (c) $\xi_1 = 0.4$. Figure 5.4 illustrates the structure of the optimal policies in three cases.

In Fig. 5.4c, to simplify the display of control actions of production and maintenance in the state space, we adopt the following codes: $0 =$ no production and no maintenance, $1 =$ production and no maintenance, $2 =$ maintenance and no production, and $3 =$ production and maintenance.

From Figs. 5.3 and 5.4, it can be observed that the optimal ordering policy is characterized by two monotonic switching curves, which is similar to the cases in Chap. 2. The optimal production and preventive maintenance policy generally has the region-switching structure. However, the monotonic property may not necessarily hold. For example, in the case (b) in Fig. 5.4, the action "production without maintenance" is taken in two separate regions in the state space. In terms of the cost, the cases under operation-dependent failures achieve lower costs than the cases under time-dependent failures, which is in agreement with intuition.

5.3   Optimal Control of Production and Preventive Maintenance in a Failure-Prone Manufacturing Supply Chain

The above section shows that the raw material ordering decisions have the similar structure to the situations without considering preventive maintenance. This indicates that the ordering decisions are loosely related to the preventive maintenance

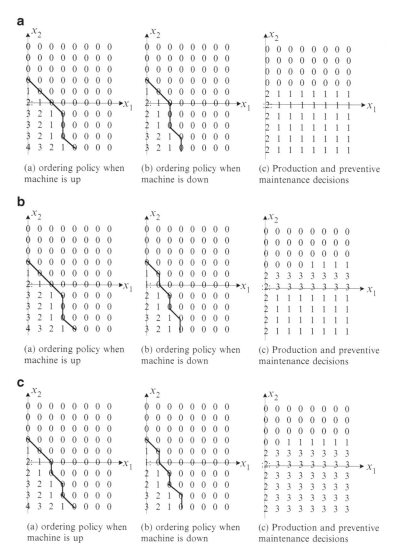

Fig. 5.4 Structure of the optimal ordering, production, and preventive maintenance policy in a time-dependent failure supply chain. (**a**) Case $\xi_1 = 0.2$. (**b**) Case $\xi_1 = 0.3$. (**c**) Case $\xi_1 = 0.4$

decisions, at least from the control structure's viewpoint. On the other hand, the production decision and the preventive maintenance decision are closely related, but they appear to be rather insensitive to the raw material inventory level. For that reason, we consider a simpler manufacturing supply chain by ignoring the raw material supply part so that we can focus more on the interaction between production and preventive maintenance.

Fig. 5.5 A failure-prone
manufacturing supply chain
with preventive maintenance
decisions

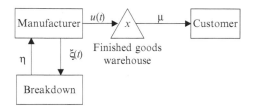

This section studies a failure-prone manufacturing supply chain with preventive
maintenance without considering the raw material supply activity. In other words, it
is assumed that raw materials are always available. We also remove the capacity
constraint of the finished goods warehouse to simplify the narrative. Our focus
is to explore the optimal control structure of the joint production and preventive
maintenance decisions.

Consider an operation-dependent failure manufacturing supply chain in
Fig. 5.5. Let $x(t)$ denote the inventory-on-hand of the finished goods (it
can take negative number to represent backlogs). The system state space
$X = \{(\alpha, x) | \alpha = 0, 1 \text{ and } x \in Z\}$, where $\alpha = 0$ represents the failure mode
and $\alpha = 1$ represents the operational mode. Let $(u(x), \xi(x))$ denote the
production rate and preventive maintenance action at system state $(1, x)$. The
admissible control set of stationary control policies is given by $\Omega = \{(u(x),$
$\xi(x)) | \xi(x) \in \{\xi_0, \xi_1\} \cdot I\{u(x) > 0\}, \xi(x) = 0 \cdot I\{u(x) = 0\}, u(x) \in (0, r_1]$
if $\xi(x) = \xi_1$, and $u(x) \in (0, r_0]$ if $\xi(x) = \xi_0, (1, x) \in X\}$.

Let $v = \mu + r_1 + \eta + \xi_1$ be the uniform transition rate . With the similar arguments to that in the previous section, the Bellman optimality equation is given as
follows:

$$J(1, x) = \frac{1}{\beta + v} \Big[c^+ x^+ + c^- x^- + \mu J(1, x - 1) + \eta J(1, x)$$

$$+ \min \{(r_1 + \xi_1) J(1, x), \xi_1 J(0, x) + r_1 J(1, x + 1),$$

$$(\xi_1 - \xi_0) c_m + \xi_0 J(0, x) + r_0 J(1, x + 1) + (\xi_1 - \xi_0 + r_1 - r_0) J(1, x)\} \Big],$$
$$(5.22)$$

$$J(0, x) = \frac{1}{\beta + v} \Big[c^+ x^+ + c^- x^- + \mu J(0, x - 1) + \eta J(1, x) + (r_1 + \xi_1) J(0, x) \Big],$$
$$(5.23)$$

where c^+, c^- are the costs per unit of product over per unit of time for finished
goods inventory and customer demand backlog, respectively.

Proposition 5.3. *The optimal joint production and preventive maintenance policy
for the operation-dependent failure manufacturing system when the machine is up,
$(\xi^*(x), u^*(x))$, is given by*

$$(\xi^*(x), u^*(x)) = \begin{cases} (0,0) & \begin{aligned} &\text{if } (\xi_1 + r_1) J(1,x) \le \xi_1 J(0,x) + r_1 J(1,x+1), \\ &(\xi_0 + r_0) J(1,x) \le (\xi - \xi_0) c_m + \xi_0 J(0,x) + r_0 J(1,x+1) \end{aligned} \\[1em] (\xi_1, r_1) & \begin{aligned} &\text{if } \xi_1 J(0,x) + r_1 J(1,x+1) < (\xi_1 + r_1) J(1,x), \\ &(\xi_1 - \xi_0) J(0,x) + (r_1 - r_0) J(1,x+1) \le (\xi - \xi_0) c_m \\ &+ (r_1 + \xi_1 - r_0 - \xi_0) J(1,x) \end{aligned} \\[1em] (\xi_0, r_0) & \begin{aligned} &\text{if } (\xi - \xi_0) c_m + \xi_0 J(0,x) + r_0 J(1,x+1) < (\xi_0 + r_0) J(1,x), \\ &(\xi - \xi_0) c_m + (r_1 + \xi_1 - r_0 - \xi_0) J(1,x) < (\xi_1 - \xi_0) J(0,x) \\ &+ (r_1 - r_0) J(1,x_1 - 1, x_2 + 1) \end{aligned} \end{cases}$$

$$(5.24)$$

The above proposition states that the optimal joint production and preventive maintenance policy consists of three actions: no production (implied no maintenance), maximum production without maintenance, and maximum production with maintenance. The optimal cost function can be approximated numerically via the value iteration algorithm. We present an example below to explore the structure of the optimal joint production and preventive maintenance policy.

Example 5.2. Consider a failure-prone manufacturing supply chain with the following setting (operation-dependent failure mode): demand rate $\mu = 0.8$; machine repair rate $\eta = 0.9$; the machine failure rate under preventive maintenance $\xi_0 = 0.1$; the maximum production rate without preventive maintenance $r_1 = 2.0$; the finished goods inventory unit cost $c^+ = 1$; the backordering cost $c- = 10$; and the preventive maintenance cost $c_m = 1.0$. The discount factor $\beta = 0.1$. The system state space is limited into a finite area with $x \in [-50, 50]$. The iterative procedure will be terminated when the number of iterations exceeds 100.

Figure 5.6 illustrates the structure of the optimal policies for five cases with different combinations of the machine failure rate without preventive maintenance and the maximum production rate under preventive maintenance, that is, case (a) – $(\xi_1 = 0.3; r_0 = 1.4)$; case (b) – $(\xi_1 = 0.3; r_0 = 1.6)$; case (c) – $(\xi_1 = 0.3; r_0 = 1.8)$; case (d) – $(\xi_1 = 0.2; r_0 = 1.6)$; and case (e) – $(\xi_1 = 0.4; r_0 = 1.6)$.

It can be seen from Fig. 5.6 that the optimal policies in the above five cases have good structural properties. For example, we can observe the following: (1) no production should be performed when the finished goods inventory level reaches a certain point and (2) three control actions in (5.24) are taken in three non-intersected intervals in the system state space, although the positions of these intervals vary in different cases. However, it should be pointed out that observation (2) is not always true. In Chap. 10, we will provide examples to show that three actions (no production, production with maintenance, production without maintenance) may be taken in four non-intersected intervals.

5.4 Discussion and Notes

When the manufacturer has failure-prone machines, preventive maintenance is an important strategy to counteract the potential disruption to the production and to the supply chain operation. Section 5.2 is an extension of the supply chain

Fig. 5.6 Structure of the optimal production and preventive maintenance policy in an operation-dependent failure manufacturing supply chain

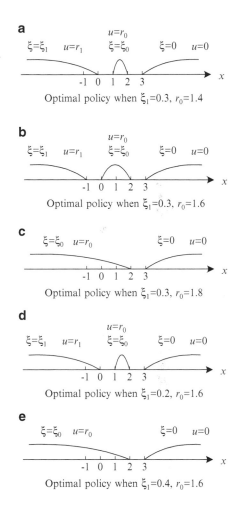

a

$u=r_0$

$\xi=\xi_1$ $u=r_1$ $\xi=\xi_0$ $\xi=0$ $u=0$

-1 0 1 2 3 x

Optimal policy when $\xi_1=0.3$, $r_0=1.4$

b

$u=r_0$

$\xi=\xi_1$ $u=r_1$ $\xi=\xi_0$ $\xi=0$ $u=0$

-1 0 1 2 3 x

Optimal policy when $\xi_1=0.3$, $r_0=1.6$

c

$\xi=\xi_0$ $u=r_0$ $\xi=0$ $u=0$

-1 0 1 2 3 x

Optimal policy when $\xi_1=0.3$, $r_0=1.8$

d

$u=r_0$

$\xi=\xi_1$ $u=r_1$ $\xi=\xi_0$ $\xi=0$ $u=0$

-1 0 1 2 3 x

Optimal policy when $\xi_1=0.2$, $r_0=1.6$

e

$\xi=\xi_0$ $u=r_0$ $\xi=0$ $u=0$

-1 0 1 2 3 x

Optimal policy when $\xi_1=0.4$, $r_0=1.6$

system in Chap. 2 to the case with preventive maintenance. Section 5.3 is partially based on Song (2009).

This chapter assumes that the preventive maintenance actions are performed without a machine stoppage . Such assumption may be interpreted in another way. For example, we may face a problem to make decisions on whether to select a more reliable machine with less production capacity or a less reliable machine with more production capacity. Nevertheless, in practice, machine stoppage maintenance is another common type of preventive maintenance. The model can be extended to this type of preventive maintenance by including additional machine modes, for example, stoppage mode for preventive maintenance, more reliable operational mode, less reliable operational mode, and failure mode. The machine may transition from the reliable mode into the less reliable mode. This problem can be formulated similarly.

Relevant literature in failure-prone manufacturing systems/supply chains with maintenance decisions may be classified into two groups (similar to Chap. 4): fluid flow models and discrete part manufacturing systems.

The first group with the maintenance decisions can be referred to Boukas and Haurie (1990), Boukas (1998), Gharbi and Kenne (2000), Boukas and Liu (2001), Kenne and Nkeungoue (2008), and the references there. In Boukas and Haurie (1990), the probability of machine failure is an increasing function of its age, while in Boukas and Liu (2001), the machine has multiple operational states and the machine failure rates are state dependent. With the assumptions of a constant demand rate and a discounted linear cost function, they established the optimal production and maintenance policies using stochastic dynamic programming. Boukas and Haurie (1990) showed the existence of hedging surfaces, in which the preventive maintenance actions are triggered by switching curves. Boukas (1998) studied the effectiveness of a hedging point policy in a production and corrective maintenance control problem. Gharbi and Kenne (2000) considered a multiple-identical-machine system and investigated the modified hedging point policy for maintenance and production using simulation-based approach. Kenne and Nkeungoue (2008) considered the control of corrective and preventive maintenance rates in a manufacturing system in which the failure rate of a machine depends on its age. Nodem et al. (2011) described a stochastic model for deteriorating failure-prone machine producing one type of product to meet constant customers' demand. The purpose is to find the optimal production, repair/replacement, and preventive maintenance policies.

The second group often takes into account the randomness in both production process and demand process. Sharafali (1984) studied the effects of a machine failure on the performance measure in which the production is controlled by an (s, S) policy. Srinivasan and Lee (1996) extended the above work by adding preventive maintenance option to the (s, S) policy. It is assumed that when the inventory level reaches S, the preventive maintenance is undertaken and the machine becomes as good as new. They obtained the optimal parameters s and S in order to minimize the total average cost. In their model, preventive maintenance and production are executed separately, and the preventive maintenance is impliedly determined by the production policy.

Van der Duyn Schouten and Vanneste (1995) considered a single machine production system with an exogenously given finite capacity buffer. The production policy is fixed and determined by the buffer level and capacity. They proposed a suboptimal preventive maintenance policy that is based on the age of the machine and the buffer level. Das and Sarkar (1999) studied a production inventory system where the unit production time, repair time, and maintenance time have general probability distributions. They presented a preventive maintenance policy based on the inventory level and the number of items produced since the last repair or maintenance operation. The production is governed by the (s, S) policy where the parameters s and S are exogenously given. Iravani and Duenyas (2002) considered a single machine system with multiple operational states and investigated how the structure of the integrated production and maintenance policy changes as the system enters different operating states. They introduced a double-threshold policy to characterize the decisions of production and maintenance and developed algorithms to implement the policy. The above three papers assumed that the unmet demands are not backordered (or partially) and therefore the system state space is finite. Yao et al. (2005) investigated the structural properties of the optimal joint preventive maintenance and production policy for an unreliable production–inventory system with constant demands and time-dependent failures. They demonstrated that the optimal production and maintenance policies have the control-limit structure and the optimal actions on the entire state space are divided into regions.

The hedging point policy can be regarded as a special type of threshold policies, and the switching-curve or switching-region policy is an extension of threshold policies. Moreover, the (s, S) type of policies (e.g., Srinivasan and Lee 1996; Das and Sarkar 1999) is a typical example of threshold policies. Therefore, threshold policies deserve more studies in production and maintenance control problems. In Chap. 10, we will investigate the effectiveness of threshold-type production and maintenance policies and analyze their steady-state performance measures.

References

Akella, R., Kumar, P.R.: Optimal control of production rate in a failure prone manufacturing system. IEEE Trans. Autom. Control **31**(2), 116–126 (1986)

Bielecki, T., Kumar, P.R.: Optimality of zero-inventory policies for unreliable manufacturing systems. Oper. Res. **36**(4), 532–541 (1988)

Boukas, E.K.: Hedging point policy improvement. J. Optim. Theory Appl. **97**(1), 47–70 (1998)

Boukas, E.K., Haurie, A.: Manufacturing flow control and preventive maintenance: a stochastic control approach. IEEE Trans. Autom. Control **35**(9), 1024–1031 (1990)

Boukas, E.K., Liu, Z.K.: Production and maintenance control for manufacturing systems. IEEE Trans. Autom. Control **46**(9), 1455–1460 (2001)

Buzacott, J.A., Hanifin, L.E.: Models of automatic transfer lines with inventory banks: a review and comparison. AIIE Trans. **10**(2), 197–207 (1978)

Dallery, Y., David, R., Xie, X.: Approximate analysis of transfer lines with unreliable machines and finite buffers. IEEE Trans. Autom. Control **34**(9), 943–953 (1989)

Das, T.K., Sarkar, S.: Optimal preventive maintenance in a production inventory system. IIE Trans. **31**, 537–551 (1999)

Fu, M., Xie, X.: Derivative estimation for buffer capacity of continuous transfer lines subject to operation-dependent failures. J. Discret. Event Dyn. Syst. Theory Appl. **12**(4), 447–469 (2002)

Gershwin, S.B.: Manufacturing Systems Engineering. Prentice-Hall, Englewood Cliffs (1994)

Gharbi, A., Kenne, J.P.: Production and preventive maintenance rates control for a manufacturing system: an experimental design approach. Intern. J. Prod. Econ. **65**, 275–287 (2000)

Hu, J.Q., Xiang, D.: Structural properties of optimal production controllers in failure prone manufacturing systems. IEEE Trans. Autom. Control **39**(3), 640–643 (1994)

Hu, J.Q., Vakili, P., Yu, G.: Optimality of hedging point policies in the production control of failure prone manufacturing systems. IEEE Trans. Autom. Control **39**, 1875–1880 (1994)

Iravani, S., Duenyas, I.: Integrated maintenance and production control of a deteriorating production system. IIE Trans. **34**, 423–435 (2002)

Kenne, J.P., Nkeungoue, L.J.: Simultaneous control of production, preventive and corrective maintenance rates of a failure-prone manufacturing system. Appl. Numer. Math. **58**(2), 180–194 (2008)

Kimemia, J.G., Gershwin, S.B.: An algorithm for the computer control of production in flexible manufacturing system. IIE Trans. **13**(4), 353–362 (1983)

Liberopoulos, G., Caramanis, M.: Production control of manufacturing systems with production rate dependent failure rates. IEEE Trans. Autom. Control **39**(4), 889–895 (1994)

Nodem, F.I.D., Kenne, J.P., Gharbi, A.: Simultaneous control of production, repair/replacement and preventive maintenance of deteriorating manufacturing systems. Int. J. Prod. Econ. **134**, 271–282 (2011)

Sethi, S., Zhang, Q.: Hierarchical Decision Making in Stochastic Manufacturing Systems. Birkhauser, Boston (1994)

Sharafali, M.: On a continuous review production–inventory problem. Oper. Res. Lett. **3**, 199–204 (1984)

Sharifnia, A.: Production control of a manufacturing system with multiple machine states. IEEE Trans. Autom. Control **33**(7), 620–625 (1988)

Song, D.P.: Production and preventive maintenance control in a stochastic manufacturing system. Int. J. Prod. Econ. **119**, 101–111 (2009)

Srinivasan, M.M., Lee, H.S.: Production/inventory system with preventive maintenance. IIE Trans. **28**, 879–890 (1996)

Van der Duyn Schouten, F.A., Vanneste, S.G.: Maintenance optimization of a production system with buffer capacity. Eur. J. Oper. Res. **82**, 323–338 (1995)

Xie, X.: Performance analysis of a transfer line with unreliable machines and finite buffers. IIE Trans. **25**(1), 99–108 (1993)

Xie, X., Mourani, I., Hennequin, S.: Performance evaluation of production lines subject to time and operation dependent failures using Petri Nets. In: The 8th International Conference on Control, Automation, Robotics and Vision, pp. 2123–2128. Kunming, 6–9 Dec 2004

Yao, X.D., Xie, X.L., Fu, M.C., Marcus, S.I.: Optimal joint preventive maintenance and production policies. Nav. Res. Logist. **52**, 668–681 (2005)

Chapter 6
Optimal Control of Supply Chain Systems with Assembly Operation

6.1 Introduction

Some members in supply chain systems often involve assembly operations, for example, manufacturers and assemblers. With the development of mass customization and agile supply chain, the concept of postponement has been applied in practice widely. For example, warehouses or distribution centers nowadays are not merely regarded as a storage place; they may also be used as an ideal place to perform final assembly operation to meet local customer requirements.

Unavailability of any type of raw materials or components disrupts the assembly operation, whereas unavailability of finished goods dissatisfies customers. On the other hand, holding too much inventory of raw materials or finished goods incurs unnecessary inventory costs. Therefore, it is essential to coordinate the raw material supply with the assembly operations and customer demands in order to manage the assembly supply chain well. However, this is challenging as the supply chain system is often subject to various types of uncertainties such as unreliable material delivery, stochastic processing time, random customer demand, and unreliable assembly machines.

This chapter considers an assembly supply chain system. The manufacturer has to make the ordering decision for n types of raw materials from n different suppliers in parallel and the production decision for assembling finished goods using all n types of raw materials. The system is subject to the uncertainties of raw material delivery lead times, assembly processing times, customer demands, and possible breakdown of the assembly machine. The objective is to minimize the expected discounted cost including raw materials inventory costs, finished goods inventory costs, and demand backlog costs.

We first formulate the optimal integrated ordering and assembly problem and derive the optimal control policy. With the assumption of maximum order size one, we then establish the structural properties of the optimal cost function and the optimal policy such as monotonicity and asymptotic behaviors. We then prove that the optimal policy can be characterized by a set of monotonic switching manifolds,

D.-P. Song, *Optimal Control and Optimization of Stochastic Supply Chain Systems*, Advances in Industrial Control, DOI 10.1007/978-1-4471-4724-4_6,
© Springer-Verlag London 2013

which lead to a set of control regions to determine the optimal control decisions. These structural properties are useful to construct simple but near-optimal control policies, which will be addressed in Chap. 11.

6.2   Problem Formulation

Consider a supply chain system with assembly operation as shown in Fig. 6.1. The manufacturer has multiple suppliers to supply n types of raw materials (or called components), which are required by the manufacturer to assemble the product (finished goods). The raw materials from suppliers are stored in raw materials warehouses. The manufacturer assembles a product by requiring one unit of components from each raw material warehouse. After the completion of the assembly operation, the finished goods may be stored in the finished goods warehouse as inventories or satisfy customer demands immediately. Customer demands arrive one at a time following a Poisson process with arriving rate μ. Unsatisfied demands are fully backlogged.

The Assumptions 2.1–2.4 stand in this chapter. In addition, it is assumed that the warehouses have sufficient capacities to simplify the problem formulation. The assembly operation cannot be performed if any of the raw materials is unavailable. The manufacturer assembles one product at a time, and the assembly time is exponentially distributed with a rate u. It is a control variable that takes 0 or r, which represents an action "not produce" or "produce at a speed r," respectively (the model can be extended easily to the case of allowing u taking more values between 0 and r, but the results remain the same). The manufacturer also makes decisions on the quantity of orders placed to the suppliers for raw materials. The quantity of an order is denoted as q_i for raw material i subject to $0 \leq q_i \leq Q_i$, where Q_i is the maximum ordering quantity in one order for raw material i. In other words, the system is subject to ordering capacity constraints instead of storage capacity constraints (cf. Sect. 3.7.4). The raw material replenishment lead time follows an independent exponential distribution with the mean $1/\lambda_i$, respectively. The lead time is assumed to be independent on actual order quantity.

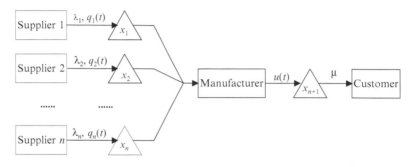

Fig. 6.1 A supply chain with assembly operation

Let $x_i(t)$ denote the on-hand inventory of raw materials at time t and $x_{n+1}(t)$ denote the on-hand inventory of finished goods at time t. Here, $x_{n+1}(t)$ could be negative, which represents the number of backordered demands. The system state space can be described by $X = \{\mathbf{x} := (x_1, x_2, \ldots, x_n, x_{n+1}) \, | \, x_i \geq 0 \text{ and } x_{n+1} \in Z\}$. Let $Z_+^n = \{(x_1, x_2, \ldots, x_n) \, | \, x_i \geq 0 \text{ and } x_i \in Z\}$. We have $X = Z_+^n \times Z$.

The manufacturer needs to make two types of decisions: the assembly rate $u(t) \in \{0, r\}$ and the raw material order quantity $q_i(t) \in [0, Q_i]$. Define an admissible control set $\Omega = \{\mathbf{u} = (u(t), q_1(t), q_2(t), \ldots, q_n(t)) \, | \, u(t) \in \{0, r\} \text{ if } x_i(t) > 0 \text{ for any } i \in \{1, 2, \ldots, n\}; u(t) = 0 \text{ if } \exists \, i \in \{1, 2, \ldots, n\} \text{ such than } x_i(t) \leq 0; q_i(t) \in [0, Q_i]; t \geq 0\}$. The problem is to find the optimal integrated policy $\mathbf{u} \in \Omega$ by minimizing the infinite horizon expected discounted cost (depending on the initial state).

$$J(\mathbf{x}_0) = \min_{\mathbf{u}} E\left[\int_0^\infty e^{-\beta t} g(\mathbf{x}(t), \mathbf{u}(t)) dt \, | \, \mathbf{x}(0) = \mathbf{x}_0 \right] \qquad (6.1)$$

where $0 < \beta < 1$ is a discount factor and $g(.)$ is a cost function to penalize raw material inventories, finished goods inventories, and backordered demands. For example, $g(.)$ can be defined as

$$g(\mathbf{x}, \mathbf{u}) = c_1 x_1 + c_2 x_2 + \cdots + c_n x_n + c_{n+1}^+ x_{n+1}^+ + c_{n+1}^- x_{n+1}^- \qquad (6.2)$$

where c_i, c_{n+1}^+ and c_{n+1}^- are the unit costs of raw material inventory, finished goods inventory, and backordered demands, respectively.

Similar to Definition 2.1, we introduce the following definition.

Definition 6.1.

- e_i be a unit vector in the state space X whose ith element is one.
- $\mathbf{e}$ be a unit vector in the state space whose each element is one, that is, $\mathbf{e} := e_1 + e_2 + \cdots + e_{n+1}$.
- $D := \{d_1, d_2, \ldots, d_n, d_{n+1}, d_{n+2}\}$ be the transition set, where $d_i = e_i$, for $i = 1, 2, \ldots, n$; $d_{n+1} = e_{n+1} - e_1 - e_2 - \cdots - e_n = 2e_{n+1} - \mathbf{e}$; and $d_{n+2} = -e_{n+1}$. Physically, d_i represents a replenishment of one item of raw material i, for $i = 1, 2, \ldots, n$; d_{n+1} represents an assembly operation, which depletes one unit of each raw material and produces one unit of finished goods; and d_{n+2} represents a customer demand arrival.
- $D(\mathbf{x}) := \{d \in D \, | \, \mathbf{x} + d \in X\}$ be the feasible transition directions from state $\mathbf{x}$.

There are three types of events in the system, that is, demand arrivals, assembly operation completions, and arrivals of placed order of any type of raw materials. Due to the memoryless properties of Poisson process and exponential distribution, unfinished assembly operation or unarrived orders interrupted by an event is statistically equivalent to that of restarting. Three types of events incur three different directions of state transition. The system state changes if and only if one of the above events occurs. The event occurring epochs comprise the set of decision epochs. At each decision epoch, a policy $\mathbf{u}$ specifies whether or not a replenishment

Fig. 6.2 State transition in a
supply chain with assembly
operation

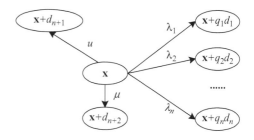

order for any of n raw materials should be placed, what the replenishment order size
is, and whether or not a product should be assembled. The state transition map at a
state $\mathbf{x}$ under the control $\mathbf{u}$ is depicted in Fig. 6.2.

6.3   Optimal Control Policy

Let $v = \mu + r + \lambda_1 + \lambda_2 + \cdots + \lambda_n$ be the uniform transition rate. By the
uniformization technique (Bertsekas 1987; Puterman 1994), the problem can be
formulated into a discrete-time Markov decision process. The one-step transition
probability function $\text{Prob}(. \mid ., \mathbf{u})$ is given by

$$\text{Prob}\{\mathbf{x} + q_i d_i \mid \mathbf{x}, \mathbf{u}\} = \lambda_i \cdot I\{q_i > 0\}/v, \quad \text{for } i = 1, 2, \ldots, n; \tag{6.3}$$

$$\text{Prob}\{\mathbf{x} + d_{n+1} \mid \mathbf{x}, \mathbf{u}\} = r \cdot I\{u > 0\}/v; \tag{6.4}$$

$$\text{Prob}\{\mathbf{x} + d_{n+2} \mid \mathbf{x}, \mathbf{u}\} = \mu/v; \tag{6.5}$$

$$\text{Prob}\{\mathbf{x} \mid \mathbf{x}, \mathbf{u}\} = \left[\sum_i (\lambda_i - \lambda_i \cdot I\{q_i > 0\}) + r - r \cdot I\{u > 0\}\right]/v. \tag{6.6}$$

Following the stochastic dynamic programming approach, the Bellman optimality equation is given as follows:

$$\begin{aligned}
J(\mathbf{x}) = \frac{1}{\beta + v}\Bigg[& g(\mathbf{x}) + \mu \cdot J(\mathbf{x} + d_{n+2}) + r \cdot J(\mathbf{x}) \cdot I\{\mathbf{x} + d_{n+1} \notin X\} \\
& + \min_{u=0,r}\{u \cdot J(\mathbf{x} + d_{n+1}) + (r - u) \cdot J(\mathbf{x})\} \cdot I\{\mathbf{x} + d_{n+1} \in X\} \\
& + \sum_i \lambda_i \cdot \min\{J(\mathbf{x} + q_i.d_i) \mid q_i \in [0, Q_i]\}\Bigg].
\end{aligned} \tag{6.7}$$

Proposition 6.1. *The optimal integrated ordering and assembly policy for the supply chain with assembly operation, $(u^*(\mathbf{x}), q_i^*(\mathbf{x})$, for $i = 1, 2, \ldots, n)$, is given by*

$$u^*(\mathbf{x}) = \begin{cases} r & J(\mathbf{x} + d_{n+1}) \leq J(\mathbf{x}), \quad \mathbf{x} + d_{n+1} \in X \\ 0 & \text{otherwise} \end{cases} ; \qquad (6.8)$$

$$q_i^*(\mathbf{x}) = j, \quad \text{if } J(\mathbf{x} + jd_i) = \min\{J(\mathbf{x}), J(\mathbf{x} + d_i), \ldots, J(\mathbf{x} + Q_i d_i)\}. \qquad (6.9)$$

The optimal discounted cost function $J(\mathbf{x})$ can be approximated numerically using the value iteration algorithm similar to (2.14) and (2.15) in Chap. 2.

To explore the explicit form of the optimal policy rigorously, we consider a special case with $Q_i = 1$ (for $i = 1, 2, \ldots, n$) in the next section.

6.4   Optimal Control Policy with Maximum Order Size One

With the assumption $Q_i = 1$ (for $i = 1, 2, \ldots, n$), the raw material ordering decision to each supplier is simplified into two options: place an order with size one, or do not place an order. To simplify the narrative, we introduce some new notation.

Definition 6.2. Define a set of mappings from $Z_+^n \times Z$ to $Z_+^n \times Z$ as follows:

- $A_i \mathbf{x} = \mathbf{x} + d_i$ for $i = 1, 2, \ldots, n$.
- $A_{n+1} \mathbf{x} = \mathbf{x} + d_{n+1}$ if $x_i > 0$ for any $i = 1, 2, \ldots, n$; $A_{n+1} \mathbf{x} = \mathbf{x}$, otherwise.
- $A_{n+2} \mathbf{x} = \mathbf{x} + d_{n+2}$.

Physically A_i $(i = 1, 2, \ldots, n)$ reflects an arrival of a unit of raw material i; A_{n+1} reflects an completion of an assembly operation if $x_i > 0$ for $i = 1, 2, \ldots, n$; and A_{n+2} reflects a departure of a finished goods in order to meet a customer demand. The Bellman optimality equation (6.7) can be simplified as

$$J(\mathbf{x}) = \frac{1}{\beta + v} \left[g(\mathbf{x}) + \mu \cdot J(A_{n+2}\mathbf{x}) + r \cdot \min\{J(A_{n+1}\mathbf{x}), J(\mathbf{x})\} \right.$$

$$\left. + \sum_i \lambda_i \cdot \min\{J(A_i\mathbf{x}), J(\mathbf{x})\} \right]. \qquad (6.10)$$

Proposition 6.2. *The optimal integrated ordering and assembly policy in the assembly supply chain with maximum order size one, $(u^*(\mathbf{x}), q_i^*(\mathbf{x}))$, for $i = 1, 2, \ldots, n$, is given by*

$$u^*(\mathbf{x}) = \begin{cases} r & J(A_{n+1}\mathbf{x}) \leq J(\mathbf{x}), x_i > 0 \,\forall i \in \{1, 2, \ldots, n\} \\ 0 & \text{otherwise} \end{cases} ; \qquad (6.11)$$

$$q_i{}^*(\mathbf{x}) = \begin{cases} 1 & J(A_i\mathbf{x}) \leq J(\mathbf{x}) \\ 0 & \text{otherwise} \end{cases}. \tag{6.12}$$

Next we aim to establish the explicit structure of the optimal policy in (6.11) and (6.12). To simplify the narrative, we assume $\beta + v = 1$ without loss of generality in the remainder of this section.

6.4.1 Structural Properties of Optimal Value Function

We first investigate the structural properties of the optimal value function. The following definitions are introduced:

Definition 6.3. $J(A_l\mathbf{x}) - J(\mathbf{x})$ is called decreasing for A_j if $J(A_l\mathbf{x}) - J(\mathbf{x}) \geq J(A_l A_j\mathbf{x}) - J(A_j\mathbf{x})$, $\forall\mathbf{x} \in X$, and increasing for A_j if the inequality is reversed.

Definition 6.4. $J(A_l\mathbf{x}) - J(\mathbf{x})$ is called decreasing in x_j if $J(A_l\mathbf{x}) - J(\mathbf{x}) \geq J(A_l(\mathbf{x} + e_j)) - J(\mathbf{x} + e_j)$, $\forall\mathbf{x} \in X$, and increasing for x_j if the inequality is reversed.

Lemma 6.1. *For every distinct pair (l, j), $J(A_l\mathbf{x}) - J(\mathbf{x})$ is decreasing for A_j, where $l, j \in \{1, 2, \ldots, n+2\}$ and $l \neq j$. In other words, $J(\mathbf{x})$ is submodular w.r.t. $D = \{d_1, d_2, \ldots, d_n, d_{n+1}, d_{n+2}\}$.*

Proof. This can be proved by the induction method on the k-stage cost function in the value iteration procedure. The assertion is obviously true when $k = 0$ and $J_0(\mathbf{x}) \equiv 0$. Suppose it holds for k, then we want to show it also holds for $k + 1$. From the value iteration algorithm, we have

$$\begin{aligned} J_{k+1}(A_l\mathbf{x}) - J_{k+1}(\mathbf{x}) &= \Delta g(l, \mathbf{x}) + \mu \cdot [J_k(A_{n+2}A_l\mathbf{x}) - J_k(A_{n+2}\mathbf{x})] \\ &+ r \cdot [\min\{J_k(A_{n+1}A_l\mathbf{x}), J_k(A_l\mathbf{x})\} - \min\{J_k(A_{n+1}\mathbf{x}), J_k(\mathbf{x})\}] \\ &+ \sum_i \lambda_i \cdot [\min\{J_k(A_i A_l\mathbf{x}), J_k(A_l\mathbf{x})\} - \min\{J_k(A_i\mathbf{x}), J_k(\mathbf{x})\}] \end{aligned} \tag{6.13}$$

where

$$\Delta g(l, \mathbf{x}) = \begin{cases} c_l & \text{if } 1 \leq l \leq n \\ -(c_1 + \cdots + c_n) + c_{n+1}^+ \cdot I\{x_{n+1} \geq 0\} - c_{n+1}^- \cdot I\{x_{n+1} < 0\} & \text{if } l = n + 1 \\ -c_{n+1}^+ \cdot I\{x_{n+1} > 0\} + c_{n+1}^- \cdot I\{x_{n+1} \leq 0\} & \text{if } l = n + 2 \end{cases}$$

where $I\{.\}$ is the indicator function. It is easy to verify that $\Delta g(l, \mathbf{x})$ is decreasing for A_j. By the induction hypothesis, the second term of the right-hand side (RHS) of (6.13) is also decreasing for A_j. Now it suffices to show that

$\forall\ i \in \{1,\ 2,\ \ldots,\ n+1\},\ \min\{J_k(A_i A_l \mathbf{x}),\ J_k(A_l \mathbf{x})\}\ -\ \min\{J_k(A_i \mathbf{x}),\ J_k(\mathbf{x})\}$ is decreasing for A_j where $j \neq l$.

Case 1: If $i = l$, then

$$\min\{J_k(A_i A_l \mathbf{x}),\ J_k(A_l \mathbf{x})\} - \min\{J_k(A_i \mathbf{x}),\ J_k(\mathbf{x})\}$$

$$= \min\{J_k(A_l A_l \mathbf{x}),\ J_k(A_l \mathbf{x})\} - \min\{J_k(A_l \mathbf{x}),\ J_k(\mathbf{x})\}$$

$$= \min\{J_k(A_l A_l \mathbf{x}) - J_k(A_l \mathbf{x}),\ 0\} + \max\{0,\ J_k(A_l \mathbf{x}) - J_k(\mathbf{x})\}.$$

Note that min{.} and max{.} preserve the monotonicity; by the induction hypothesis, it is decreasing for A_j.

Case 2: If $i \neq l$, then it suffices to show that

$$\min\{J_k(A_i A_l \mathbf{x}),\ J_k(A_l \mathbf{x})\} - \min\{J_k(A_i \mathbf{x}),\ J_k(\mathbf{x})\}$$

$$\geq \min\{J_k(A_i A_l A_j \mathbf{x}),\ J_k(A_l A_j \mathbf{x})\} - \min\{J_k(A_i A_j \mathbf{x}),\ J_k(A_j \mathbf{x})\}.$$

That is,

$$\min\{J_k(A_i A_l \mathbf{x}),\ J_k(A_l \mathbf{x})\} + \min\{J_k(A_i A_j \mathbf{x}),\ J_k(A_j \mathbf{x})\}$$

$$\geq \min\{J_k(A_i A_l A_j \mathbf{x}),\ J_k(A_l A_j \mathbf{x})\} + \min\{J_k(A_i \mathbf{x}),\ J_k(\mathbf{x})\}. \tag{6.14}$$

If $i = j$, then the above inequality is equivalent to

$$\min\{0,\ J_k(A_l \mathbf{x}) - J_k(A_i A_l \mathbf{x})\} + \min\{J_k(A_i A_i \mathbf{x}) - J_k(A_i \mathbf{x}),\ 0\}$$

$$\geq \min\{J_k(A_i A_l A_i \mathbf{x}) - J_k(A_l A_i \mathbf{x}),\ 0\} + \min\{0,\ J_k(\mathbf{x}) - J_k(A_i \mathbf{x})\}. \tag{6.15}$$

By the induction hypothesis, we have

$$J_k(A_l \mathbf{x}) - J_k(A_i A_l \mathbf{x}) \geq J_k(\mathbf{x}) - J_k(A_i \mathbf{x});$$

$$J_k(A_i A_i \mathbf{x}) - J_k(A_i \mathbf{x}) \geq J_k(A_i A_l A_i \mathbf{x}) - J_k(A_l A_i \mathbf{x}).$$

The above two inequalities yield directly the inequality (6.15).

If $i \neq j$, then we have

$$J_k(A_i A_l \mathbf{x}) + J_k(A_i A_j \mathbf{x}) \geq J_k(A_i A_l A_j \mathbf{x}) + J_k(A_i \mathbf{x}) \geq \text{RHS of (6.14)}; \tag{6.16}$$

$$J_k(A_i A_l \mathbf{x}) + J_k(A_j \mathbf{x}) = J_k(A_i A_l \mathbf{x}) - J_k(A_l \mathbf{x}) + J_k(A_l \mathbf{x}) + J_k(A_j \mathbf{x})$$

$$\geq J_k(A_i A_l A_j \mathbf{x}) - J_k(A_l A_j \mathbf{x}) + J_k(A_l \mathbf{x}) + J_k(A_j \mathbf{x})$$

$$\geq J_k(A_i A_l A_j \mathbf{x}) + J_k(\mathbf{x}) \geq \text{RHS of (6.14)}; \tag{6.17}$$

$$J_k(A_l\mathbf{x}) + J_k(A_j\mathbf{x}) \geq J_k(A_l A_j\mathbf{x}) + J_k(\mathbf{x}) \geq \text{RHS of } (6.14); \qquad (6.18)$$

$$J_k(A_l\mathbf{x}) + J_k(A_i A_j\mathbf{x}) = J_k(A_i A_j\mathbf{x}) - J_k(A_j\mathbf{x}) + J_k(A_j\mathbf{x}) + J_k(A_l\mathbf{x})$$

$$\geq J_k(A_i A_l A_j\mathbf{x}) - J_k(A_l A_j\mathbf{x}) + J_k(A_j\mathbf{x}) + J_k(A_l\mathbf{x})$$

$$\geq J_k(A_i A_l A_j\mathbf{x}) + J_k(\mathbf{x}) \geq \text{RHS of } (6.14). \qquad (6.19)$$

The inequalities (6.16), (6.17), (6.18), and (6.19) lead to the inequality (6.14). Thus, the assertion holds for $k + 1$. This completes the induction proof.   □

Proposition 6.3. *For the supply chain with assembly operation in Fig. 6.1, we have:*

(i) $J(A_i\mathbf{x}) - J(\mathbf{x})$ is increasing in x_i and x_{n+1} and is decreasing in x_j ($j \neq i$, $j \neq n + 1$), where $i = 1, 2, \ldots, n$.
(ii) $J(A_{n+1}\mathbf{x}) - J(\mathbf{x})$ is decreasing in x_j ($j \neq n + 1$) and increasing in x_{n+1}.
(iii) $J(A_{n+1}\mathbf{x}) - J(\mathbf{x}) \leq 0$, for $\forall \mathbf{x} \in \{\mathbf{x} \in X \,|\, x_{n+1} < 0\}$.

Proof. Assertion (i). From Lemma 6.1, for every distinct pair (i, j) such that $i, j \in \{1, 2, \ldots, n + 2\}$, we have

$$J(A_i\mathbf{x}) - J(\mathbf{x}) \geq J(A_i A_j\mathbf{x}) - J(A_j\mathbf{x}).$$

Recall the definition of A_j that yields $J(A_i\mathbf{x}) - J(\mathbf{x})$ is decreasing in x_j ($j \neq i$, $j \neq n + 1$) and is increasing in x_{n+1} (by letting $j = n + 2$). Moreover, we have

$$J(A_i A_i\mathbf{x}) - J(A_i\mathbf{x}) \geq J(A_i A_1 A_i\mathbf{x}) - J(A_1 A_i\mathbf{x}) \geq \ldots$$

$$\geq J(A_i A_{i-1} \ldots A_1 A_i\mathbf{x}) - J(A_i A_{i-1} \ldots A_1 A_i\mathbf{x})$$

$$= J(A_i A_i \ldots A_1\mathbf{x}) - J(A_i A_i \ldots A_1\mathbf{x})$$

$$\geq J(A_i A_{i+1} \ldots A_1\mathbf{x}) - J(A_{i+1} \ldots A_1\mathbf{x}) \geq \ldots$$

$$\geq J(A_i A_{n+2} \ldots A_1\mathbf{x}) - J(A_{n+2} \ldots A_1\mathbf{x})$$

$$= J(A_i\mathbf{x}) - J(\mathbf{x}).$$

The last equation is obtained by applying $A_{n+2}A_{n+1} \ldots A_1\mathbf{x} = \mathbf{x}$. So $J(A_i\mathbf{x}) - J(\mathbf{x})$ is increasing in x_i.

Assertion (ii). From Lemma 6.1, we have $J(A_{n+1}\mathbf{x}) - J(\mathbf{x}) \geq J(A_{n+1}A_j\mathbf{x}) - J(A_j\mathbf{x})$ for $j \neq n + 1$. This implies that $J(A_{n+1}\mathbf{x}) - J(\mathbf{x})$ is decreasing in x_j ($j \neq n + 1$) and is increasing in x_{n+1} (by letting $j = n + 2$).

Assertion (iii). Define $\underline{X} := \{\mathbf{x} \in X \mid x_{n+1} < 0\}$. We can prove this assertion by induction on k-stage value function. Suppose it holds for k, then we want to show it also holds for $k + 1$. From the value iteration algorithm, we have (for $\mathbf{x} \in \underline{X}$)

$$J_{k+1}(A_{n+1}\mathbf{x}) - J_{k+1}(\mathbf{x}) = -(c_1 + \cdots + c_n + c_{n+1}^-)$$

$$+ \mu \cdot [J_k(A_{n+2}A_{n+1}\mathbf{x}) - J_k(A_{n+2}\mathbf{x})]$$

$$+ r \cdot [\min\{J_k(A_{n+1}A_{n+1}\mathbf{x}), J_k(A_{n+1}\mathbf{x})\} - \min\{J_k(A_{n+1}\mathbf{x}), J_k(\mathbf{x})\}] \quad (6.20)$$

$$+ \sum_i \lambda_i \cdot [\min\{J_k(A_i A_{n+1}\mathbf{x}), J_k(A_{n+1}\mathbf{x})\} - \min\{J_k(A_i\mathbf{x}), J_k(\mathbf{x})\}].$$

Clearly, the first term on the RHS of (6.20) is negative. If $\mathbf{x} \in \underline{X}$, then $A_{n+2}\mathbf{x}$ certainly belongs to $\underline{X}$. By the induction hypothesis and $A_{n+2}A_{n+1}\mathbf{x} = A_{n+1}A_{n+2}\mathbf{x}$, the second term on the RHS of (6.20) is not greater than zero. Besides, the induction hypothesis yields

$$J_k(A_{n+1}\mathbf{x}) \le J_k(\mathbf{x}), \quad \text{for } \mathbf{x} \in \underline{X};$$

$$J_k(A_i A_{n+1}\mathbf{x}) \le J_k(A_i\mathbf{x}), \quad \text{for } \mathbf{x} \in \underline{X} \text{ and } i = 1, 2, \ldots, n.$$

So the third and the forth term on the RHS of (6.20) is also not greater than zero. Thus, $J_{k+1}(A_{n+1}\mathbf{x}) - J_{k+1}(\mathbf{x}) \le 0$. This completes the induction proof. $\qquad\square$

Recall (6.12), Proposition 6.3(i) implies that if it is the optimal decision to order type i raw materials at the state $\mathbf{x} = (x_1, x_2, \ldots, x_{n+1})$, then so is it at any of the following three states: $(x_1, \ldots, x_{i-1}, y, x_{i+1}, \ldots, x_{n+1})$, $(x_1, \ldots, x_n, z)$, and $(x_1, \ldots, x_{j-1}, w, x_{j+1}, \ldots, x_{n+1})$, where y, z and w are any reasonable integers subject to $y \le x_i$, $z \le x_{n+1}$, and $w \ge x_j$, $(j \ne i, j \ne n+1)$. Physically, the first case means that we should order raw material i if its inventory level is lower; the second case means that we should order raw material i if the finished goods inventory level is lower; and the third case means that we should order raw material i if the inventory level of any other type of raw materials is higher.

Recall (6.11), Proposition 6.3(ii) indicates that if it is the optimal decision for the manufacturer to perform assembly operation with the maximum rate r at the state $\mathbf{x} = (x_1, x_2, \ldots, x_{n+1})$, then so is it at $(x_1, \ldots, x_{j-1}, w, x_{j+1}, \ldots, x_{n+1})$ and $(x_1, \ldots, x_n, z)$, where w and z are any reasonable integers s.t. $w \ge x_j$, $(j \ne n+1)$ and $z \le x_{n+1}$. Physically, the first case indicates that we should assemble finished goods if the inventory level of any raw material is higher; the second case indicates that we should assemble finished goods if the finished goods inventory level is lower.

Proposition 6.3(iii) implies that the manufacturer should always assemble products at the maximum speed if the current state $\mathbf{x} = (x_1, x_2, \ldots, x_{n+1})$ satisfies $x_{n+1} < 0$. This is intuitively true since backlogs should be avoided if possible.

6.4.2   Characterization of the Optimal Policy

To better describe the structure of the optimal policy, we introduce the concept of switching manifolds as follows.

Definition 6.5. Define the switching manifolds as follows:

- $S_i(\mathbf{x} \backslash x_i) := S_i(x_1, \ldots, x_{i-1}, x_{i+1}, \ldots, x_{n+1}) := \min\{x_i \mid J(A_i\mathbf{x}) - J(\mathbf{x}) > 0,$ $x_i \geq 0\}$ for $i = 1, 2, \ldots, n$.
- $S_{n+1}(\mathbf{x} \backslash x_{n+1}) := S_{n+1}(x_1, x_2, \ldots, x_n) := \min\{x_{n+1} \mid J(A_{n+1}\mathbf{x}) - J(\mathbf{x}) > 0,$ $x_{n+1} \in Z\}$; if $J(A_{n+1}\mathbf{x}) - J(\mathbf{x}) \leq 0$ for any $x_{n+1} \in Z$, then we define $S_{n+1}(\mathbf{x} \backslash x_{n+1}) = +\infty$.

Proposition 6.4.

(i) The switching manifold $S_i(\mathbf{x} \backslash x_i)$ is decreasing in x_{n+1} and is increasing in x_j $(j \neq i, j \neq n+1)$, where $i = 1, 2, \ldots, n$;
(ii) $S_{n+1}(\mathbf{x} \backslash x_{n+1})$ is increasing in $x_j, j = 1, 2, \ldots, n$.

Proof. We only give the proof the assertion (i) here, because the proof of the assertion (ii) is similar. The definition of $S_i(.)$ yields

$$S_i(x_1, \ldots, x_{i-1}, x_{i+1}, \ldots, x_{n+1}) = \min\{x_i \mid J(A_i\mathbf{x}) - J(\mathbf{x}) > 0\},$$

$$S_i(x_1, \ldots, x_{i-1}, x_{i+1}, \ldots, x_{n+1} - 1) = \min\{x_i \mid J(A_i A_{n+2}\mathbf{x}) - J(A_{n+2}\mathbf{x}) > 0\}.$$

From Lemma 6.1, we obtain $J(A_i\mathbf{x}) - J(\mathbf{x}) \geq J(A_i A_{n+2}\mathbf{x}) - J(A_{n+2}\mathbf{x})$. In addition, $J(A_i\mathbf{x}) - J(\mathbf{x})$ is increasing in x_i by Proposition 6.3. Thus,

$$S_i(x_1, \ldots, x_{i-1}, x_{i+1}, \ldots, x_{n+1}) \leq S_i(x_1, \ldots, x_{i-1}, x_{i+1}, \ldots, x_{n+1} - 1).$$

Therefore, $S_i(\mathbf{x} \backslash x_i)$ is decreasing in x_{n+1}. Similarly, we can show that $S_i(\mathbf{x} \backslash x_i)$ is increasing in x_j $(j \neq i, j \neq n+1)$. This completes the proof. $\square$

Proposition 6.5.

(i) The switching manifold $S_{n+1}(\mathbf{x} \backslash x_{n+1}) \geq 0$. Moreover, $S_{n+1}(\mathbf{x} \backslash x_{n+1}) \equiv +\infty$ if $c_1 + \cdots + c_n \geq c_{n+1}^+$, and it converges to a finite asymptote as x_i tends to positive infinity if $c_1 + \cdots + c_n < c_{n+1}^+, i = 1, 2, \ldots, n$.
(ii) $S_i(\mathbf{x} \backslash x_i)$ converges to zero as x_{n+1} tends to positive infinity and converges to a finite asymptote as x_{n+1} tends to negative infinity.

Proof. Assertion (i). From the definition of $S_{n+1}(\mathbf{x} \backslash x_{n+1})$ and Proposition 6.3 (iii), we have $S_{n+1}(\mathbf{x} \backslash x_{n+1}) \geq 0$. This is the first part of (i). Moreover, if $c_1 + \cdots + c_n \geq c_{n+1}^+$, following the induction approach similar to the proof of Lemma 6.1, it can be shown that $J(A_{n+1}\mathbf{x}) - J(\mathbf{x}) \leq 0$ for any $x_{n+1} \in Z$, which leads to $S_{n+1}(\mathbf{x} \backslash x_{n+1}) \equiv +\infty$.

Now assume $c_1 + \cdots + c_n < c_{n+1}^+$, and $x_i = +\infty$ for $i = 1, 2, \ldots, n$. The system is reduced to be a single-stage manufacturing system with random customer demands. For a single-stage manufacturing system, it is easy to show that there exists a finite number $x_{n+1}^* < +\infty$ such that $S_{n+1}(+\infty, +\infty, \ldots, +\infty) = x_{n+1}^*$. That is,

$$\lim_{(x_1, \ldots, x_n) \to \infty} S_{n+1}(x_1, x_2, \ldots, x_n) = x_{n+1}^*.$$

By Proposition 6.4(ii), we know that $S_{n+1}(\mathbf{x}\backslash x_{n+1})$ is increasing in x_j, $j = 1, 2,$ $\ldots, n$. Thus, there exists a finite integer $x_{n+1}^{(i)}$ such that $x_{n+1}^{(i)} \leq x_{n+1}^{*}$ and

$$\lim_{x_i \to \infty} S_{n+1}(x_1, x_2, \ldots, x_n) = x_{n+1}^{(i)}, \quad \text{for } i = 1, 2, \ldots, n.$$

Assertion (ii). If $x_{n+1} = +\infty$, that is, there are infinite inventories of finished goods, so it is unnecessary to order any raw materials. That means $S_i(x_1, \ldots, x_{i-1},$ $x_{i+1}, \ldots, x_n, +\infty) = 0$. By Proposition 6.4(i), we know that $S_i(\mathbf{x}\backslash x_i)$ is decreasing in x_{n+1}. It follows

$$\lim_{x_{n+1} \to +\infty} S_i(x_1, \ldots, x_{i-1}, x_{i+1}, \ldots x_{n+1}) = 0.$$

On the other hand, if $x_{n+1} = -\infty$, Proposition 6.3 (iii) indicates that the assembly operation should be performed whenever possible. Suppose $x_j = +\infty$ for $j \neq i$, $j \neq n+1$. Then the ordering process for raw material i is reduced to be a one-stage ordering system with random demands. For one-stage ordering system, there exists a finite integer number $x_i^{*} < +\infty$ such that $S_i(+\infty, +\infty, \ldots, +\infty, -\infty) = x_i^{*}$. That is,

$$\lim_{(x_1, \ldots, x_{i-1}, x_{i+1}, \ldots, x_n) \to \infty} S_i(x_1, \ldots, x_{i-1}, x_{i+1}, \ldots, x_n, -\infty) = x_i^{*}.$$

By Proposition 6.4(i), we know that $S_i(\mathbf{x}\backslash x_i)$ is increasing in x_j, for $j \neq i$, $j \neq n+1$. Thus, there exists a finite integer $x_i^{(n+1)}$ such that $x_i^{(n+1)} \leq x_i^{*}$ and

$$\lim_{x_{n+1} \to -\infty} S_i(x_1, \ldots, x_{i-1}, x_{i+1}, \ldots x_{n+1}) = x_i^{(n+1)}.$$

This completes the proof.   □

Physically, $c_1 + \cdots + c_n \geq c_{n+1}^{+}$ means the cost of holding one unit of finished goods inventory is not greater than the sum of holding one unit of each raw material inventory. Proposition 6.5 (i) indicates that we should always perform the assembly operation whenever there are available raw materials. On the other hand, if $c_1 + \cdots + c_n < c_{n+1}^{+}$, we should stop assembling finished goods when its inventory level reaches to a certain point.

Note that $S_i(\mathbf{x}\backslash x_i)$ and $S_{n+1}(\mathbf{x}\backslash x_{n+1})$ are monotone in their any component and $x_1, x_2, \ldots, x_{n+1}$ are all integers; Proposition 6.5 (i) means that there exists finite integers $\bar{x}_i$ s.t. $S_{n+1}(\mathbf{x}\backslash x_{n+1}) = x_{n+1}^{(i)}$ when $x_i \geq \bar{x}_i$, $i = 1, 2, \ldots, n$. Proposition 6.5 (ii) means that there exist two finite integers $\bar{x}_{n+1}, \underline{x}_{n+1}$ s.t. $S_i(\mathbf{x}\backslash x_i) = 0$ when $x_{n+1} \geq \bar{x}_{n+1}$, and $S_i(\mathbf{x}\backslash x_i) = x_i^{(n+1)}$ when $x_{n+1} \leq \underline{x}_{n+1}$.

From the switching manifolds and their properties, we can define control regions in the state space X.

Definition 6.6. The control region R_i is defined as follows: $R_i := \{\mathbf{x} = (x_1, x_2, \ldots, x_{n+1}) \in X | x_i < S_i(\mathbf{x}\backslash x_i)\}$ for $i = 1, 2, \ldots, n$ and $R_{n+1} := \{\mathbf{x} = (x_1, x_2, \ldots, x_{n+1}) \in X | x_{n+1} < S_{n+1}(\mathbf{x}\backslash x_{n+1}), \text{and } x_i > 0 \text{ for } i = 1, 2, \ldots, n\}$.

Fig. 6.3 Illustration of control regions of the optimal policy. (**a**) Ordering policy for raw material 1. (**b**) Ordering policy for raw material 2. (**c**) Assembly operation policy

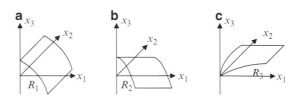

Now we are able to characterize the optimal control policy in (6.11) and (6.12) explicitly using the defined control regions.

Proposition 6.6. *The optimal integrated ordering and assembly policy in the assembly supply chain with maximum order size one, $(u^*(\mathbf{x}), q_i^*(\mathbf{x}))$, for $i = 1, 2,$ . . . , n, has the switching structure and can be determined by a set of control regions,*

$$u^*(\mathbf{x}) = \begin{cases} r & \mathbf{x} \in R_{n+1} \\ 0 & \mathbf{x} \notin R_{n+1} \end{cases}, \quad \text{and} \quad q_i^*(\mathbf{x}) = \begin{cases} 1 & \mathbf{x} \in R_i \\ 0 & \mathbf{x} \notin R_i \end{cases}, \quad \text{for } i = 1, 2, \ldots, n .$$

(6.21)

Physically, the optimal policy (6.21) states that we should order raw material i if and only if the system state locates below the switching manifold $S_i(\mathbf{x} \backslash x_i)$; we should perform assembly operation at the maximum rate if and only if the current state locates below the switching manifold $S_{n+1}(\mathbf{x} \backslash x_{n+1})$.

To visualize the control regions in the system state space intuitively, let $n = 2$. Then, R_1, R_2 and R_3 can be illustrated in Fig. 6.3.

6.5 The Failure-Prone Assembly Supply Chain

This section extends the assembly supply chain in the above section (with $Q_i = 1$ for $i = 1, 2, \ldots, n$) to the situation where the assembly machine is failure-prone. More specifically, we consider the case in which the assembly processes or machines are subject to Markovian failures and repairs with failure rate ξ and repair rate η.

The failure process is assumed to be time dependent. Let $\alpha(t)$ denote the machine state, that is, $\alpha(t) = 1$ if the machine is up and $\alpha(t) = 0$ if the machine is down. The system state space is redefined as $X = \{(\alpha, \mathbf{x}) \mid \alpha \in Z_2 \text{ and } \mathbf{x} \in Z_+^n \times Z\}$. The assembly operation can be performed only if the assembly machine is up, whereas the raw material ordering decisions can be placed regardless the status of the assembly machine.

Redefine the uniform transition rate $v := \xi + \eta + \mu + r + \lambda_1 + \lambda_2 + \cdots + \lambda_n$. Following the uniformization technique and stochastic dynamic programming theory, the Bellman optimality equations are given by

$$J(1, \mathbf{x}) = \frac{1}{\beta + \upsilon} \Bigg[g(\mathbf{x}) + \xi J(0, \mathbf{x}) + \eta J(1, \mathbf{x}) + \mu \cdot J(1, A_{n+2}\mathbf{x})$$

$$+ r \cdot \min \{ J(1, A_{n+1}\mathbf{x}), J(1, \mathbf{x}) \} + \sum_i \lambda_i \cdot \min \{ J(1, A_i\mathbf{x}), J(1, \mathbf{x}) \} \Bigg];$$

$$(6.22)$$

$$J(0, \mathbf{x}) = \frac{1}{\beta + \upsilon} \Bigg[g(\mathbf{x}) + (\xi + r) J(0, \mathbf{x}) + \eta J(1, \mathbf{x}) + \mu \cdot J(0, A_{n+2}\mathbf{x})$$

$$+ \sum_i \lambda_i \cdot \min \{ J(0, A_i\mathbf{x}), J(0, \mathbf{x}) \} \Bigg].$$

$$(6.23)$$

Proposition 6.7. *The optimal integrated ordering and assembly policy* $(u^*(\alpha, \mathbf{x}),$ $q_i^*(\alpha, \mathbf{x})),$ *for* $\alpha = 0, 1,$ *and* $i = 1, 2, \ldots, n,$ *in a failure-prone assembly supply chain with maximum order size one is given by*

$$u^*(\alpha, \mathbf{x}) = \begin{cases} r & J(\alpha, A_{n+1}\mathbf{x}) \leq J(\alpha, \mathbf{x}), \alpha = 1, x_i > 0, \forall i \in \{1, 2, \ldots n\} \\ 0 & \text{otherwise} \end{cases}; \quad (6.24)$$

$$q_i^*(\alpha, \mathbf{x}) = \begin{cases} 1 & J(\alpha, A_i\mathbf{x}) \leq J(\alpha, \mathbf{x}) \\ 0 & \text{otherwise} \end{cases}. \quad (6.25)$$

Follow the same arguments as that in Sect. 6.4, the similar structural properties such as Proposition 6.3–6.6 for the failure-prone assembly supply chain system can also be established, which are summarized below.

Proposition 6.8. *For the supply chain with failure-prone assembly operation, we have (for* $\alpha = 0, 1$*):*

(iv) $J(\alpha, A_i\mathbf{x}) - J(\alpha, \mathbf{x})$ is increasing in x_i and x_{n+1} and is decreasing in x_j $(j \neq i,$ $j \neq n + 1)$, where $i = 1, 2, \ldots, n$.
(v) $J(\alpha, A_{n+1}\mathbf{x}) - J(\alpha, \mathbf{x})$ is decreasing in x_j $(j \neq n + 1)$ and increasing in x_{n+1}.
(vi) $J(\alpha, A_{n+1}\mathbf{x}) - J(\alpha, \mathbf{x}) \leq 0$, for $\forall \ \mathbf{x} \in \{ \mathbf{x} \in X \mid x_{n+1} < 0 \}$.

Definition 6.7. Define the switching manifolds as follows:

• $S_i(\alpha, \mathbf{x} \backslash x_i) := S_i(\alpha, x_1, \ldots, x_{i-1}, x_{i+1}, \ldots, x_{n+1}) := \min\{x_i \mid J(\alpha, A_i\mathbf{x})$ $- J(\alpha, \mathbf{x}) > 0, x_i \geq 0\}$ for $i = 1, 2, \ldots, n$; and $\alpha = 0, 1$.
• $S_{n+1}(1, \mathbf{x} \backslash x_{n+1}) := S_{n+1}(1, x_1, x_3, \ldots, x_n) := \min\{x_{n+1} \mid J(1, A_{n+1}\mathbf{x}) - J(1,$ $\mathbf{x}) > 0, x_{n+1} \in Z\}$; if $J(1, A_{n+1}\mathbf{x}) - J(1, \mathbf{x}) \leq 0$ for any $x_{n+1} \in Z$, then we define $S_{n+1}(1, \mathbf{x} \backslash x_{n+1}) = +\infty$.

Proposition 6.9.

(i) The switching manifold $S_i(\alpha, \mathbf{x} \backslash x_i)$ is decreasing in x_{n+1} and is increasing in x_j $(j \neq i, j \neq n + 1)$, where $i = 1, 2, \ldots, n$; and $\alpha = 0, 1$;
(ii) $S_{n+1}(1, \mathbf{x} \backslash x_{n+1})$ is increasing in $x_j, j = 1, 2, \ldots, n$.

Proposition 6.10.

(i) The switching manifold $S_{n+1}(1, \mathbf{x}\backslash x_{n+1}) \geq 0$. Moreover, $S_{n+1}(1, \mathbf{x}\backslash x_{n+1})$
  $\equiv +\infty$ if $c_1 + \cdots + c_n \geq c_{n+1}{}^+$, and it converges to a finite asymptote as x_i
  tends to positive infinity if $c_1 + \cdots + c_n < c_{n+1}{}^+$, $i = 1, 2, \ldots, n$.
(ii) $S_i(\alpha, \mathbf{x}\backslash x_i)$ converges to zero as x_{n+1} tends to positive infinity and converges to
  a finite asymptote as x_{n+1} tends to negative infinity.

Definition 6.8. The control region $R_{i,\alpha}$ is defined as follows: $R_{i,\alpha} := \{\mathbf{x} = (x_1,$
$x_2, \ldots, x_{n+1}) \in X \mid x_i < S_i(\alpha, \mathbf{x}\backslash x_i)\}$ for $i = 1, 2, \ldots, n$, and $\alpha = 0, 1$; and
$R_{n+1} := \{\mathbf{x} = (x_1, x_2, \ldots, x_{n+1}) \in X \mid x_{n+1} < S_{n+1}(1, \mathbf{x}\backslash x_{n+1}),$ and $x_i > 0$ for
$i = 1, 2, \ldots, n\}$.

Proposition 6.11. *The optimal integrated ordering and assembly policy in the
failure-prone assembly supply chain with maximum order size one, $(u^*(\alpha, \mathbf{x}), q_i^*(\alpha,$
$\mathbf{x}))$, for $i = 1, 2, \ldots, n$, has the switching structure and can be determined by a set
of control regions,*

$$u^*(\alpha, \mathbf{x}) = \begin{cases} r & \mathbf{x} \in R_{n+1}, \alpha = 1 \\ 0 & \text{otherwise} \end{cases}, \quad q_i^*(\alpha, \mathbf{x}) = \begin{cases} 1 & \mathbf{x} \in R_{i,\alpha} \\ 0 & \mathbf{x} \notin R_{i,\alpha} \end{cases}, \quad \text{for } i = 1, 2, \ldots, n.$$

$$(6.26)$$

6.6   Discussion and Notes

Sections 6.4 and 6.5 are mainly based on Song et al. (1998). Assembly operation
is one of the most common processes for manufacturers. It has also become an
important value-adding activity for warehouses or distribution centers to implement
the postponement concept, which can improve supply chain flexibility and achieve
quick response to customer demand changes. Zipkin (2000) and Axsater (2006)
included chapters and sections to discuss assembly systems. However, there has
been relatively limited literature examining the optimal control policies for assembly
systems (Benjaafar et al. 2011).

Rosling (1989) generalized the serial model in Clark and Scarf (1960) to a
multistage assembly system with fixed lead times on a periodic review basis.
He characterized the optimal policy as a set of simple reorder policies, which is
equivalent to the case of a serial system, based on some assumptions on the initial
inventory levels. Chen and Zheng (1994) further extended the model to a multistage
network with setup costs at all stages and compound Poisson demand. The network
can represent serial, assembly, and one-warehouse multi-retailer systems. They
provided the performance bounds and showed the optimality of the echelon base-
stock based policies. The above papers assumed deterministic lead times.

A number of studies considered the assembly-to-order (ATO) systems in which
no inventory of finished goods is kept. Zhao and Simchi-Levi (2006) analyzed
the performance of a base-stock ATO policy and a batch-ordering ATO policy of

component inventory control in which replenishment lead times of the components are stochastic. Benjaafar and ElHafsi (2006) characterized the structure of an optimal production policy for components, in which unsatisfied demands are lost and the finished goods assembly time is neglected.

Benjaafar et al. (2011) extended the results in Benjaafar and ElhAFSI (2006) to a multistage assembly system with variable batch sizes. With the assumption that unsatisfied customer demands are lost, they showed that the optimal production policy for each stage is a state-dependent base-stock policy with the base-stock level nonincreasing in the inventory level of items that are downstream and nondecreasing in the inventory level of all other items. This work complements the results of the backlog cases in this book.

Instead of focusing on the optimal control of assembly systems, some researchers investigated the effectiveness of specific type of control policies. For example, Chaouiya et al. (2000) presented and compared pull-type production control policies, which are the combination of base-stock and Kanban control, for the production coordination of assembly manufacturing systems. Zhao (2008) considered a general class of supply chains including assembly, distribution, tree, and two-level networks as special cases. Assuming each stage is controlled by a base-stock policy with unsatisfied demands fully backordered, he focused on evaluating the stock-out delay for each unite of demand at each stage of the supply chain. Huh and Janakiraman (2010) studied an assembly system with capacity constraints under an echelon base-stock policy. In Chap. 11, we will discuss near-optimal threshold-type control policies and their effectiveness for assembly supply chain systems in detail.

References

Axsater, S.: Inventory Control, 2nd edn. Springer, New York (2006)

Benjaafar, S., ElHafsi, M.: Production and inventory control of a single product assemble-to-order system with multiple customer classes. Manage. Sci. **52**(12), 1896–1912 (2006)

Benjaafar, S., ElHafsi, M., Lee, C.Y., Zhou, W.H.: Optimal control of an assembly system with multiple stages and multiple demand classes. Oper. Res. **59**(2), 522–529 (2011)

Bertsekas, D.P.: Dynamic Programming: Deterministic and Stochastic Models. Prentice-Hall, Englewood Cliffs (1987)

Chaouiya, C., Liberopoulos, G., Dallery, Y.: The extended Kanban control system for production coordination of assembly manufacturing systems. IIE Trans. **32**(10), 999–1012 (2000)

Chen, F., Zheng, Y.S.: Lower bounds for multi-echelon stochastic inventory systems. Manage. Sci. **40**(11), 1426–1443 (1994)

Clark, A.J., Scarf, H.E.: Optimal policies for a multi-echelon inventory problem. Manage. Sci. **6**, 475–490 (1960)

Huh, W.T., Janakiraman, G.: Base-stock policies in capacitated assembly systems: convexity properties. Nav. Res. Logist. **57**(2), 109–118 (2010)

Puterman, M.L.: Markov Decision Processes: Discrete Stochastic Dynamic Programming. Wiley, New York (1994)

Rosling, K.: Optimal inventory policies for assembly systems under random demands. Oper. Res. **37**(4), 565–579 (1989)

Song, D.P., Sun, Y.X., Xing, W.: Optimal control of a stochastic assembly production line. J. Optim. Theory Appl. **98**(3), 681–700 (1998)

Zhao, Y.: Evaluation and optimization of installation base-stock policies in supply chains with compound Poisson demand. Oper. Res. **56**(2), 437–452 (2008)

Zhao, Y., Simchi-Levi, D.: Performance analysis and evaluation of assemble-to-order systems with stochastic sequential lead times. Oper. Res. **54**(4), 706–724 (2006)

Zipkin, P.H.: Foundations of Inventory Management. McGraw-Hill, New York (2000)

Chapter 7
Optimal Control of Supply Chain Systems with Multiple Products

7.1 Introduction

Internationalization or globalization has been rapidly developed in the last few decades. One of the important approaches toward internationalization is the focused factory, in which manufacturers attempt to concentrate their production by focusing on a relatively few standard products but serving the global market. Nevertheless, it is rare in reality that a business organization only produces one type of products, even using the same type of raw materials. It is therefore interesting to investigate how supply chain systems coordinate the production processes over multiple types of products in stochastic situations.

When producing multiple products, there is a competing need for finite resources, which requires a trade-off in allocating finite production capacity over multiple products/customers. For example, allocating too much capacity on one product can sufficiently satisfy one type of customers, but may incur backlog costs of other types of products. When the systems are subject to uncertainties such as random customer demands, stochastic processing times, uncertain raw material supply, and unreliable machine, it requires an intelligent trade-off.

This chapter tackles two problems. The first is the joint optimal control problem for raw material ordering and production capacity allocation over n part-types. We consider the production capacity cycling, in which only one product can be produced at any time. We will present the mathematical formulation and explore its optimal control structure. This is an extension of the problem in Chap. 2 to the situation with multiple products.

The second is the optimal production rate allocation problem in a failure-prone manufacturing supply chain producing two part-types. Two types of products are allowed to be produced at the same time. Apart from the system being subject to random customer demand and stochastic processing times, it is also subject to machine unreliability, which may force the manufacturer to stop production

D.-P. Song, *Optimal Control and Optimization of Stochastic Supply Chain Systems*, Advances in Industrial Control, DOI 10.1007/978-1-4471-4724-4_7,
© Springer-Verlag London 2013

unpredictably. We are able to establish the explicit structural properties of the optimal production capacity allocation policy, which will be further utilized to develop near-optimal threshold-type policies in Chap. 12.

7.2 Optimal Ordering and Production Control in a Supply Chain with Multiple Products

Consider a supply chain system with multiple products shown in Fig. 7.1. The manufacturer produces n types of products (finished goods) to meet different customers. The raw materials are stored in raw materials warehouses. The finished goods may be stored in the finished goods warehouse as inventories or satisfy customer demands immediately. Unsatisfied demands are fully backlogged.

The Assumptions 2.1–2.4 stand in this chapter. In addition, it is assumed that warehouses have sufficient capacities. However, the raw material order size is constrained by the ordering capacity. The manufacturer produces one product at a time using one unit of raw materials, and the processing time is exponentially distributed with a rate u_i for producing product type i. It is a control variable that takes a value between 0 and r, subject to $0 \leq \sum_i I\{u_i > 0\} \leq 1$ and $0 \leq u_i \leq r$ for $i = 1$, $2, \ldots, n$, where r represents the maximum production capacity. The manufacturer makes decisions on the production capacity allocation over multiple products (i.e., which product to produce and in what production rate, but at most, one product can be produced at any time), and also makes decisions on the quantity of orders placed to the supplier for raw materials. The quantity of an order is denoted as q for raw materials subject to $0 \leq q \leq Q$, where Q is the maximum ordering quantity in one order. The raw material replenishment lead time follows an exponential distribution with the mean $1/\lambda$. The lead time is assumed to be independent on actual order quantity. The product demands are random and the time between two demand events for product type i is exponentially distributed with average time $1/\mu_i$ for $i = 1, 2, \ldots, n$.

Let $x_0(t)$ denote the on-hand inventory of raw materials at time t and $x_i(t)$ denote the on-hand inventory of product type i at time t. Here $x_i(t)$ $(i > 0)$ could

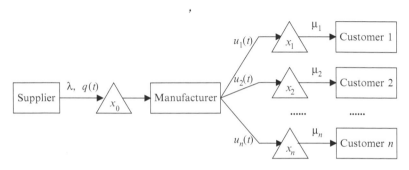

Fig. 7.1 A supply chain with multiple products

Fig. 7.2 State transition in a
supply chain with multiple
products

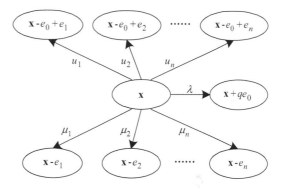

be negative, which represents the number of backlogged demands of product
type i. The system state space can be described by $X = \{\mathbf{x} := (x_0, x_1, \ldots, x_n) \mid x_0 \geq 0$ and $x_i \in Z$ for $i > 0\}$, namely, $X = Z_+ \times Z^n$.

Define an admissible control set $\Omega := \{\mathbf{u} = (u_1(t), u_2(t), \ldots, u_n(t), q(t)) \mid 0 \leq \sum_i I\{u_i(t) > 0\} \leq 1$ and $0 \leq u_i(t) \leq r$ for $i = 1, 2, \ldots, n; u_i(t) = 0$ if $x_0(t) \leq 0$; $q(t) \in [0, Q]; t \geq 0\}$. The problem is to find the optimal integrated policy $\mathbf{u} \in \Omega$ by minimizing the infinite horizon expected discounted cost (depending on the initial state):

$$J(\mathbf{x}_0) = \min_{\mathbf{u}} E \left[\int_0^\infty e^{-\beta t} g(\mathbf{x}(t), \mathbf{u}(t)) dt \mid \mathbf{x}(0) = \mathbf{x}_0 \right] \qquad (7.1)$$

where $0 < \beta < 1$ is a discount factor and $g(.)$ is a cost function to penalize raw material inventories, finished goods inventories, and backlogged demands. For example, $g(.)$ can be defined as

$$g(\mathbf{x}, \mathbf{u}) = c_0 x_0 + c_1^+ x_1^+ + c_1^- x_1^- + \cdots + c_n^+ x_n^+ + c_n^- x_n^- \qquad (7.2)$$

where c_0, c_i^+ and c_i^- are the unit costs of raw material inventory, finished goods inventory for product type i, and backlogged demands for product type i, respectively. To simplify the narrative, define e_i as a unit vector in the state space X whose $(i+1)$th element is one.

There are three types of events in the system, that is, demand arrivals for any type of products, production completions, and arrivals of placed orders for raw materials. Due to the memoryless properties of exponential distributions, unfinished production or unarrived orders interrupted by an event is statistically equivalent to that of restarting. The system state changes if and only if one of the above events occurs. The event occurring epochs comprise the set of decision epochs. At each decision epoch, a policy $\mathbf{u}$ specifies whether or not a replenishment order for the raw material should be placed, how much of the order size is, whether or not a product should be produced, which product to be produced, and what the production rate should be. The state transition map at a state $\mathbf{x}$ under the control $\mathbf{u}$ is depicted in Fig. 7.2.

Let $v = \lambda + r + \mu_1 + \mu_2 + \cdots + \mu_n$ be the uniform transition rate. By the uniformization technique (Bertsekas 1987; Puterman 1994), the problem can be formulated into a discrete-time Markov decision process. The one-step transition probability function Prob$(. \mid ., \mathbf{u})$ is given

$$\text{Prob}\,\{\mathbf{x} + qe_0|\mathbf{x}, \mathbf{u}\} = \lambda \cdot I \,\{q > 0\}\,/v, \tag{7.3}$$

$$\text{Prob}\,\{\mathbf{x} - e_0 + e_i|\mathbf{x}, \mathbf{u}\} = u_i \cdot I \,\{u_i > 0\}\,/v, \quad \text{for } i = 1, 2, \ldots, n, \tag{7.4}$$

$$\text{Prob}\,\{\mathbf{x} - e_i|\mathbf{x}, \mathbf{u}\} = \mu_i/v, \quad \text{for } i = 1, 2, \ldots, n, \tag{7.5}$$

$$\text{Prob}\,\{\mathbf{x}|\mathbf{x}, \mathbf{u}\} = \left[(\lambda - \lambda \cdot I \,\{q > 0\}) + r - \sum_i u_i \cdot I \,\{u_i > 0\} \right]/v. \tag{7.6}$$

Following the stochastic dynamic programming approach, the Bellman optimality equation is given as follows:

$$\begin{aligned} J\,(\mathbf{x}) = \frac{1}{\beta + v} \Big[&g\,(\mathbf{x}) + \sum_i \mu_i \cdot J\,(\mathbf{x} - e_i) \\ &+ r \cdot \min\{J\,(\mathbf{x}), J\,(\mathbf{x} - e_0 + e_1), \ldots, J\,(\mathbf{x} - e_0 + e_n)\} \\ &+ \lambda \cdot \min\{J\,(\mathbf{x} + q \cdot e_0)\,|q \in [0, Q]\}\,. \end{aligned} \tag{7.7}$$

Proposition 7.1. *The optimal stationary ordering and production policy for the supply chain with multiple products, $(u_i{}^*(\mathbf{x}), q^*(\mathbf{x})$, for $i = 1, 2, \ldots, n)$ is given by*

$$u_i{}^*\,(\mathbf{x}) = \begin{cases} r & \begin{aligned} &J\,(\mathbf{x} - e_0 + e_i) = \min\{J\,(\mathbf{x}), J\,(\mathbf{x} - e_0 + e_j), j = 1, 2, \ldots, n\}, \\ &J\,(x - e_0 + e_i) < J\,(x - e_0 + e_{i-1}) \quad \text{if } i > 1 \end{aligned} \\ 0 & \qquad\qquad\qquad\qquad\qquad\quad \text{otherwise} \end{cases}$$
$$\tag{7.8}$$

$$q^*\,(\mathbf{x}) = j, \quad \text{if } J\,(\mathbf{x} + je_0) = \min\{J\,(\mathbf{x}), J\,(\mathbf{x} + e_0), \ldots, J\,(\mathbf{x} + Qe_0)\}\,. \tag{7.9}$$

The above proposition states that the optimal policies are of switching structure and multiple optimal choices may be made only if the system state is on the switching surfaces. Proposition 7.1 gives priority to the products with smaller codes to break the ties. Note that there are finite control actions for both production capacity allocation decision and material ordering decision. The optimal discounted cost function $J(\mathbf{x})$ can therefore be approximated numerically using the value iteration algorithm similar to (2.14) and (2.15) in Chap. 2.

Proposition 7.2. $J(x - e_0 + e_j) - J(x) \leq 0$ *for any $x_0 > 0$ and $x_j < 0$.*

Proof. This can be proved by the induction method on the k-stage cost function in the value iteration procedure. Suppose the assertion is true for k and we want to show it also holds for $k + 1$. Consider (assume $\beta + \nu = 1$):

$$
\begin{aligned}
J_{k+1}\left(\mathbf{x} - e_0 + e_j\right) - J_{k+1}\left(\mathbf{x}\right) &= \left[g\left(\mathbf{x} - e_0 + e_j\right) - g\left(\mathbf{x}\right)\right] \\
&+ \sum_i \mu_i \cdot \left[J_k\left(\mathbf{x} - e_i - e_0 + e_j\right) - J_k\left(\mathbf{x} - e_i\right)\right] \\
&+ r \cdot \left[\min\left\{J_k\left(\mathbf{x} - e_0 + e_j\right), J_k\left(\mathbf{x} - e_0 + e_j - e_0 + e_1\right), \dots,\right.\right. \\
&\quad J_k\left(\mathbf{x} - e_0 + e_j - e_0 + e_n\right) - \min\left\{J_k\left(\mathbf{x}\right), J_k\left(\mathbf{x} - e_0 + e_1\right), \dots,\right. \\
&\quad \left[J_k\left(\mathbf{x} - e_0 + e_n\right)\right] + \lambda \cdot \left[\min\left\{J_k\left(\mathbf{x} - e_0 + e_j + q \times e_0\right) \mid q \in [0, Q]\right\}\right. \\
&\quad \left.\left. - \min\left\{J_k\left(\mathbf{x} + q \cdot e_0\right) \mid q \in [0, Q]\right\}\right]
\end{aligned}
$$

By the induction hypothesis, it can be seen that each term on the RHS of the above equation is not greater than zero for any $x_0 > 0$ and $x_j < 0$. This completes the induction proof. $\qquad\square$

To explore the explicit form of the optimal policy theoretically, we consider a special case by ignoring the raw material ordering process (i.e., assuming the raw materials are sufficient), but considering the unreliability in manufacturing process producing two types of products (i.e., $n = 2$) in next section.

7.3 Optimal Production Rate Allocation in a Failure-Prone Manufacturing Supply Chain Producing Two Part-Types

Consider a failure-prone manufacturing supply chain producing two types of products, in which the raw material ordering process is neglected. The manufacturing process is subject to Markovian failures and repairs with failure rate ξ and repair rate η. The failure process is time-dependent, that is, the probability of a failure occurring during a time interval does not depend on how much the machine has been used, but only on the length of the interval being up.

Let $\alpha(t)$ denote the machine state, that is, $\alpha(t) = 1$ if the machine is up and $\alpha(t) = 0$ if the machine is down. The system state space is redefined as $X = \{(\alpha, \mathbf{x}) \mid \alpha \in Z_2 \text{ and } \mathbf{x} \in Z^2\}$. The manufacturing operation can only be performed when the machine is up.

At each decision epoch, a policy $\mathbf{u} = (u_1(t), u_2(t))$ specifies whether or not a product should be produced, and the production rate allocation over two types of products subject to $0 \le u_1(t) + u_2(t) \le r$ and $u_i(t) \ge 0$. Different from the problem in Sect. 7.2, here we allow two products to be produced at same time but the total production capacity is constrained by r. The state transition map at a state $(\alpha, \mathbf{x})$ under the control $\mathbf{u}$ is depicted in Fig. 7.3.

Fig. 7.3 State transition in a
failure-prone manufacturing
supply chain with two
products

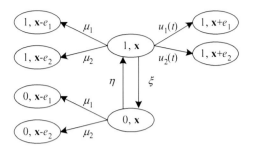

When the raw material ordering activity is ignored, the inventory holding cost
and demand backlog cost with two types of products can be simplified as

$$g\left(\mathbf{x}\right) = c_1^+ x_1^+ + c_1^- x_1^- + c_2^+ x_2^+ + c_2^- x_2^-. \tag{7.10}$$

Redefine the uniform transition rate $v := \xi + \eta + r + \mu_1 + \mu_2$. Following the
uniformization technique and stochastic dynamic programming theory, the Bellman
optimality equations are given by

$$J\left(1, \mathbf{x}\right) = \frac{1}{\beta + v} \left[g\left(\mathbf{x}\right) + \xi J\left(0, \mathbf{x}\right) + \eta J\left(1, \mathbf{x}\right) + \mu_1 \cdot J\left(1, \mathbf{x} - e_1\right)\right.$$

$$\left. + \mu_2 \cdot J\left(1, \mathbf{x} - e_2\right) + r \cdot \min\left\{J\left(1, \mathbf{x}\right), J\left(1, \mathbf{x} + e_1\right), J\left(1, \mathbf{x} + e_2\right)\right\}\right], \tag{7.11}$$

$$J\left(0, \mathbf{x}\right) = \frac{1}{\beta + v} \left[g\left(\mathbf{x}\right) + \xi J\left(0, \mathbf{x}\right) + \eta J\left(1, \mathbf{x}\right)\right.$$

$$\left. + \mu_1 \cdot J\left(0, \mathbf{x} - e_1\right) + \mu_2 \cdot J\left(0, \mathbf{x} - e_2\right) + r \cdot J\left(0, \mathbf{x}\right)\right]. \tag{7.12}$$

Proposition 7.3. *The optimal production rate allocation policy in the failure-prone
manufacturing supply chain producing two-type products when machine is up,*
$(u_1^*(1, \mathbf{x}), u_2^*(1, \mathbf{x}))$ *is given by*

$$\left(u_1^*, u_2^*\right) = \begin{cases} (r, 0) & J\left(1, \mathbf{x} + e_1\right) < J\left(1, \mathbf{x}\right), J\left(1, \mathbf{x} + e_1\right) \leq J\left(1, \mathbf{x} + e_2\right) \\ (0, r) & J\left(1, \mathbf{x} + e_2\right) < J\left(1, \mathbf{x}\right), J\left(1, \mathbf{x} + e_2\right) < J\left(1, \mathbf{x} + e_1\right) s. \\ (0, 0) & \text{otherwise} \end{cases}$$

$$\tag{7.13}$$

The above proposition gives only one type of the optimal policies for production
rate allocation, called bang-bang type. In fact, if $J(\mathbf{x} - e_0 + e_1) = J(\mathbf{x} - e_0 + e_2) <$
$J(\mathbf{x})$, then any element in $\{(u_1^*, u_2^*) : u_1^* + u_2^* = r, u_1^* \geq 0, u_2^* \geq 0\}$ is optimal.
However, the optimal policies are of switching structure, and multiple optimal
choices of rate allocation are possible only when the system state is on the switching
surfaces. If we limit the optimal policies in the form of bang-bang type, namely, the
manufacturer will either stop producing or allocate all the production capacity to

one of the product types. This implies that there are finite control actions for the production rate allocation, and the problem is equivalent to the production capacity cycling (i.e., only allow producing one type at a time).

Next we aim to explore the structural properties of the optimal value function first, and then to establish the explicit structure of the optimal policy in (7.13). To simplify the narrative, we assume $\beta + v = 1$ without loss of generality in the remainder of this section.

7.3.1 Structural Properties of the Optimal Value Function

Proposition 7.4. *For the failure-pone manufacturing supply chain producing two-type products, we have the following (for $\alpha = 0, 1$):*

(i) $J(\alpha, \mathbf{x} + e_1) - J(\alpha, \mathbf{x})$ is increasing in x_1 and x_2.
(ii) $J(\alpha, \mathbf{x} + e_2) - J(\alpha, \mathbf{x})$ is increasing in x_1 and x_2.
(iii) $J(\alpha, \mathbf{x} + e_1) - J(\alpha, \mathbf{x} + e_2)$ is increasing in x_1 and decreasing in x_2.

Proof. The proof can be shown by the induction method on k using the value iteration equation. Since $J_0(\alpha, \mathbf{x}) \equiv 0$, the assertions are true for $k = 0$. Suppose they hold for k, we want to prove they also hold for $k + 1$.

For assertion (i): Consider the case $\alpha = 1$. We have

$$
\begin{aligned}
J_{k+1}\left(1, \mathbf{x} + e_1\right) - J_{k+1}\left(1, \mathbf{x}\right) = {}& c_1{}^+ \cdot I\left\{x_1 \geq 0\right\} - c_1{}^- \cdot I\left\{x_1 < 0\right\} \\
& + \xi \left(J_k\left(0, \mathbf{x} + e_1\right) - J_k\left(0, \mathbf{x}\right)\right) + \eta \left(J_k\left(1, \mathbf{x} + e_1\right) - J_k\left(1, \mathbf{x}\right)\right) \\
& + \mu_1 \cdot \left(J_k\left(1, \mathbf{x}\right) - J_k\left(1, \mathbf{x} - e_1\right)\right) + \mu_2 \cdot \left(J_k\left(1, \mathbf{x} + e_1 - e_2\right) - J_k\left(1, \mathbf{x} - e_2\right)\right) \\
& + r \cdot \min\left(\left\{J_k\left(1, \mathbf{x} + e_1\right), J_k\left(1, \mathbf{x} + 2e_1\right), J_k\left(1, \mathbf{x} + e_1 + e_2\right)\right\}\right. \\
& \left. - \min\left\{J_k\left(1, \mathbf{x}\right), J_k\left(1, \mathbf{x} + e_1\right), J_k\left(1, \mathbf{x} + e_2\right)\right\}\right).
\end{aligned}
$$

In order to prove $J_{k+1}(1, \mathbf{x} + e_1) - J_{k+1}(1, \mathbf{x})$ is increasing in x_1 and x_2, it suffices to show that TERM (defined below) is increasing in x_1 and x_2:

$$
\begin{aligned}
\text{TERM} := {}& \min\left\{J_k\left(1, \mathbf{x} + e_1\right), J_k\left(1, \mathbf{x} + 2e_1\right), J_k\left(1, \mathbf{x} + e_1 + e_2\right)\right\} \\
& - \min\left\{J_k\left(1, \mathbf{x}\right), J_k\left(1, \mathbf{x} + e_1\right), J_k\left(1, \mathbf{x} + e_2\right)\right\}.
\end{aligned} \tag{7.14}
$$

First, rewrite the above expression as follows:

$$
\begin{aligned}
\text{TERM} = {}& \min\left\{0, J_k\left(1, \mathbf{x} + 2e_1\right) - J_k\left(1, \mathbf{x} + e_1\right), J_k\left(1, \mathbf{x} + e_1 + e_2\right)\right. \\
& \left. - J_k\left(1, \mathbf{x} + e_1\right)\right\} + \max\left\{J_k\left(1, \mathbf{x} + e_1\right) - J_k\left(1, \mathbf{x}\right), 0, J_k\left(1, \mathbf{x} + e_1\right)\right. \\
& \left. - J_k\left(1, \mathbf{x} + e_2\right)\right\}.
\end{aligned} \tag{7.15}
$$

Clearly, by the induction hypotheses (i)–(iii), each term of the right-hand side of (7.15) is increasing in x_1 and the operations min and max preserve the monotonicity, that means TERM is increasing in x_1. Secondly, to prove TERM is increasing in x_2, it suffices to show that

$$\text{TERM} \leq \min\{J_k(1, \mathbf{x} + e_1 + e_2), J_k(1, \mathbf{x} + 2e_1 + e_2), J_k(1, \mathbf{x} + e_1 + 2e_2)\}$$
$$- \min\{J_k(1, \mathbf{x} + e_2), J_k(1, \mathbf{x} + e_1 + e_2), J_k(1, \mathbf{x} + 2e_2)\}. \qquad (7.16)$$

It is not difficult to show that

$$\min\{J_k(1, \mathbf{x} + e_1), J_k(1, \mathbf{x} + 2e_1), J_k(1, \mathbf{x} + e_1 + e_2)\} - J_k(1, \mathbf{x})$$
$$\leq \text{RHS of (7.16)}$$
$$\min\{J_k(1, \mathbf{x} + e_1), J_k(1, \mathbf{x} + 2e_1), J_k(1, \mathbf{x} + e_1 + e_2)\} - J_k(1, \mathbf{x} + e_1)$$
$$\leq \text{RHS of (7.16)}.$$

Thus, in order to prove inequality (7.16), it suffices to prove

$$\min(J_k(1, \mathbf{x} + e_1), J_k(1, \mathbf{x} + 2e_1), J_k(1, \mathbf{x} + e_1 + e_2)) - J_k(1, \mathbf{x} + e_2)$$
$$\leq \text{RHS of (7.16)}. \qquad (7.17)$$

Note that

$$\text{LHS of (7.17)} \leq J_k(1, \mathbf{x} + e_1 + e_2) - J_k(1, \mathbf{x} + e_2); \qquad (7.18)$$

$$J_k(1, \mathbf{x} + e_1 + e_2) - J_k(1, \mathbf{x} + e_2)$$
$$\leq J_k(1, \mathbf{x} + e_1 + e_2) - \min\{J_k(1, \mathbf{x} + e_2), J_k(1, \mathbf{x} + e_1 + e_2), J_k(1, \mathbf{x} + 2e_2)\}; \qquad (7.19)$$

$$J_k(1, \mathbf{x} + e_1 + e_2) - J_k(1, \mathbf{x} + e_2) \leq J_k(1, \mathbf{x} + 2e_1 + e_2) - J_k(1, \mathbf{x} + e_1 + e_2)$$
$$\leq J_k(1, \mathbf{x} + 2e_1 + e_2) - \min\{J_k(1, \mathbf{x} + e_2), J_k(1, \mathbf{x} + e_1 + e_2), J_k(1, \mathbf{x} + 2e_2)\}; \qquad (7.20)$$

$$J_k(1, \mathbf{x} + e_1 + e_2) - J_k(1, \mathbf{x} + e_2) \leq J_k(1, \mathbf{x} + e_1 + 2e_2) - J_k(1, \mathbf{x} + 2e_2)$$
$$\leq J_k(1, \mathbf{x} + e_1 + 2e_2) - \min\{J_k(1, \mathbf{x} + e_2), J_k(1, \mathbf{x} + e_1 + e_2), J_k(1, \mathbf{x} + 2e_2)\}. \qquad (7.21)$$

From (7.19), (7.20), and 7.21), inequality (7.16) holds. For the case $\alpha = 0$, the proof is straightforward.

For assertion (ii): Due to the symmetry of x_1 and x_2. It can be shown using the same argument.

For assertion (iii): Consider the case $\alpha = 1$ (for $\alpha = 0$, the proof is simple). The value iteration equation yields

$$
\begin{aligned}
J_{k+1}\left(1, \mathbf{x} + e_1\right) - J_{k+1}\left(1, \mathbf{x} + e_2\right) &= \left[c_1{}^+ \cdot I\left\{x_1 \geq 0\right\} - c_1{}^- \cdot I\left\{x_1 < 0\right\}\right] \\
&\quad - \left[c_2{}^+ \cdot I\left\{x_2 \geq 0\right\} - c_2{}^- \cdot I\left\{x_2 < 0\right\}\right] \\
&\quad + \xi\left[J_k\left(0, \mathbf{x} + e_1\right) - J_k\left(0, \mathbf{x} + e_2\right)\right] + \eta\left[J_k\left(1, \mathbf{x} + e_1\right) - J_k\left(1, \mathbf{x} + e_2\right)\right] \\
&\quad + \mu_1 \cdot \left[J_k\left(1, \mathbf{x}\right) - J_k\left(1, \mathbf{x} - e_1 + e_2\right)\right] + \mu_2 \cdot \left[J_k\left(1, \mathbf{x} + e_1 - e_2\right) - J_k\left(1, \mathbf{x}\right)\right] \\
&\quad + r \cdot \left[\min\left\{J_k\left(1, \mathbf{x} + e_1\right), J_k\left(1, \mathbf{x} + 2e_1\right), J_k\left(1, \mathbf{x} + e_1 + e_2\right)\right\}\right. \\
&\quad \left. - \min\left\{J_k\left(1, \mathbf{x} + e_2\right), J_k\left(1, \mathbf{x} + e_1 + e_2\right), J_k\left(1, \mathbf{x} + 2e_2\right)\right\}\right].
\end{aligned}
$$

From the induction hypotheses, in order to prove $J_{k+1}(1, \mathbf{x} + e_1) - J_{k+1}(1, \mathbf{x} + e_2)$ is increasing in x_1 and decreasing in x_2, it suffices to show that TERM1 (defined below) is increasing in x_1 and decreasing in x_2:

$$
\begin{aligned}
\text{TERM1} := &\min\left\{J_k\left(1, \mathbf{x} + e_1\right), J_k\left(1, \mathbf{x} + 2e_1\right), J_k\left(1, \mathbf{x} + e_1 + e_2\right)\right\} \\
&- \min\left\{J_k\left(1, \mathbf{x} + e_2\right), J_k\left(1, \mathbf{x} + e_1 + e_2\right), J_k\left(1, \mathbf{x} + 2e_2\right)\right\}. \quad (7.22)
\end{aligned}
$$

We only show that TERM1 is increasing in x_1 (for x_2, the proof is similar). Define TERM2 as follows:

$$
\begin{aligned}
\text{TERM2} := &\min\left\{J_k\left(1, \mathbf{x} + 2e_1\right), J_k\left(1, \mathbf{x} + 3e_1\right), J_k\left(1, \mathbf{x} + 2e_1 + e_2\right)\right\} \\
&- \min\left\{J_k\left(1, \mathbf{x} + e_1 + e_2\right), J_k\left(1, \mathbf{x} + 2e_1 + e_2\right), J_k\left(1, \mathbf{x} + e_1 + 2e_2\right)\right\}. \\
&\hspace{10cm} (7.23)
\end{aligned}
$$

Then, TERM1 is increasing in x_1 is equivalent to TERM1 $\leq$ TERM2. The proof will be divided into three parts: (iii-1), (iii-2), and (iii-3) as follows.

(iii-1): Note that

$$
\begin{aligned}
J_k\left(1, \mathbf{x} + 2e_1\right) &- J_k\left(1, \mathbf{x} + e_2\right) \\
&= J_k\left(1, \mathbf{x} + 2e_1\right) - J_k\left(1, \mathbf{x} + e_1\right) + J_k\left(1, \mathbf{x} + e_1\right) - J_k\left(1, \mathbf{x} + e_2\right) \\
&\leq J_k\left(1, \mathbf{x} + 3e_1\right) - J_k\left(1, \mathbf{x} + 2e_1\right) + J_k\left(1, \mathbf{x} + 2e_1\right) - J_k\left(1, \mathbf{x} + e_1 + e_2\right) \\
&= J_k\left(1, \mathbf{x} + 3e_1\right) - J_k\left(1, \mathbf{x} + e_1 + e_2\right), \hspace{3cm} (7.24)
\end{aligned}
$$

$$
J_k\left(1, \mathbf{x} + e_1 + e_2\right) - J_k\left(1, \mathbf{x} + e_2\right) \leq J_k\left(1, \mathbf{x} + 2e_1 + e_2\right) - J_k\left(1, \mathbf{x} + e_1 + e_2\right), \quad (7.25)
$$

$$
J_k\left(1, \mathbf{x} + e_1\right) - J_k\left(1, \mathbf{x} + e_2\right) \leq J_k\left(1, \mathbf{x} + 2e_1\right) - J_k\left(1, \mathbf{x} + e_1 + e_2\right). \quad (7.26)
$$

From (7.24), (7.25), and (7.26), we have

$$\min\left\{J_k\left(1,\mathbf{x}+e_1\right), J_k\left(1,\mathbf{x}+2e_1\right), J_k\left(1,\mathbf{x}+e_1+e_2\right)\right\} - J_k\left(1,\mathbf{x}+e_2\right)$$
$$\leq \min\left\{J_k\left(1,\mathbf{x}+2e_1\right), J_k\left(1,\mathbf{x}+3e_1\right), J_k\left(1,\mathbf{x}+2e_1+e_2\right)\right\} - J_k\left(1,\mathbf{x}+e_1+e_2\right).$$
$$(7.27)$$

It follows

$$\min\left\{J_k\left(1,\mathbf{x}+e_1\right), J_k\left(1,\mathbf{x}+2e_1\right), J_k\left(1,\mathbf{x}+e_1+e_2\right)\right\} - J_k\left(1,\mathbf{x}+e_2\right)$$
$$\leq \text{TERM2}. \qquad\qquad (7.28)$$

(iii-2): Note that

$$\min\left\{J_k\left(1,\mathbf{x}+e_1\right), J_k\left(1,\mathbf{x}+2e_1\right), J_k\left(1,\mathbf{x}+e_1+e_2\right)\right\} - J_k\left(1,\mathbf{x}+e_1+e_2\right)$$
$$\leq J_k\left(1,\mathbf{x}+2e_1\right) - J_k\left(1,\mathbf{x}+e_1+e_2\right) \leq J_k\left(1,\mathbf{x}+3e_1\right) - J_k\left(1,\mathbf{x}+2e_1+e_2\right)$$
$$\leq J_k\left(1,\mathbf{x}+3e_1\right) - \min\left\{J_k\left(1,\mathbf{x}+e_1+e_2\right), J_k\left(1,\mathbf{x}+2e_1+e_2\right),\right.$$
$$\left. J_k\left(1,\mathbf{x}+e_1+2e_2\right)\right\}; \qquad\qquad (7.29)$$

$$\min\left\{J_k\left(1,\mathbf{x}+e_1\right), J_k\left(1,\mathbf{x}+2e_1\right), J_k\left(1,\mathbf{x}+e_1+e_2\right)\right\} - J_k\left(1,\mathbf{x}+e_1+e_2\right)$$
$$\leq 0 \leq J_k\left(1,\mathbf{x}+2e_1+e_2\right) - \min\left\{J_k\left(1,\mathbf{x}+e_1+e_2\right), J_k\left(1,\mathbf{x}+2e_1+e_2\right),\right.$$
$$\left. J_k\left(1,\mathbf{x}+e_1+2e_2\right)\right\}; \qquad\qquad (7.30)$$

$$\min\left\{J_k\left(1,\mathbf{x}+e_1\right), J_k\left(1,\mathbf{x}+2e_1\right), J_k\left(1,\mathbf{x}+e_1+e_2\right)\right\} - J_k\left(1,\mathbf{x}+e_1+e_2\right)$$
$$\leq J_k\left(1,\mathbf{x}+2e_1\right) - J_k\left(1,\mathbf{x}+e_1+e_2\right)$$
$$\leq J_k\left(1,\mathbf{x}+2e_1\right) - \min\left\{J_k\left(1,\mathbf{x}+e_1+e_2\right), J_k\left(1,\mathbf{x}+2e_1+e_2\right),\right.$$
$$\left. J_k\left(1,\mathbf{x}+e_1+2e_2\right)\right\}. \qquad\qquad (7.31)$$

The inequalities (7.29), (7.30), and (7.31) yield

$$\min\left\{J_k\left(1,\mathbf{x}+e_1\right), J_k\left(1,\mathbf{x}+2e_1\right), J_k\left(1,\mathbf{x}+e_1+e_2\right)\right\} - J_k\left(1,\mathbf{x}+e_1+e_2\right)$$
$$\leq \text{TERM2}. \qquad\qquad (7.32)$$

(iii-3): Note that

$$\min\left\{J_k\left(1,\mathbf{x}+e_1\right), J_k\left(1,\mathbf{x}+2e_1\right), J_k\left(1,\mathbf{x}+e_1+e_2\right)\right\} - J_k\left(1,\mathbf{x}+2e_2\right)$$
$$\leq J_k\left(1,\mathbf{x}+e_1+e_2\right) - J_k\left(1,\mathbf{x}+2e_2\right)$$
$$\leq J_k\left(1,\mathbf{x}+2e_1+e_2\right) - J_k\left(1,\mathbf{x}+e_1+2e_2\right)$$
$$\leq J_k\left(1,\mathbf{x}+2e_1+e_2\right) - \min\left\{J_k\left(1,\mathbf{x}+e_1+e_2\right), J_k\left(1,\mathbf{x}+2e_1+e_2\right),\right.$$
$$\left. J_k\left(1,\mathbf{x}+e_1+2e_2\right)\right\}; \qquad\qquad (7.33)$$

$$\min \{J_k (1, \mathbf{x} + e_1), J_k (1, \mathbf{x} + 2e_1), J_k (1, \mathbf{x} + e_1 + e_2)\} - J_k (1, \mathbf{x} + 2e_2)$$

$$\leq J_k (1, \mathbf{x} + 2e_1) - J_k (1, \mathbf{x} + 2e_2) \leq J_k (1, \mathbf{x} + 3e_1) - J_k (1, \mathbf{x} + e_1 + 2e_2)$$

$$\leq J_k (1, \mathbf{x} + 3e_1) - \min \{J_k (1, \mathbf{x} + e_1 + e_2), J_k (1, \mathbf{x} + 2e_1 + e_2),$$

$$J_k (1, \mathbf{x} + e_1 + 2e_2)\}; \tag{7.34}$$

$$\min \{J_k (1, \mathbf{x} + e_1), J_k (1, \mathbf{x} + 2e_1), J_k (1, \mathbf{x} + e_1 + e_2)\} - J_k (1, \mathbf{x} + 2e_2)$$

$$\leq J_k (1, \mathbf{x} + e_1 + e_2) - J_k (1, \mathbf{x} + 2e_2) \leq J_k (1, \mathbf{x} + 2e_1) - J_k (1, \mathbf{x} + e_1 + e_2)$$

$$\leq J_k (1, \mathbf{x} + 2e_1) - \min \{J_k (1, \mathbf{x} + e_1 + e_2), J_k (1, \mathbf{x} + 2e_1 + e_2),$$

$$J_k (1, \mathbf{x} + e_1 + 2e_2)\}. \tag{7.35}$$

The inequalities (7.32), (7.33), and (7.34) yield

$$\min \{J_k (1, \mathbf{x} + e_1), J_k (1, \mathbf{x} + 2e_1), J_k (1, \mathbf{x} + e_1 + e_2)\} - J_k (1, \mathbf{x} + 2e_2)$$

$$\leq \text{TERM2}. \tag{7.36}$$

From (7.28), (7.32), and (7.36), we obtain TERM1 $\leq$ TERM2. Thus, by (iii-1), (iii-2), and (iii-3), assertion (iii) is true. This completes the induction proof. $\square$

Proposition 7.5. *(i)* $J(\alpha, \mathbf{x} + e_1) - J_k(\alpha, \mathbf{x}) \leq 0$ *for any* $x_1 < 0$, $x_2 \in Z$; *(ii)* $J(\alpha, \mathbf{x} + e_2) - J_k(\alpha, \mathbf{x}) \leq 0$ *for any* $x_2 < 0$, $x_1 \in Z$.

Proof. With the similar argument to Proposition 7.4, the results can be established by the induction approach. $\square$

Proposition 7.6. *(i)* $J(\alpha, \mathbf{x} + e_1) - J_k(\alpha, \mathbf{x} + e_2) \leq 0$ *for any* $x_1 < 0$, $x_2 < 0$, *if* $c_1^- \geq c_2^-$. *(ii)* $J(\alpha, \mathbf{x} + e_2) - J_k(\alpha, \mathbf{x} + e_1) \leq 0$ *for any* $x_1 < 0$, $x_2 < 0$, *if* $c_2^- \geq c_1^-$.

Proof. We only need consider assertion (i). Use the induction approach on k in the value iteration equation. Since $J_0(\alpha, \mathbf{x}) \equiv 0$, the assertions are true for $k = 0$. Suppose they hold for k, we want to prove they also hold for $k + 1$. Consider the case $\alpha = 1$:

$$J_{k+1} (1, \mathbf{x} + e_1) - J_{k+1} (1, \mathbf{x} + e_2) = \left[(c_1^+ \cdot I \{x_1 \geq 0\} - c_1^- \cdot I \{x_1 < 0\}) \right.$$

$$- (c_2^+ \cdot I \{x_2 \geq 0\} - c_2^- \cdot I \{x_2 < 0\}) \right]$$

$$+ \xi [J_k (0, \mathbf{x} + e_1) - J_k (0, \mathbf{x} + e_2)] + \eta [J_k (1, \mathbf{x} + e_1) - J_k (1, \mathbf{x} + e_2)]$$

$$+ \mu_1 \cdot [J_k (1, \mathbf{x}) - J_k (1, \mathbf{x} - e_1 + e_2)] + \mu_2 \cdot [J_k (1, \mathbf{x} + e_1 - e_2) - J_k (1, \mathbf{x})]$$

$$+ r \cdot [\min \{J_k (1, \mathbf{x} + e_1), J_k (1, \mathbf{x} + 2e_1), J_k (1, \mathbf{x} + e_1 + e_2)\}$$

$$- \min \{J_k (1, \mathbf{x} + e_2), J_k (1, \mathbf{x} + e_1 + e_2), J_k (1, \mathbf{x} + 2e_2)\}]. \tag{7.37}$$

From the condition $c_1^- \geq c_2^-$ and the induction hypotheses, it is clear that the first four terms on the RHS of (7.37) is nonpositive when $x_1 < 0$ and $x_2 < 0$. It suffices to show that, for any $x_1 < 0, x_2 < 0$,

$$\min \{J_k (1, \mathbf{x} + e_1), J_k (1, \mathbf{x} + 2e_1), J_k (1, \mathbf{x} + e_1 + e_2)\}$$
$$- \min \{J_k (1, \mathbf{x} + e_2), J_k (1, \mathbf{x} + e_1 + e_2), J_k (1, \mathbf{x} + 2e_2)\} \leq 0. \quad (7.38)$$

From the induction hypothesis and Proposition 7.4(iii), we have $0 \geq J_k (1, \mathbf{x} + e_1) - J_k (1, \mathbf{x} + e_2) \geq J_k (1, \mathbf{x} + e_1 + e_2) - J_k (1, \mathbf{x} + 2e_2)$. This implies that the second term on the LHS of (7.38) is equal to $\min \{J_k (1, \mathbf{x} + e_2), J_k (1, \mathbf{x} + e_1 + e_2)\}$. From Proposition 7.5(i), we have $\min \{J_k (1, \mathbf{x} + e_2), J_k (1, \mathbf{x} + e_1 + e_2)\} = J_k (1, \mathbf{x} + e_1 + e_2)$. It is therefore clear that (7.38) holds. This completes the proof. $\qquad \square$

Physically, Proposition 7.5 indicates that we should always produce the product type that is backlogged, and Proposition 7.6 further reveals that the production priority should be given to the type of products with larger backlog cost when both of them are backlogged.

7.3.2 Characterization of the Optimal Policy

First, we define three switching curves as follows:

Definition 7.2. Define the switching curves as follows:

- $S_1 (x_2) := \max \{x_1 | J(1, \mathbf{x}) - J (1, \mathbf{x} + e_1) \geq 0\}$.
- $S_2 (x_1) := \max \{x_2 | J(1, \mathbf{x}) - J (1, \mathbf{x} + e_2) \geq 0\}$.
- $S_3 (x_2) := \max \{x_1 | J (1, \mathbf{x} + e_2) - J (1, \mathbf{x} + e_1) \geq 0\}$.

From Proposition 7.5, it is clear that $S_1(x_2) \geq -1$ for any x_2 and $S_2(x_1) \geq -1$ for any x_1.

Proposition 7.7. *The switching curves $S_1(x_2)$ and $S_2(x_1)$ are monotonic decreasing, and the $S_3(x_2)$ is monotonic increasing.*

Proof. From the definition of $S_1(x_2)$, it follows

$$S_1 (x_2) := \max \{x_1 | J (1, \mathbf{x}) - J (1, \mathbf{x} + e_1) \geq 0\},$$
$$S_1 (x_2 - 1) := \max \{x_1 | J (1, \mathbf{x} - e_1) - J (1, \mathbf{x}) \geq 0\}.$$

By Proposition 7.5(i), we have

$$J (1, \mathbf{x} - e_1) - J (1, \mathbf{x}) \geq J (1, \mathbf{x}) - J (1, \mathbf{x} + e_1).$$

This implies that $S_1(x_2 - 1) \geq S_1(x_2)$, namely, $S_1(x_2)$ is monotonic decreasing. With the similar arguments, we can prove that $S_2(x_1)$ are monotonic decreasing, and the $S_3(x_2)$ is monotonic increasing. This completes the proof.   □

Proposition 7.8. *(i) There exist finite integers y_1 and y_2 such that $J(\alpha, y_1 + 1, y_2) - J(\alpha, y_1, y_2) > 0$; (ii) there exist finite integers z_1 and z_2 such that $J(\alpha, z_1, z_2 + 1) - J(\alpha, z_1, z_2) > 0$.*

Proof. We only need to verify the assertion (i). Suppose the assertion (i) does not hold. That means, for all $x_1 \in Z$ and $x_2 \in Z$, we have

$$J(\alpha, x_1 + 1, x_2) - J(\alpha, x_1, x_2) \leq 0.$$

Especially, for any $x_1 > 0$ and $x_2 > 0$, we have (by Proposition 7.4(i))

$$J(\alpha, x_1, x_2) - J(\alpha, 0, x_2) \leq 0.$$

From the Bellman optimality equation, it follows (for any $x_1 > 0$ and $x_2 > 0$)

$$J(\alpha, \mathbf{x}) = \frac{1}{\beta + v} \left[c_1{}^+ x_1 + c_2{}^+ x_2 + \xi J(0, \mathbf{x}) + \eta J(1, \mathbf{x}) \right.$$
$$+ \mu_1 \cdot J(\alpha, \mathbf{x} - e_1) + \mu_2 \cdot J(\alpha, \mathbf{x} - e_2)$$
$$+ r \cdot \min\{J(\alpha, \mathbf{x}), J(\alpha, \mathbf{x} + e_1), J(\alpha, \mathbf{x} + e_2)\}$$
$$\left. \cdot I\{\alpha = 1\} + r \cdot J(a, \mathbf{x}) \cdot I\{\alpha = 0\} \right]. \qquad (7.39)$$

In the above equation, let $x_2 > 0$ be a fixed positive integer and x_1 tend to infinity. Then the LHS of (7.39) is not greater than $J(\alpha, 0, x_2)$. On the other hand, each term of the RHS of (7.39) is nonnegative. In addition, $c_1{}^+ x_1$ tends to be infinity, which means the LHS of (7.39) is infinite. This results in contradiction. Therefore, the assertion (i) holds. This completes the proof.   □

Proposition 7.9. *(i) The switching curve $S_1(x_2)$ converges to a nonnegative finite asymptote as x_2 tends to infinity; (ii) the switching curve $S_2(x_1)$ converges to a nonnegative finite asymptote as x_1 tends to infinity.*

Proof. We only need to verify the assertion (i). From Proposition 7.8(i), there exist some finite integers y_1 and y_2 such that $J(\alpha, y_1 + 1, y_2) - J(\alpha, y_1, y_2) > 0$. Together with Proposition 7.4(i) and the definition of $S_1(x_2)$, we have $-1 \leq S_1(y_2) < \infty$. This implies that $S_1(y_2)$ is a finite integer. From Proposition 7.8, it follows

$$-1 \leq S_1(x_2) \leq S_1(y_2), \qquad \text{for any } x_2 \geq y_2.$$

Since $S_1(x_2)$ is a monotonic decreasing integer-valued function, the assertion (i) is obvious. This completes the proof.   □

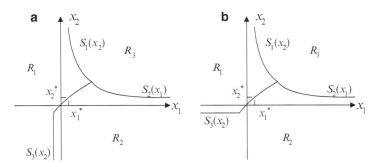

Fig. 7.4 The optimal control regions for the failure-prone manufacturing supply chain with two products. (**a**) When $c_1 > c_2^-$. (**b**) When $c_1^- < c_2^-$

Intuitively, if $x_2 = +\infty$, that means, there are infinite inventories of the second product and it is unnecessary to produce the second product type anymore. At this time, the control decision is made only for producing the first part-type. Hence, this case is equivalent to the one part-type system. The results in Proposition 7.9 are in agreement with the intuition.

Note that $S_1(x_2)$ and $S_2(x_1)$ both are decreasing and bounded, and x_2 and x_1 both are integers, thus, there exist $\bar{x}_1$ and $\bar{x}_2$ such that

$$S_1(x_2) \equiv x_1^* \quad \text{for any } x_2 \geq \bar{x}_2 \text{ and } S_2(x_1) \equiv x_2^* \quad \text{for any } x_1 \geq \bar{x}_1.$$

From Propositions 7.5 and 7.6, the following results are obvious:

Proposition 7.10. *The switching curve $S_3(x_2)$ has the following properties: (i) $S_1(x_2) \equiv -1$ for $x_2 < 0$, if $c_1^- \geq c_2^-$. (ii) $S_1(x_2) \equiv -\infty$ for $x_2 < 0$, if $c_2^- > c_1^-$.*

Proposition 7.11. *For the failure-prone manufacturing supply chain producing two-type products, the optimal production rate allocation policy $u^* = (u_1^*(1, x), u_2^*(1, x))$ is given by*

$$\left(u_1^*(1, x), u_2^*(1, x)\right) = \begin{cases} (r, 0) & x \in R_1 \\ (0, r) & x \in R_2 \\ (0, 0) & x \in R_3 \end{cases}$$

where $R_1 := \{x \mid x_1 \leq S_1(x_2) \text{ and } x_1 \geq S_3(x_2)\}$, $R_2 := \{x \mid x_2 \leq S_2(x_1) \text{ and } x_1 < S_3(x_2)\}$, and $R_3 := \{x \mid x_2 > S_2(x_1) \text{ and } x_1 \geq S_1(x_2)\}$. The control regions R_1, R_2, and R_3 are illustrated in Fig. 7.4a when $c_1^- \geq c_2^-$ and Fig. 7.4b when $c_1^- < c_2^-$.

From Proposition 7.11, we know that the optimal control for two part-types case is of region switching structure and can be described by three monotone curves. Moreover, $S_1(x_2)$ and $S_2(x_1)$ converge to nonnegative finite asymptotes. Based on these properties, we are able to obtain optimal or near-optimal policies, which are easy to operate and compute in practice.

7.4 Discussion and Notes

This chapter is based on Song and Sun (1999) with an extension to the supply chain context. The results in Sect. 7.3 indicate that the optimal production rate allocation policy is actually equivalent to the case of production capacity cycling among multiple products. For the failure-prone manufacturing supply chain producing two types of products, the optimal production control policy can be characterized by three monotonic switching curves, which is similar to the case with reliable manufacturing process in Ha (1997c).

Relevant literature may be classified into three groups. The first group focuses on stochastic production control/scheduling problem for multiple products. There may be different types of finished goods inventories. The decisions include when and which type of products should be produced. The second group considered the stochastic production/inventory systems with two types of products/customers in which one type has a higher priority than the other. A common problem in this group is to determine the appropriate stock rationing level for the low-priority product/customer. The third group focuses on production control and rationing of a single product among multiple classes of customers. There is only one type of finished goods inventories. The decisions include when a product should be produced and which class of customers should be satisfied. The priority rule is often an endogenous result in this group rather than an exogenous specification in the second group.

In the first group of literature, Graves (1980) considered the multiproduct production cycling problem for a capacitated facility with stochastic demand. A heuristic rule is proposed and shown to be effective using simulation. Zheng and Zipkin (1990) considered a symmetric model producing two types of products with Poisson demands, where two products have the same parameters. They applied the $(S - 1, S)$ base-stock policy to determine the on–off decision, and stated that whenever an allocation decision is required, the manufacturer should produce the product with the lower net inventory (inventory minus backorders). Wein (1992) considered the production scheduling problem with multiple products. Each product has its own general service time distribution and demand for each product is a random process. Under the heavy traffic conditions, a parametric scheduling policy is proposed, which is an aggregated base-stock on–off policy in associated with cost and service rate based priority rules. Pena-Perez and Zipkin (1997) developed a simple but effective heuristic policy for the multiple product scheduling problem based on the results in the above two papers and provided a numerical study to compare the myopic, static-priority, and optimal policies.

Ha (1997c) was among the first to establish the characteristics of the optimal policy for the two-product dynamic scheduling problem. With the assumption of Poisson demands and exponential production times, it is shown that the optimal policy should produce the product with larger $b\mu$ index when there is a backlog for this product regardless the inventory state of the other product, where b is the backlog cost and μ is the production rate. If the production times are identically distributed, the optimal policy can be further characterized by monotonic switching curves. De Vericourt et al. (2000) used fluid flow models to study a similar system to Ha (1997c), in which two products have different production rates. They showed that the monotonic switching curve that separates the priority regions of the two products is actually a straight line whose position can be expressed by a simple equation. Arruda and Do Val (2008) presented a discrete event model for a multistage multiproduct production and storage problem with random lead times and demands. They introduced the notion of stochastic stability and truncated the state space to circumvent the computational burden and obtain the suboptimal policies. Zhao and Lian (2011) considered a queuing-inventory system with two classes of customers requiring different service rates consuming one unit of raw materials supplied by an external supplier with exponentially distributed lead times. The system is more like a two-stage production system to meet two classes of customers on a make-to-order basis. The raw material ordering policy follows a continuous review (r, Q) policy. They showed that the $b\mu$ index rule is optimal to determine which class of customers should be served, and obtained the steady-state probability distribution using quasi-birth-and-death method.

Choia et al. (2005) considered a stochastic periodic-review multiproduct inventory system with unequal replenishment lead times for different products and limited warehouse capacity. They presented a few heuristic methods to minimize warehouse inventory-related costs.

In the second group, the system produces two types of products (or serves two types of customers) in which one type has a higher priority than the other. Such priority relationship is exogenously specified and is often static. Nahmias and Demmy (1981) considered an inventory system which maintains stock to meet both high- and low-priority demands, and analyzed a specific threshold control policy with or without stock rationing. Cohen et al. (1988) presented a model of an (s, S) inventory system in which there are two priority classes of customers. Melchiors et al. (2000) analyzed a continuous review (s, Q) model with lost sales and two prioritized demand classes. A single-threshold level policy is applied to ration the inventory among the two demand classes. For Poisson demand and deterministic lead times, they presented an exact formulation of the average inventory cost and optimized the threshold value.

Ha (1997b) considered a make-to-stock production system with two priority classes of customers. The type one customer has a high priority and will be backordered if no inventory is available, whereas type two may be satisfied from inventory, or backordered/rejected when inventory is low. With Poisson arriving demands and exponential production times, he is able to shown that the optimal production control and the inventory rationing policies can be characterized by a single monotonic switching curve. Carr and Duenyas (2000) considered the joint admission and sequencing decisions for a stochastic production system producing two types of products. The first type has the high priority on a make-to-stock (MTS) basis with unmet demands equivalently lost, whereas the second type has the low priority on a make-to-order (MTO) basis with an option to reject the orders. They found an optimal policy characterized by monotonic nonlinear switching curves.

Isotupa (2006) considered a lost-sales Markovian inventory system with two types of customers (a priority type and an ordinary type). Under the (s, Q) ordering policy, they obtained the expression for the long-run expected cost and provided efficient algorithm to determine the optimal values for the reorder level and reorder quantity. Iravani et al. (2007) considered a capacitated manufacturing system producing a single product with two prioritized customers, in which the advance demand information regarding the order size of the primary customer is available to the manufacturer. They showed that the manufacturer's optimal production and stock reservation policies are threshold-type policies.

Song (2009) considered a prioritized base-stock control for a stochastic production system to meet two priority demand classes, and obtained the explicit steady-state probability distribution using the spectral expansion approach. Iravani et al. (2012) extended the work in Carr and Duenyas (2000) and studied a joint optimization problem for production and admission control in stochastic manufacturing systems with a high-priority MTS product and a low-priority MTO product. They characterized the optimal production and admission policies with a partial-linear structure.

In the periodic-review situation, Frank et al. (2003) considered a multi-period inventory system with two priority demand classes, in which type one is deterministic and must be satisfied in each period, whereas type two is stochastic and the firm has the option to ration inventory to type two customers. They showed that the optimal order quantity and rationing policy are state dependent but do not have a simple structure. Teunter and Haneveld (2008) studied the dynamic inventory rationing strategies for stochastic inventory systems with two demand classes (critical and noncritical) in a single-period situation. They focused on determining the number of items to be reserved for the critical demand depending on the remaining time until the next order arrives. They also provided a comprehensive review on inventory rationing literature in various contexts.

In the third group of literature, the main concern is the stock rationing among multiple classes of customers for a single type of products, in which the priority of customers is determined endogenously. Because the economic parameters for different classes of customers are often different, the manufacturer has to make decisions not only on which class to serve first, but also on whether to conserve the inventory for more important customers later on. The problem has been studied in various contexts since the late 1960s. For example, Topkis (1968) considered the optimal ordering and rationing policies in a dynamic inventory model with multiple demand classes. He provided conditions under which the optimal rationing policy between successive procurements of new stock can be characterized by a set of critical rationing levels. Ha (1997a) studied the inventory rationing in a make-to-stock production system with several demand classes with Poisson arrivals. He showed that the optimal inventory rationing policy is characterized by a set of threshold parameters. The above results were further extended to a MTS production system with Erlang processing times in Ha (2000).

De Vericourt et al. (2002) considered a capacitated supply system producing a single item with several classes of customers. They obtained the intuitive structure of the optimal inventory allocation policy and presented an efficient algorithm to calculate the control parameters. Huang and Iravani (2008) characterized the manufacturer's optimal production and inventory rationing policies in a production system producing a single product serving multiple customer classes with batch ordering and lose sales.

Benjaafar et al. (2010) consider the optimal control of a production inventory system with a single product and two customer classes and showed that the optimal policy can be described by three state-dependent thresholds: A production base-stock level and two order-admission levels, one for each class. Benjaafar and ElHafsi (2006) considered an assemble-to-order system producing a single end product for multiple classes of customers. They showed that a state-dependent base-stock production policy is optimal, and the optimal inventory allocation policy is characterized by a set of state-dependent rationing levels. Benjaafar et al. (2011) further investigated the structure of the optimal policy for a multistage assembly system producing a single product with multiple classes of customers.

Mollering and Thonemann (2008) considered a periodic-review inventory system with two demand classes under the threshold control policy. They focused on finding the optimal parameters of the policy based on a sample path approach.

References

Arruda, E.F., Do Val, J.B.R.: Stability and optimality of a multi-product production and storage system under demand uncertainty. Eur. J. Oper. Res. **188**(2), 406–427 (2008)

Benjaafar, S., ElHafsi, M.: Production and inventory control of a single product assemble-to-order system with multiple customer classes. Manage. Sci. **52**(12), 1896–1912 (2006)

Benjaafar, S., ElHafsi, M., Huang, T.: Optimal control of a production-inventory system with both backorders and lost sales. Nav. Res. Logist. **57**(3), 252–265 (2010)

Benjaafar, S., ElHafsi, M., Lee, C.Y., Zhou, W.H.: Optimal control of an assembly system with multiple stages and multiple demand classes. Oper. Res. **59**(2), 522–529 (2011)

Bertsekas, D.P.: Dynamic Programming: Deterministic and Stochastic Models. Prentice-Hall, Englewood Cliffs (1987)

Carr, S., Duenyas, I.: Optimal admission control and sequencing in a make-to-stock/make-to-order production system. Oper. Res. **48**(5), 709–720 (2000)

Choia, J., Caob, J.J., Romeijnb, H.E., Geunesb, J., Baib, S.X.: A stochastic multi-item inventory model with unequal replenishment intervals and limited warehouse capacity. IIE Trans. **37**(12), 1129–1141 (2005)

Cohen, M.A., Kleindorfer, P.R., Lee, H.L.: Service constrained (s, S) inventory systems with priority demand classes and lost sales. Manage. Sci. **34**(4), 482–499 (1988)

De Vericourt, F., Karaesmen, F., Dallery, Y.: Dynamic scheduling in a make-to-stock system: a partial characterization of optimal policies. Oper. Res. **48**(5), 811–819 (2000)

De Vericourt, F., Karaesmen, F., Dallery, Y.: Optimal stock allocation for a capacitated supply system. Manage. Sci. **48**(11), 1486–1501 (2002)

Frank, K.C., Zhang, R.Q., Duenyas, I.: Optimal policies for inventory systems with priority demand classes. Oper. Res. **51**(6), 993–1002 (2003)

Graves, S.C.: The multi-product production cycling problem. AIIE Trans. **12**(3), 233–240 (1980)

Ha, A.: Inventory rationing in a make-to-stock production system with several demand classes and lost sales. Manage. Sci. **43**(8), 1093–1103 (1997a)

Ha, A.: Stock-rationing policy for a make-to-stock production system with two priority classes and backordering. Nav. Res. Logist. **44**(5), 458–472 (1997b)

Ha, A.: Optimal dynamic scheduling policy for a make-to-stock production system. Oper. Res. **45**(1), 42–53 (1997c)

Ha, A.: Stock-rationing in an M/Ek/1 make-to-stock queue. Manage. Sci. **46**(1), 77–87 (2000)

Huang, B., Iravani, S.M.R.: Technical note—A make-to-stock system with multiple customer classes and batch ordering. Oper. Res. **56**(5), 1312–1320 (2008)

Iravani, S.M.R., Liu, T., Luangkesorn, K.L., Simchi-Levi, D.: A produce-to-stock system with advance demand information and secondary customers. Nav. Res. Logist. **54**(3), 331–345 (2007)

Iravani, S.M.R., Liu, T., Simchi-Levi, D.: Optimal production and admission policies in make-to-stock/make-to-order manufacturing systems. Prod. Oper. Manage. **21**(2), 224–235 (2012)

Isotupa, K.P.S.: An Markovian inventory system with lost sales and two demand classes. Math. Comput. Model. **43**(7–8), 687–694 (2006)

Melchiors, P., Dekker, R., Kleijn, M.J.: Inventory rationing in an (s, Q) inventory model with lost sales and two demand classes. J. Oper. Res. Soc. **51**(1), 111–122 (2000)

Mollering, K.T., Thonemann, U.W.: An optimal critical level policy for inventory systems with two demand classes. Nav. Res. Logist. **55**(7), 632–642 (2008)

Nahmias, S., Demmy, W.S.: Operating characteristics of an inventory system with rationing. Manage. Sci. **27**(11), 1236–1245 (1981)

Pena-Perez, A., Zipkin, P.: Dynamic scheduling rules for a multi-product make-to-stock queue. Oper. Res. **45**(6), 919–930 (1997)

Puterman, M.L.: Markov Decision Processes: Discrete Stochastic Dynamic Programming. Wiley, New York (1994)

Song, D.P.: Stability and optimization of a production inventory system under prioritized base-stock control. IMA J. Manage. Math. **20**(1), 59–79 (2009)

Song, D.P., Sun, Y.X.: Optimal control structure of an unreliable manufacturing system with random demands. IEEE Trans. Autom. Control. **44**(3), 619–622 (1999)

Teunter, R.H., Haneveld, W.K.K.: Dynamic inventory rationing strategies for inventory systems with two demand classes, Poisson demand and backordering. Eur. J. Oper. Res. **190**(1), 156–178 (2008)

Topkis, D.M.: Optimal ordering and rationing policies in a nonstationary dynamic inventory model with n demand classes. Manage. Sci. **15**(3), 160–176 (1968)

Wein, L.M.: Dynamic scheduling of a multiclass make-to-stock queue. Oper. Res. **40**(4), 724–735 (1992)

Zhao, N., Lian, Z.T.: A queueing-inventory system with two classes of customers. Int. J. Prod. Econ. **129**(1), 225–231 (2011)

Zheng, Y., Zipkin, P.: A queueing model to analyze the value of centralized inventory information. Oper. Res. **38**(2), 296–307 (1990)

Chapter 8
Threshold-Type Control Policies and System Stability for Serial Supply Chain Systems

8.1 Introduction

An important issue in stochastic dynamic systems is the system stability, in particular, when the long-run average cost is considered. To some extent, system stability is more fundamental than the optimal control problem. This chapter will establish the sufficient and necessary conditions to ensure the stability of the basic stochastic supply chain using the matrix analytic method and then extend the results in Chap. 2 to the long-run average cost case.

Although Chaps. 2 and 3 have illustrated the detailed structural properties of the optimal integrated ordering and production control policies such as the switching structures, it is still difficult to implement them in reality because the shapes of these switching curves and their locations in the state space could vary significantly in different scenarios. One way to overcome this drawback is to simplify and parameterize the control policies. Such treatment is common in practice for two reasons. Firstly, practical policies emphasize on the operationability. It is vital that control policies should be easy to understand and easy to operate. If a policy's logic is hidden from the operators and/or it requires a complicated set of input data to implement the policy, then human errors are more likely to occur that result in worse performance. Secondly, simple and parameterized policies are often more robust to the environment. They could offer fairly good performance in a wide range of system settings. A typical example is the application of just-in-time (JIT) or Kanban systems, which has rather simple control structure but has produced good performance in the automotive industry. How to construct an appropriate form of the parameterized policy for stochastic serial supply chain systems is one of the key issues in this chapter.

D.-P. Song, *Optimal Control and Optimization of Stochastic Supply Chain Systems*, Advances in Industrial Control, DOI 10.1007/978-1-4471-4724-4_8,
© Springer-Verlag London 2013

8.2 Stability Conditions and the Long-Run Average Cost Case

The system stability issue in our problem concerns whether the backordered demands may increasingly go to infinity. Mathematically, the induced Markov chain does not have a stationary distribution. When the long-run average cost is considered, the cost function will not converge if the system is unstable. Therefore, we must ensure the stability of the system before investigating the characteristics of the optimal integrated policies for the long-run average cost case.

Physically, the effective production rate should be greater than the demand arrival rate so that unsatisfied demands will not be increasingly backordered. If the raw material warehouse has infinite capacity, it is obvious that the conditions $\lambda > \mu$ and $r > \mu$ would guarantee the stability of the system. However, when the raw material (RM) warehouse has finite capacity, this may not be sufficient because finite warehouse capacity will affect the actual production capacity. We will apply the matrix analytic method (Latouche and Ramaswami 1999) to establish the sufficient and necessary condition for the stability of the stochastic supply chain in Fig. 2.1.

The maximum potential production capacity of the system can be achieved by ordering raw materials as much as possible and producing finished goods whenever possible. This policy, denoted by (u_N, q_M), can be described by $u(x_1, x_2) = r$ if $x_1 > 0$ and $x_2 < N$, $u(x_1, x_2) = 0$ otherwise; $q(x_1, x_2) = \max\{M - x_1, 0\}$. Therefore, to find the condition to ensure the stability of the system is equivalent to find condition to ensure that the induced Markov chain under (u_N, q_M) is positive recurrent.

We sequence the system state as follows: $(0, N)$, $(1, N)$, $\ldots$, (M, N), $(0, N-1)$, $(1, N-1)$, $\ldots$, $(M, N-1)$, $\ldots$, in which the raw material state (i.e., x_1) is treated as different phases from 0 to M and finished goods state x_2 is treated as different stages from N to $-\infty$. This is a quasi-birth–death (QBD) Markov chain. The state transition map of the QBD can be partially depicted in Fig. 8.1.

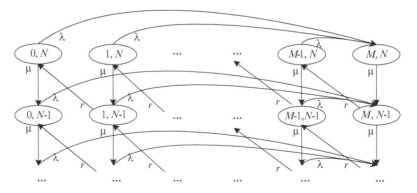

Fig. 8.1 State transition map of QBD Markov chain in the basic supply chain

For the QBD Markov chain in Fig. 8.1, its infinitesimal generator Q is given by

$$Q = \begin{pmatrix} B_0 & A_0 & 0 & 0 & \cdots \\ A_2 & A_1 & A_0 & 0 & \cdots \\ 0 & A_2 & A_1 & A_0 & \cdots \\ 0 & 0 & A_2 & A_1 & \cdots \\ \cdots & \cdots & \cdots & \cdots & \cdots \end{pmatrix} \qquad (8.1)$$

where B_0, A_1, A_2, and A_0 are $(M+1) \times (M+1)$ matrices given by

$$B_0 = \begin{pmatrix} -\lambda - \mu & 0 & & & \lambda \\ 0 & -\lambda - \mu & & & \lambda \\ & & \cdots & & \cdots \\ & & & -\lambda - \mu & \lambda \\ 0 & & 0 & & -\mu \end{pmatrix},$$

$$A_0 = \begin{pmatrix} \mu & & & 0 \\ & \mu & & \\ & & \cdots & \\ & & & \mu \\ 0 & & & \mu \end{pmatrix}, \qquad A_2 = \begin{pmatrix} 0 & & & 0 \\ r & 0 & & \\ & r & \cdots & \\ & & \cdots & 0 \\ 0 & & & r & 0 \end{pmatrix},$$

$$A_1 = \begin{pmatrix} -\lambda - \mu & 0 & & & \lambda \\ 0 & -r - \lambda - \mu & & & \lambda \\ & & \cdots & & \cdots \\ & & & -r - \lambda - \mu & \lambda \\ 0 & & 0 & & -r - \mu \end{pmatrix}.$$

Lemma 8.1. *(Latouche and Ramaswami 1999). Consider an irreducible continuous time quasi-birth–death process and assume M is finite and the matrix $A := A_1 + A_2 + A_0$ is irreducible; then the QBD is positive recurrent if and only if $\boldsymbol{\pi}(A_2 - A_0)\boldsymbol{e} > 0$ where $\boldsymbol{e}$ is a column vector of dimension $M+1$ consisting entirely of ones and $\boldsymbol{\pi} := (\pi_0, \pi_1, \ldots, \pi_M)$ is the solution to the equations $\boldsymbol{\pi} A = 0$ and $\boldsymbol{\pi e} = 1$.*

Proposition 8.1. *The sufficient and necessary condition for the stability of the basic stochastic supply chain system is*

$$r \cdot \left(1 - \left(\frac{r}{r + \lambda}\right)^M\right) > \mu. \qquad (8.2)$$

Proof. Matrix $A = A_1 + A_2 + A_0$ is irreducible since the associated diagraph is strongly connected. Solve the equations $\pi A = 0$ and $\pi e = 1$, which gives the unique solution

$$\pi_0 = (r/(r+1))^M. \tag{8.3}$$

From Lemma 8.1, the induced Markov chain under (u_N, q_M) is positive recurrent if and only if $r(1 - \pi_0) - \mu > 0$. This completes the proof. $\qquad\square$

Proposition 8.1 indicates that $r \cdot (r/(r+\lambda))^M$ is the amount of production capacity that is reduced by the impact of the finite RM warehouse capacity. It can be seen that as the RM warehouse capacity tends to infinity (i.e., $M \to \infty$) or the order replenishment lead time tends to be zero (i.e., $\lambda \to \infty$), the condition becomes $r > \mu$, which is intuitively true.

From the viewpoint of designing or selecting the RM warehouse capacity, the condition (8.2) provides an insight into the minimum RM warehouse capacity required in order to meet the customer demands in long term.

Definition 8.1. The system stability index is defined as

$$\rho := \mu \Big/ \left(r \cdot \left(1 - (r/(r+\lambda))^M \right) \right). \tag{8.4}$$

From now on, we assume that the system always satisfies the stability condition (8.2), that is, $\rho < 1$. Consider the long-run average cost (independent of the initial state) in (8.3), where $g(.)$ is defined in (2.2),

$$J^* = \min_{u,q} \lim_{T \to \infty} \frac{1}{T} E \int_0^T g(\mathbf{x}(t), u(t), q(t)) dt. \tag{8.5}$$

Now we are able to apply the average cost optimality theory developed by Sennott (1999) to our problem to derive the optimal ordering and production policy. Consider the admissible policies such that the induced Markov chain is positive recurrent. From the discounted cost case in Chap. 2, we know that it is always optimal to produce finished goods if $x_1 = 1$ and $x_2 = -1$. The system is then transitioning into the state $(0, 0)$ with rate r. Therefore, it is reasonable to assume that the induced Markov chain includes the state $(0, 0)$. The average cost optimality equation is given by

$$w(x_1, x_2) + \frac{J^*}{v} = \frac{1}{v} [g(x_1, x_2) + \mu \cdot w(x_1, x_2 - 1)$$

$$+ r \cdot w(x_1, x_2) \cdot I\{x_1 = 0 \text{ or } x_2 = N\}$$

$$+ r \cdot \min\{w(x_1 - 1, x_2 + 1), w(x_1, x_2)\} \cdot I\{x_1 > 0 \text{ and } x_2 < N\}$$

$$+ \lambda \cdot \min\{w(x_1 + q, x_2) | q \in [0, M - x_1]\}] \tag{8.6}$$

where J^* is the optimal average cost defined in (8.5) and $w(x_1, x_2)$ is a finite function.

Proposition 8.2. *For the long-run average cost case, we have:*

(i) $w(x_1 + 1, x_2) - w(x_1, x_2)$ is increasing in x_1 and x_2.
(ii) $w(x_1, x_2 + 1) - w(x_1, x_2)$ is increasing in x_1 and x_2.
(iii) $w(x_1 - 1, x_2 + 1) - w(x_1, x_2)$ is decreasing in x_1 and increasing in x_2.
(iv) $w(x_1 - 1, x_2 + 1) - w(x_1, x_2) \leq 0$ for any $x_1 > 0$ and $x_2 < 0$.

Proof. Treat $(0, 0)$ as the distinguished state in the induced Markov Chain. Note that $w(x_1, x_2) = \lim_{\beta \to 0} \left(J^\beta (x_1, x_2) - J^\beta (0, 0) \right)$ by Sennott (1999), where $J^\beta(x_1, x_2)$ represents the discounted cost case objective function in (2.1). Recalling the results in Lemma 2.1 and Proposition 2.1, the assertions are true. This completes the proof.   □

Similar to the discounted cost case in Chap. 2, we can define two switching curves $S_p(x_1)$ and $S_o(x_2)$ based on Proposition 8.2. The following results hold for the long-run average cost case.

Proposition 8.3. *The optimal integrated ordering and production policy for the long-run average cost can be characterized by two switching curves $S_p(x_1)$ and $S_o(x_2)$ and:*

(i) $S_p(x_1)$ is monotonic increasing in x_1.
(ii) $S_o(x_2)$ is skip-free monotonic decreasing in x_2.
(iii) $\lim_{x_2 \to -\infty} (w(i, x_2 + 1) - w(i, x_2)) = -\infty$ for $i \in [0, M]$.
(iv) $\lim_{x_2 \to -\infty} (w(i - 1, x_2 + 1) - w(i, x_2)) = -\infty$ for $i \in [1, M]$.
(v) $\lim_{x_2 \to -\infty} (w(i + 1, x_2 + 1) - w(i, x_2)) = -\infty$ for $i \in [0, M - 1]$.
(vi)   There exists a finite negative integer $\underline{x_2}$, such $S_o(x_2) = M$ for $x_2 \leq \underline{x_2}$.

From Proposition 8.3, we can see that the asymptotic behaviors in the long-run average cost case are different from that in the discounted cost case. Particularly, $S_o(x_2)$ may be bounded by an integer that is less than M for the discounted cost case, while $S_o(x_2)$ definitely converges to M in the average cost case. Intuitively, for the average cost case, it is preferable to have more RM warehouse capacity and RM inventory when there are a large number of backordered demands.

8.3   Threshold Control Policies in the Basic Supply Chain System

Based on the established structural properties of the switching curves, we propose a linear switching threshold policy as shown in Fig. 8.2, which is obtained by replacing piecewise increasing/decreasing parts of two switching curves $S_o(x_2)$ and $S_p(x_2)$ with linear line segments.

Fig. 8.2 Linear switching threshold control policy in the basic supply chain

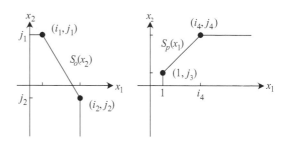

Mathematically, the linear switching threshold (LST) policy can be expressed as follows:

$$
u(x) = \begin{cases} r & x_1 > 0, x_2 \le j_4, (x_2 - j_3)\,(i_4 - 1) \le (j_4 - j_3)\,(x_1 - 1) \\ 0 & \text{otherwise} \end{cases}, \quad (8.7)
$$

$$
q(x) = \begin{cases} i_2 - x_1 & x_2 \le j_2, x_1 \le i_2 \\ \max\left\{ \left(\frac{(i_2 - i_1)(x_2 - j_1)}{j_2 - j_1} + i_1 \right) - x_1, 0 \right\} & j_2 < x_2 \le j_1, x_1 \le i_2 \\ 0 & \text{otherwise} \end{cases}. \quad (8.8)
$$

Clearly, the linear switching threshold policy is determined by four points: (i_1, j_1), (i_2, j_2), $(1, j_3)$, and (i_4, j_4). The advantage of the linear switching threshold policy is twofold: firstly, it can closely approximate the shape of the optimal integrated policy; secondly, it is easy to operate and implement. From Proposition 2.7, we can take $i_2 = x''_1$, which can be analytically determined. In addition, the following relationships between the parameters, for example, $0 \le x'_1 \le x''_1, \underline{x}_2 \le \bar{x}_2 \le N, 1 \le \bar{x}_1 \le M$, and $0 \le x'_2 \le x''_2 \le N$, can be utilized to simplify the procedure of seeking the optimal parameters in (i_1, j_1), (i_2, j_2), $(1, j_3)$, and (i_4, j_4).

In the long-run average cost case, the above LST policy can be further simplified, for example, we have $j_2 = M$ from Proposition 8.3.

In the remainder of this section, numerical examples are given to verify the analytical results in Chap. 2 and to demonstrate the effectiveness of the proposed LST policy. More specifically, there are three purposes of the numerical examples: (1) examine the structural properties of the optimal integrated ordering and production policy, (2) compare the performance of the linear switching threshold policy with that of the optimal policy and evaluate their sensitivity to system parameters, and (3) assess the impacts of raw material warehouse capacity on system stability and performance.

We use the value iteration algorithm to evaluate the performance (Bertsekas 1987; Sennott 1999). The system state space is limited into a finite area with $x_2 \in [N - 50, N]$. The iterative procedure will be terminated when the value difference is less than 10^{-3} or the number of iterations exceeds 5,000.

Ten cases with different combinations of system parameters (see Table 8.1) are tested, in which the lead-time rate λ takes 0.5, 1.0, and 2.0; demand rate μ takes

Table 8.1 Parameter setting
for different cases

Case	λ	r	μ	c_1	c_2^+	c_2^-
1	0.5	1.0	0.5	1	2	10
2	0.5	1.0	0.7	1	2	10
3	0.5	1.0	0.9	1	2	10
4	1.0	1.0	0.5	1	2	10
5	1.0	1.0	0.7	1	2	10
6	1.0	1.0	0.9	1	2	10
7	2.0	1.0	0.5	1	2	10
8	2.0	1.0	0.7	1	2	10
9	2.0	1.0	0.9	1	2	10
10	2.0	1.0	0.9	1	2	20

Table 8.2 Optimal policy and LST policy for different case

Case	ρ	$J_{optimal}$	J_{LST}	% above	$(i_1, j_1)\ (i_2, j_2)\ (1, j_3)\ (i_4, j_4)$
Disc.					
1	0.548	12.855	12.855	0.00%	$(0, 4)\ (4, -4)\ (1, 0)\ (1, 0)$
2	0.767	18.598	18.618	0.11%	$(0, 3)\ (4, -3)\ (1, 1)\ (1, 1)$
3	0.987	24.920	24.920	0.00%	$(0, 5)\ (4, -3)\ (1, 1)\ (1, 1)$
4	0.508	10.733	10.746	0.12%	$(0, 1)\ (3, -2)\ (1, 0)\ (1, 0)$
5	0.711	15.747	15.787	0.26%	$(0, 4)\ (3, -2)\ (1, 1)\ (1, 1)$
6	0.914	21.309	21.309	0.00%	$(0, 4)\ (3, -2)\ (1, 1)\ (1, 1)$
7	0.501	9.257	9.292	0.38%	$(0, 2)\ (3, -4)\ (1, 0)\ (1, 0)$
8	0.701	13.617	13.621	0.03%	$(0, 3)\ (3, -3)\ (1, 1)\ (1, 1)$
9	0.901	18.678	18.679	0.01%	$(0, 3)\ (3, -3)\ (1, 1)\ (1, 1)$
10	0.901	33.726	33.726	0.00%	$(0, 4)\ (3, -2)\ (1, 2)\ (1, 2)$
Ave.					
1	0.548	8.649	8.651	0.03%	$(2, 4)\ (6, -2)\ (1, 2)\ (1, 2)$
2	0.767	16.700	16.700	0.00%	$(3, 6)\ (6, 3)\ (1, 5)\ (1, 5)$
3	0.987	168.63	168.63	0.00%	$(6, 6)\ (6, 6)\ (1, 5)\ (1, 5)$
4	0.508	7.080	7.083	0.05%	$(0, 4)\ (6, -4)\ (1, 2)\ (1, 2)$
5	0.711	13.085	13.101	0.12%	$(1, 6)\ (6, -1)\ (1, 5)\ (1, 5)$
6	0.914	62.954	62.954	0.00%	$(4, 6)\ (6, 4)\ (1, 5)\ (1, 5)$
7	0.501	6.341	6.345	0.06%	$(0, 4)\ (6, -7)\ (1, 2)\ (1, 2)$
8	0.701	11.989	12.030	0.34%	$(1, 6)\ (6, -6)\ (1, 4)\ (1, 4)$
9	0.901	54.280	54.340	0.11%	$(3, 6)\ (6, -3)\ (1, 5)\ (1, 5)$
10	0.901	100.683	100.683	0.00%	$(3, 6)\ (6, 0)\ (1, 5)\ (1, 5)$

0.5, 0.7, and 0.9; and the backordering cost c_2^- takes 10 and 20. The warehouse
capacities are set as $M = N = 6$, and the discount factor $\beta = 0.5$.

Table 8.2 gives the stability index, the optimal cost, the cost under the best linear
switching threshold (LST) policy, and the control parameters of the best LST policy
for all ten cases in both discounted and average cost situations, respectively. The
system initial state is assumed to be $(0, 0)$ for the discounted cost case. In Table 8.2,
the fifth column is the percentage of the cost under the best LST policy above the
cost under the optimal policy.

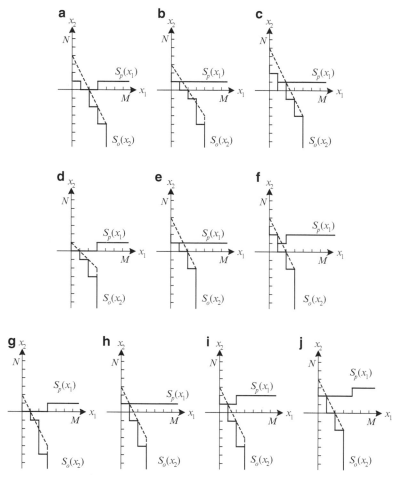

Fig. 8.3 Optimal and LST policies for discounted costs in the basic supply chain. (**a**) Case 1, (**b**) case 2, (**c**) case 3, (**d**) case 4, (**e**) case 5, (**f**) case 6, (**g**) case 7, (**h**) case 8, (**i**) case 9, (**j**) case 10

The optimal ordering and production policies and the best linear switching threshold policies for cases 1–10 are shown in Fig. 8.3 for the discounted cost cases and in Fig. 8.4 for the long-run average cost cases. The ordering decisions of the best LST polices are described by the dash lines. The production decisions of the best LST policy are very simple (see Table 8.2) and therefore omitted in Figs. 8.3 and 8.4.

Based on the computational results in Table 8.2 and Figs. 8.3 and 8.4, it can be seen that:

1. The optimal integrated ordering and production policy can be characterized by two monotonic switching curves $S_p(x_1)$ and $S_o(x_2)$. In addition, $S_o(x_2)$ is a skip-free piecewise curve when $S_o(x_2) > 0$.

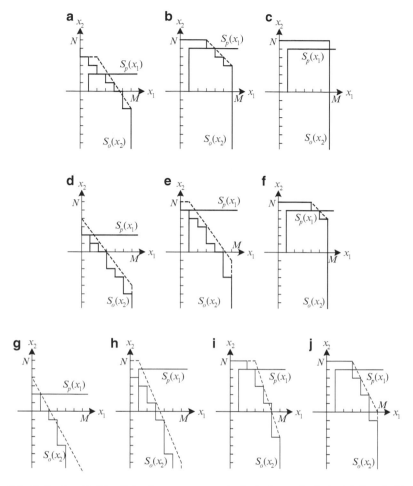

Fig. 8.4 Optimal and LST policies for average costs in the basic supply chain. (**a**) Case 1, (**b**) case 2, (**c**) case 3, (**d**) case 4, (**e**) case 5, (**f**) case 6, (**g**) case 7, (**h**) case 8, (**i**) case 9, (**j**) case 10

2. With the fixed inventory level of finished goods (i.e., x_2), the optimal ordering policy is an order-up-to-point policy, but the order-up-to-point (i.e., $S_o(x_2)$) is increasing and converging to a finite number as x_2 decreases to negative infinity. For example, $\lim_{x_2 \to -\infty} S_o(x_2) = 4$ for cases 1–3, and $\lim_{x_2 \to -\infty} S_o(x_2) = 3$ for cases 4–10 in the discounted cost situations, while in the average cost situations, $\lim_{x_2 \to -\infty} S_o(x_2) = M$ for all cases, which verifies Proposition 8.3(vi). Moreover, the order-up-to-point could vary substantially for different x_2, which implies that the FG inventory level has significant impact on the ordering decision. This supports the use of vendor-managed inventory (VMI) approach under which the supplier should be responsible for managing the stock levels and availability for the downstream entities based on the information of demand and inventory provided by the downstream entities.

Table 8.3 Parameter setting
for different cases

M	λ	r	μ	c_2^-
6	0.5	1.0	0.4	10
8	1.0	2.0	0.6	20
10	2.0		0.8	40

3. With the fixed inventory level of raw materials (i.e., x_1), the optimal production policy is a base-stock policy, but the base-stock level (i.e., $S_p(x_1)$) is increasing as x_1 increases. However, we observed that the base-stock level is a constant in most cases, particularly in the average cost situations. This indicates that the production decision is much less dependent of the raw material inventory level compared with the ordering decision on the FG inventory level.
4. The linear switching threshold (LST) policy performs extremely well in all cases. It is only 0.38% above the optimal cost in the worst case (i.e., case 7 with discounted cost).
5. Both switching curves $S_p(x_1)$ and $S_o(x_2)$ move upward as demand rate μ increases and move downward as lead-time rate λ increases. For the control parameters of the best LST policy, i_1 always takes 0 in the discounted cost situations; $i_2 = \max\{S_o(x_2)\}$, which can be analytically determined by Proposition 2.7 for the discounted cost situations and $i_2 = M$ for the average cost situations; j_1 and j_2 are generally increasing in demand rate μ and decreasing in lead-time rate λ; $j_3 = j_4$ is increasing in demand rate μ and decreasing in lead-time rate λ. Therefore, in terms of implementing the LST policy, we only need to find the appropriate values of j_1, j_2, and j_3 for the discounted cost situations and i_1, j_1, j_2, and j_3 for the average cost situations. These values have the relationships: $0 \leq i_1 \leq \max\{S_o(x_2)\}, 0 \leq j_1 \leq N, j_2 \leq j_1$, and $0 \leq j_3 \leq N$.

The above ten cases illustrate the structural properties of the optimal integrated ordering and production policies and visualize the closeness of the linear switching threshold (LST) policies to the optimal policies. To further demonstrate the effectiveness of the LST policies, a greater range of values for the system parameters and their combinations are tested. Consider the long-run average cost situations with $N = 10$, $c_1 = 1$, and $c_2^+ = 2$ as examples. Other parameters vary in ranges as shown in Table 8.3, which gives total $3 \times 3 \times 2 \times 3 \times 3 = 162$ cases. The results show that the average percentage of the cost under the best LST policy above the optimal cost is 0.08% with a standard deviation 0.0014. The worst case is 0.82% above the optimal cost, corresponding to the parameter setting: $M = 10$, $\lambda = 2$, $r = 1$, $\mu = 0.8$, and $c_2^- = 10$ in Table 8.3. This confirms the effectiveness of the LST policy.

Next we investigate the effects of the raw material warehouse capacity M on the stability index and the discounted/average costs under the optimal and LST policies. As an example, we only consider cases 1 and 3 in Table 8.1. Let the RM warehouse capacity M take different values from 2 to 16. Table 8.4 gives the stability index, the optimal cost, the percentage of the cost under the best LST policy above the optimal cost, and the control parameters of the best LST policy.

Table 8.4 Optimal policy and LST policy with different M

M	ρ	J_{optimal}	J_{LST} % above	$(i_1, j_1)\,(i_2, j_2)\,(1, j_3)\,(i_4, j_4)$
Case 1 with discounted cost				
2	0.900	12.963	0.00	$(1, 1)\,(2, 0)\,(1, 0)\,(1, 0)$
4	0.623	12.855	0.00	$(1, 2)\,(4, -4)\,(1, 0)\,(1, 0)$
>5	0.548	12.855	0.00	$(0, 4)\,(4, -4)\,(1, 0)\,(1, 0)$
Case 1 with average cost				
2	0.900	40.757	0.00	$(2, 6)\,(2, 6)\,(1, 5)\,(1, 5)$
4	0.623	9.266	0.00	$(0, 5)\,(4, 1)\,(1, 2)\,(1, 2)$
6	0.548	8.649	0.03	$(2, 4)\,(6, -2)\,(1, 2)\,(1, 2)$
8	0.520	8.602	0.01	$(2, 4)\,(8, -4)\,(1, 2)\,(1, 2)$
10	0.509	8.599	0.01	$(2, 4)\,(10, -7)\,(1, 2)\,(1, 2)$
12	0.504	8.599	0.01	$(2, 4)\,(12, -10)\,(1, 2)\,(1, 2)$
14	0.502	8.599	0.02	$(1, 5)\,(14, -12)\,(1, 2)\,(1, 2)$
16	0.501	8.599	0.02	$(2, 4)\,(16, -15)\,(1, 2)\,(1, 2)$
Case 3 with discounted cost				
2	1.620	25.432	0.00	$(0, 3)\,(2, 1)\,(1, 1)\,(1, 1)$
4	1.122	24.920	0.00	$(0, 5)\,(4, -3)\,(1, 1)\,(1, 1)$
>5	0.987	24.920	0.00	$(0, 5)\,(4, -3)\,(1, 1)\,(1, 1)$
Case 3 with average cost				
2	1.620	–	–	–
4	1.122	–	–	–
6	0.987	168.63	0.00	$(6, 6)\,(6, 6)\,(1, 5)\,(1, 5)$
8	0.937	86.379	0.00	$(8, 6)\,(8, 6)\,(1, 5)\,(1, 5)$
10	0.916	67.711	0.00	$(8, 6)\,(10, 4)\,(1, 5)\,(1, 5)$
12	0.907	62.731	0.01	$(8, 6)\,(12, 0)\,(1, 5)\,(1, 5)$
14	0.903	61.452	0.03	$(8, 6)\,(14, -3)\,(1, 5)\,(1, 5)$
16	0.901	61.237	0.10	$(9, 6)\,(16, -9)\,(1, 5)\,(1, 5)$

From Table 8.4, we can observe that:

1. The stability index ρ is decreasing and tends to be μ/r as the RM warehouse capacity M increases.
2. The cost under the best LST policy is very close to the optimal cost (only 0.10% above in the worst scenario in Table 8.4).
3. The optimal cost is decreasing in M and appears to be very sensitive to M when it is small, particularly in the average cost situations. In the discounted cost situations, the cost converges quickly and the impact of the RM warehouse capacity is not significant.
4. In case 3, the system is unstable when $M \leq 5$ (e.g., $\rho = 1.0365 > 1$ if $M = 5$). The discounted cost can still be calculated due to its quick convergence rate, but the average cost tends to be infinity.

It should be pointed out that the discounted costs in Tables 8.2 and 8.4 are sufficiently accurate using the specified finite state space and the maximum iteration number, but the average costs in Tables 8.2 and 8.4 for high stability indices (e.g., $\rho = 0.987$ and 0.937) are not the accurate converged values. This is because when the system is close to unstable, a much larger state space and a much larger number of iterations are required in order to achieve the convergence of the cost function. However, we believe that these results will not affect the qualitative comparison between different scenarios since they are based on the same maximum number of iterations.

8.4 Threshold Control Policies in More General Supply Chain Systems

In Chap. 3, we have illustrated the characteristics of the optimal integrated policies in more general supply chain systems. In this section, we aim to construct threshold control policies in those cases based on the characteristics we observed in Chap. 3 and provide numerical examples to demonstrate their effectiveness compared to the optimal policies. We follow the notation defined in Chap. 3.

8.4.1 Supply Chain Systems Subject to One Non-changeable Outstanding Order

Based on Figs. 3.2 and 3.7, and the discussions there for the case of at most one outstanding order with its size not changeable once issued, we present a truncated linear switching threshold (TLST) policy as follows:

$$
u(x_1, x_2, y) = \begin{cases} r & x_1 > 0, x_2 \leq j_4, (x_2 - j_3)(i_4 - 1) \leq (j_4 - j_3)(x_1 - 1) \\ 0 & \text{otherwise} \end{cases},
$$

$$(8.9)$$

$$
q(x_1, x_2, 0) = \begin{cases} i_2 - x_1 & x_2 \leq j_2, x_1 \leq i_2, x_1 \leq K \\ \max\left\{ \left(\frac{(i_2 - i_1)(x_2 - j_1)}{j_2 - j_1} + i_1 \right) - x_1, 0 \right\} & j_2 < x_2 \leq j_1, x_1 \leq i_2, x_1 \leq K \\ 0 & \text{otherwise} \end{cases}
$$

$$(8.10)$$

The truncated linear switching threshold policy is determined by four points, (i_1, j_1), (i_2, j_2), $(1, j_3)$, and (i_4, j_4), and a constant K. The production policy is the same as (8.7), whereas the ordering policy is slightly different from (8.8). Here we only place orders when $x_1 \leq K$. Namely, if we have had a certain amount of raw materials, we would not place a small order. This truncation mechanism partially reflects the above intuition and observation. It should be pointed out that the parameter K is often less than i_2. However, determining the best K is not trivial. Simulation-based optimization methods could be applied for this purpose.

8.4.2 Supply Chain Systems with Two Parallel Outstanding Orders

For the cases that there are at most two outstanding orders but with the order sizes that cannot be modified once they were issued, the control structure of the optimal integrated policy is illustrated in Figs. 3.4 and 3.8. A similar truncated linear switching threshold (TLST) policy can be constructed:

$$u(x_1, x_2, y_1, y_2) = \begin{cases} r & x_1 > 0, x_2 \le j_4, (x_2 - j_3)(i_4 - 1) \le (j_4 - j_3)(x_1 - 1) \\ 0 & \text{otherwise} \end{cases},$$

(8.11)

$$q(x_1, x_2, 0, y_2)$$

$$= \begin{cases} \max\{i_2 - x_1 - y_2, 0\} & x_2 \le j_2, x_1 \le i_2, x_1 \le K \\ \left(\min\left\{\frac{(i_2 - i_1)(x_2 - j_1)}{j_2 - j_1} + i_1 - x_1, i_2 - x_1 - y_2\right\}\right)^+ & \begin{array}{l} j_2 < x_2 \le j_1, x_1 \le i_2, \\ x_1 \le K \end{array} \\ 0 & \text{otherwise} \end{cases}.$$

(8.12)

The ordering policy at state $(x_1, x_2, y_1, 0)$ is the same as (8.12) by replacing y_2 with y_1. The ordering policy is slightly different from (8.10) because we now have to take into account the effect of the other outstanding order.

8.4.3 Supply Chain Systems Subject to Erlang Distributed Lead Times

For the case that there is at most one outstanding order with its size not changeable and an Erlang distributed lead time with L stages and a rate $L\lambda$ at each stage, its control structure of the optimal integrated policy is illustrated in Figs. 3.6 and 3.9. A truncated linear switching threshold (TLST) policy similar to (8.9) and (8.10) can be constructed:

$$u(x_1, x_2, y, l) = \begin{cases} r & x_1 > 0, x_2 \le j_4, (x_2 - j_3)(i_4 - 1) \le (j_4 - j_3)(x_1 - 1) \\ 0 & \text{otherwise} \end{cases},$$

(8.13)

$$q(x_1, x_2, 0, 0) = \begin{cases} i_2 - x_1 & x_2 \le j_2, x_1 \le i_2, x_1 \le K \\ \max\left\{\left(\frac{(i_2 - i_1)(x_2 - j_1)}{j_2 - j_1} + i_1\right) - x_1, 0\right\} & j_2 < x_2 \le j_1, x_1 \le i_2, x_1 \le K. \\ 0 & \text{otherwise} \end{cases}$$

(8.14)

Table 8.5 Optimal policy and truncated linear switching threshold policy in generalized cases

	Case	J_{optimal}	J_{TLST}	% above	$(i_1, j_1)\, (i_2, j_2)\, (1, j_3)\, (i_4, j_4);\, K$
One order	4	10.967	10.972	0.04	$(0, 1)\, (4, -3)\, (1, 0)\, (1, 0);\, 2$
Non-changeable	5	15.839	15.842	0.02	$(0, 3)\, (4, -3)\, (1, 1)\, (1, 1);\, 2$
	6	21.472	21.498	0.12	$(0, 4)\, (4, -3)\, (1, 2)\, (1, 2);\, 2$
Two orders	4	9.7345	9.8212	0.89	$(0, 2)\, (3, -3)\, (1, 0)\, (1, 0);\, 2$
Non-changeable	5	14.215	14.394	1.25	$(0, 3)\, (3, -3)\, (1, 1)\, (1, 1);\, 2$
	6	19.439	19.654	1.11	$(0, 3)\, (3, -3)\, (1, 1)\, (1, 1);\, 2$
One order	4	10.701	10.703	0.01	$(0, 3)\, (3, -2)\, (1, 0)\, (1, 0);\, 2$
Non-changeable	5	15.566	15.570	0.03	$(0, 3)\, (4, -3)\, (1, 1)\, (1, 1);\, 2$
Erlang distr.	6	21.213	21.218	0.02	$(0, 4)\, (4, -2)\, (1, 1)\, (1, 1);\, 2$

8.4.4   Numerical Examples in More General Supply Chain Systems

We take cases 5–7 with discounted costs in Table 8.1 as examples to compare the truncated linear switching threshold (TLST) policies with the optimal policies in three generalized cases (with $L = 5$). The results are given in Table 8.5.

It can be seen from Table 8.5 that the TLST policy performs very well in all three generalized cases. The worst scenario is case 5 under the second relaxation, in which TLST has a cost 1.25% above the optimal policy. It appears that TLST is more robust to the first relaxation (order size becomes not changeable) and the third relaxation (lead time follows more centralized distributions) but slightly sensitive to the second relaxation (allowing parallel orders). This may be explained by the control structure shown in Figs. 3.7, 3.8, and 3.9, in which the optimal ordering policy under the second relaxation is characterized by many switching curves, while TLST uses a single switching curve to approximate them. The last column in Table 8.5 shows that four points and the truncation value K that determine the TLST policy appear quite similar for the cases under study, which indicates that the TLST policy is fairly robust.

Comparing Table 8.5 with Table 8.2, the optimal costs of cases 4–6 increase in the first generalization and decrease in the second and third generalizations. This is in agreement with the intuitions that the first generalization reduces the flexibility of modifying order sizes and therefore incurs more cost. The second generalization allows parallel outstanding orders and therefore achieves quicker responses. The third generalization has a more centralized distribution of lead times and becomes more reliable in raw material replenishment. It appears that the second and third relaxations offset the effect of the relaxation of "outstanding order not changeable" and lead to the cost reduction. Allowing parallel orders appears more beneficial compared with improving lead-time reliability, but this depends on the degree of the improvements.

8.5 Threshold Control Policy for Multistage Serial Supply Chains

In the multistage supply chain systems, the proposed linear switching threshold policies tend to be too complicated to implement. In order to reduce the number threshold parameters, it is necessary to further simplify the structure of the LST control policies. From the results in Tables 8.4 and 8.5, it can be observed that the production decisions are fairly insensitive to the state of raw material inventory. On the other hand, the decisions of the ordering decisions for raw materials are quite sensitive to the state of the finished goods inventory. Extending this observation to the multistage serial supply chain, it is reasonable to make ordering decisions by considering the inventory level at the current stage and the echelon base-stock for all the downstream stages (i.e., the cumulative inventory level of all stages from the next downstream stage to the final stage serving customers). In this way, we are able to extend the LST policy to multistage serial supply chain systems.

Consider the multistage supply chain in Fig. 3.10. Let $x^e_{i+1} := x_{i+1} + x_{i+2} + \cdots + x_n$ be the echelon base-stock from stage $i + 1$ to the final stage n. The LST policy is a kind of order-up-to-point policies that can be determined by a set of control parameters $\{(L_{i1}, H_{i1}, L_{i2}, H_{i2}) \mid L_{i1} \leq L_{i2}, H_{i2} \leq H_{i1}$, for $i = 1, 2, \ldots, n$, and $L_{n1} = H_{n1} = L_{n2} = H_{n2}\}$. In other words, except the final stage n, each stage is characterized by four parameters, whereas the final stage is characterized only by one parameter. Assuming $Q_i \geq L_{i2}$, (i.e., the stage ordering capacity is greater than the corresponding order-up-to-point), the mathematical formula of the LST policy is given below (for the situation of $Q_i < L_{i2}$, similar threshold policies can be constructed):

$$
q_i(\mathbf{x}) = \begin{cases} L_{i2} - x_i & x^e_{i+1} \leq H_{i2}, x_i \leq L_{i2} \\ \max\left\{ \left(\frac{(L_{i2}-L_{i1})(x^e_{i+1}-H_{i1})}{H_{i2}-H_{i1}} + L_{i1} \right) - x_i, 0 \right\} & H_{i2} < x^e_{i+1} \leq H_{i1}, x_i \leq L_{i2} \\ 0 & \text{otherwise} \end{cases}
$$

(8.15)

The LST policy in (8.15) can be further simplified into an order-up-to-point base-stock Kanban control (BKC) policy, in which each stage (except the final stage) is characterized by two parameters, one for the current stage's inventory level (i.e., Kanban) and the other for the echelon base-stock level from the current stage to the final stage. Let the threshold parameters be denoted as $\{(L_i, H_i) \mid i = 1, 2, \ldots, n$ and $L_n = H_n\}$; the mathematical formula of the BKC policy is given as follows:

$$
q_i(\mathbf{x}) = \begin{cases} L_i - x_i & x^e_{i+1} \leq H_i - L_i, x_i \leq L_i \\ \max\{H_i - x^e_{i+1} - x_i, 0\} & H_i - L_i < x^e_{i+1} \leq H_i, x_i \leq L_i \\ 0 & \text{otherwise} \end{cases} \quad (8.16)
$$

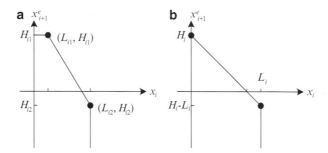

Fig. 8.5 Threshold control policies for multistage serial supply chains. (**a**) LST policy. (**b**) KBT policy

To have an intuitive view of the threshold policies in (8.15) and (8.16), they are illustrated in Fig. 8.5a, b.

It can be seen that the traditional echelon base-stock policy, in which each stage is controlled by an echelon inventory target level, can be regarded as a special form of the BKC policy. The above results can also be further extended to multistage serial supply chains with failure-prone machines (cf. Song and Sun 1998).

8.6   Discussion and Notes

This chapter is partially based on Song (2009). We have addressed the stability conditions for the basic stochastic supply chain in Chap. 2 and extended the main results to the long-run average cost case. More importantly, based on the structural characteristics presented in Chaps. 2 and 3, we have constructed linear switching threshold-type control policies for both the basic supply chain and more general supply chains and extended the threshold policies to multistage serial supply chains.

In the literature, various forms of threshold-type policies have been proposed for stochastic production/inventory systems. According to the inventory information being utilized to determine the control actions, there are three common types of threshold policies: base-stock (or echelon base-stock) policy, Kanban policy, and CONWIP policy (constant work in process).

The base-stock policy tries to maintain a target level of the echelon base-stock for each stage. Here the stage's echelon base-stock is the sum of all downstream stages' inventories minus the backlogged finished goods demand. Under the base-stock control, it is necessary that customer demand information is transmitted to all upstream supply chain entities immediately. For example, in the serial manufacturing system, the base-stock policy states whenever a customer demand for finished goods arrives, the information is immediately transmitted to every stage where it authorizes the processing of a new part (Duri et al. 2000). The policy operates on a pull mechanism. The advantage of this policy is the timely information sharing of customer demand and downstream stages' inventory. The disadvantage is that there is no guarantee for the upper bound of the local inventories at upstream stages due to the possible severe backlogged demands.

Kanban systems were initially developed in Japan to manage multistage manufacturing systems. Every stage has a target inventory level of the local inventories. The customer demand information is transmitted upward stage by stage. More specifically, it states that whenever an item is withdrawn from the local inventory, a card is issued to the corresponding production unit to signal that a new part should be produced (Duri et al. 2000). This policy also operates on a pull mechanism. The advantage of this policy is that the number of parts at each stage (i.e., inventory-on-hand) is limited by the number of Kanbans of that stage. The disadvantages are the indirect flow of demand information from customers to suppliers and the lack of utilizing inventory information at downstream stages.

CONWIP system can be viewed as a single stage Kanban system, in which a single type of cards is used to control the total amount of inventories permitted in the entire system (Spearman et al. 1990). Essentially, there is only one target inventory level for the whole system to trigger whether the most upstream production unit should produce a new part (i.e., pull the raw material into the first stage of the system), whereas other stages follow a simple push mechanism, that is, producing parts whenever possible. Spearman and Zazanis (1992) provided more discussions of push and pull control mechanisms.

The comparison of the above three control policies and the combination or extension of them has been extensively performed (e.g., Muckstadt and Tayur 1995; Gstettner and Kuhn 1996; Bonvik et al. 1997; Chaouiya et al. 2000; Dallery and Liberopoulos 2000; Duri et al. 2000; Karaesmen and Dallery 2000; Yang 2000; Ovalle and Marquez 2003; Takahashi et al. 2005; Pettersen and Segerstedt 2009; Lavoie et al. 2010; Khojasteh-Ghamari 2012). More relevant literature can be found in the following review papers: Framinan et al. (2003) and Junior and Filho (2010).

The proposed linear switching threshold (LST) policy is more general than many base-stock or Kanban-based policies in the literature. For example, the three-parameter extended Kanban-type policy in Karaesmen and Dallery (2000) can be regarded as a special form of the LST policy with one linear segment is determined by $x_1 + x_2 = parameter$. However, as supply chain systems involve more stages, it is often necessary to further simplify the LST policies in order to reduce the number of threshold parameters as in Sect. 8.4. The main reason is the challenge of determining those threshold parameters because an inappropriately designed threshold values may result in poor supply chain performance. In later chapters, we will discuss different ways to optimize threshold parameters.

References

Bertsekas, D.P.: Dynamic Programming: Deterministic and Stochastic Models. Prentice-Hall, Englewood Cliffs (1987)
Bonvik, A.M., Couch, C.E., Gershwin, S.B.: A comparison of production-line control mechanisms. Int. J. Prod. Res. **35**(3), 789–804 (1997)

Chaouiya, C., Liberopoulos, G., Dallery, Y.: The extended Kanban control system for production coordination of assembly manufacturing systems. IIE Trans. **32**, 999–1012 (2000)

Dallery, Y., Liberopoulos, G.: Extended Kanban control system: combining Kanban and base stock. IIE Trans. **32**(4), 369–386 (2000)

Duri, D., Frein, Y., Di Mascolo, M.: Comparison among three pull control policies: Kanban, base tock, and generalized Kanban. Ann. Oper. Res. **93**, 41–69 (2000)

Framinan, J.M., Gonzalez, P.L., Ruiz-Usano, R.: The CONWIP production control system: review and research issues. Prod. Plan. Control **14**, 255–265 (2003)

Gstettner, S., Kuhn, H.: Analysis of production control systems Kanban and CONWIP. Int. J. Prod. Res. **34**(11), 3253–3274 (1996)

Junior, M.L., Filho, M.G.: Variations of the Kanban system: literature review and classification. Int. J. Prod. Econ. **125**(1), 13–21 (2010)

Karaesmen, F., Dallery, Y.: A performance comparison of pull type control mechanisms for multi-stage manufacturing. Int. J. Prod. Econ. **68**(1), 59–71 (2000)

Khojasteh-Ghamari, Y.: Developing a framework for performance analysis of a production process controlled by Kanban and CONWIP. J. Intell. Manuf. **23**(1), 61–71 (2012)

Latouche, G., Ramaswami, V.: Introduction to Matrix Analytic Methods in Stochastic Modelling. SIAM, Philadelphia (1999)

Lavoie, P., Gharbi, A., Kenne, J.P.: A comparative study of pull control mechanisms for unreliable homogenous transfer lines. Int. J. Prod. Econ. **124**(1), 241–251 (2010)

Muckstadt, J.A., Tayur, S.R.: A comparison of alternative Kanban control mechanisms: I, background and structural results. IIE Trans. **27**(1), 140–150 (1995)

Ovalle, O.R., Marquez, A.C.: Exploring the utilization of a CONWIP system for supply chain management. A comparison with fully integrated supply chains. Int. J. Prod. Econ. **83**(2), 195–215 (2003)

Pettersen, J.A., Segerstedt, A.: Restricted work-in-process: a study of differences between Kanban and CONWIP. Int. J. Prod. Econ. **118**, 199–207 (2009)

Sennott, L.I.: Stochastic Dynamic Programming and the Control of Queueing Systems. Wiley, New York (1999)

Song, D.P.: Optimal integrated ordering and production policy in a supply chain with stochastic lead-time, processing-time and demand. IEEE Trans. Autom. Control **54**(9), 2027–2041 (2009)

Song, D.P., Sun, Y.X.: Optimal service control of a serial production line with unreliable workstations and random demand. Automatica **34**(9), 1047–1060 (1998)

Spearman, M.L., Zazanis, M.A.: Push and pull production systems: issues and comparisons. Oper. Res. **40**, 521–532 (1992)

Spearman, M.L., Woodruff, D.L., Hopp, W.J.: CONWIP: a pull alternative to Kanban. Int. J. Prod. Res. **23**, 879–894 (1990)

Takahashi, K., Myreshka, Hirotani, D.: Comparing CONWIP, synchronized CONWIP, and Kanban in complex supply chains. Int. J. Prod. Econ. **93–94**, 25–40 (2005)

Yang, K.K.: Managing a flow line with single-Kanban, dual-Kanban or CONWIP. Prod. Oper. Manage. **9**(4), 349–366 (2000)

Chapter 9
Threshold-Type Control of Supply Chain Systems with Backordering Decisions

9.1 Introduction

In Chap. 4, we have investigated the optimal control policies in supply chain systems with backordering decision. Two types of supply chains were discussed: the basic stochastic supply chain and a failure-prone manufacturing supply chain. This chapter attempts to develop relative simple but near-optimal policies and extend some analytical results to the average cost case.

For the basic stochastic supply chain system in Sect. 4.2, similar to Chap. 8, we will use the linear switching threshold (LST) policies to approximate the optimal policy. A range of experiments will be conducted to explore the sensitivity of the system performance to the decision types and the system parameters.

For the failure-prone manufacturing supply chain in Sect. 4.3, we will first discuss the system stability and the long-run average cost. The similar structural properties of the optimal average cost and its optimal policy will be established. It will be shown that the optimal policy is actually of threshold type. We then analytically derive the stationary distribution of the induced Markov chain under a general threshold control policy by using the flow balance equations and the characteristic equation method. Steady-state performance measures can then be calculated from the stationary distribution.

9.2 Threshold Control in the Basic Serial Supply Chain with Backordering Decisions

The system stability of the basic supply chain with backordering would be the same as the one in Chap. 8 under the completely backordering policy. If a finite backordering policy is applied, the system is obviously stable due to the finite state

Table 9.1 Parameter setting for different cases

Case	λ	r	μ	c_1	$c_2{}^+$	$c_2{}^-$	c_r	$\mu c_2{}^-/\beta$
1	0.5	1.0	0.5	1	2	10	12	10
2	0.5	1.0	0.7	1	2	10	12	14
3	0.5	1.0	0.9	1	2	10	12	18
4	1.0	1.0	0.5	1	2	10	12	10
5	1.0	1.0	0.7	1	2	10	12	14
6	1.0	1.0	0.9	1	2	10	12	18
7	2.0	1.0	0.5	1	2	10	12	10
8	2.0	1.0	0.7	1	2	10	12	14
9	2.0	1.0	0.9	1	2	10	12	18

space. We would assume that the condition in Proposition 8.1 holds so that the completely backordering policy is also feasible.

We are interested in constructing relatively simple but near-optimal threshold control policies. Based on the structural properties of the switching curves in Chap. 4, we are able to propose a linear switching threshold (LST) policy. Its ordering decision and production decision parts are the same as (8.7) and (8.8) as shown in Fig. 8.2. Its backordering (i.e., order accepting) decision is approximated by an additional control parameter j_5, which is independent of the raw material inventory level. Mathematically,

$$a\,(x_1, x_2) = \begin{cases} 0 & x_2 < j_5 \\ 1 & \text{otherwise} \end{cases}. \tag{9.1}$$

Therefore, the LST policy consists of three parts, that is, (8.7), (8.8), and (9.1), which are determined by five points: (i_1, j_1), (i_2, j_2), $(1, j_3)$, (i_4, j_4), and $(0, j_5)$. In the remainder of this section, numerical examples are given to demonstrate the effectiveness of the proposed LST policy and examine the impacts of the backordering decision on the system performance.

We use the value iteration algorithm to evaluate the performance. The system state space is limited into a finite area with $x_2 \in [N - 50, N]$. The iterative procedure will be terminated when the value difference is less than 10^{-3} or the number of iterations exceeds 5,000. Nine cases with different combinations of system parameters (see Table 9.1) are tested. The last column gives the value of $\mu c_2{}^-/\beta$, which indicates that the completely backordering policy is optimal if $c_r \geq \mu c_2{}^-/\beta$. The warehouse capacities are set as $M = N = 6$ and the discount factor $\beta = 0.5$.

Table 9.2 gives the optimal cost, the percentage of the cost under the best linear switching threshold (LST) policy above the optimal policy, the control parameters of the best LST policy, and the optimal cost with completely backordering (denoted as $J_{\text{Opt\_CB}}$) for all nine cases, respectively. The system initial state is assumed to be (0, 0) for the discounted cost case. Here the last column $J_{\text{Opt\_CB}}$ is the same as the optimal cost in Table 8.2.

In Table 9.2, $j_5 = -9$ means that the maximum backordering boundary in the state space has been reached, namely, the completely backordering policy should be

Table 9.2 Optimal policy and LST policy for different case

Case	J_{optimal}	% above of J_{LST}	$(i_1, j_1)\,(i_2, j_2)\,(1, j_3)\,(i_4, j_4)\,(j_5)$	$J_{\text{Opt\_CB}}$
1	12.855	0.00%	(0; 4) (4; −4) (1; 0) (1; 0) (−9)	12.855
2	18.122	0.44%	(0; 3) (4; −3) (1; 1) (1; 1) (−1)	18.598
3	20.081	4.84%	(0; 4) (4; −4) (1; 1) (1; 1) (1)	24.920
4	10.733	0.12%	(0; 1) (3; −2) (1; 0) (1; 0) (−9)	10.733
5	15.553	0.29%	(0; 6) (3;−3) (1; 1) (1; 1) (−1)	15.747
6	18.465	1.38%	(0; 4) (3; −2) (1; 1) (1; 1) (0)	21.309
7	9.257	0.38%	(0; 2) (3; −4) (1; 0) (1; 0) (−9)	9.257
8	13.500	0.02%	(0; 3) (3; −3) (1; 1) (1; 1) (−1)	13.617
9	16.730	0.00%	(0; 3) (3; −3) (1; 1) (1; 1) (0)	18.678

used in the best LST policy. This verifies the analytical results in Chap. 4, that is, the completely backordering is optimal for the cases with $\mu c_2^- / \beta > c_r$.

It can be observed that the best LST is very close to the optimal policy (less than 1.4%) in all cases except Case 3, which has 4.84% difference. The switching curves for ordering decisions and production decisions in the best LST policy are close to those in Table 8.2. Comparing the optimal cost to the one in the completely back-ordering situations (the last column), it can be seen that rejecting customer demands could lead to significant cost reduction in some cases, that is, cases 3, 6, and 9.

As case 3 has the largest difference between the optimal policy and the best LST policy, it may be interesting to take a closer look at their control structure, which are shown in Fig. 9.1.

From Fig. 9.1, it can be observed that two policies have the same decisions for raw material ordering, a slight difference of decisions for production and backordering. The optimal cost is 20.0810, and the best LST policy has a cost 21.0522. By adjusting the decisions for production to be the same as the optimal policy, the LST policy would incur a cost of 21.2378, which is worse than the best LST policy as we expected. By keeping the production policy the same as that in the best LST policy but adjust the backordering policy to be the same as the one in the optimal policy (i.e., make the backordering policy characterized by two threshold values rather than a single parameter), the LST policy would incur a cost of 20.0810, which is the same as the optimal cost. The above two additional experiments indicate that the system performance appears to be more sensitive to the backordering decision than to the production decision in this case. Another implication is that the approximation of the backordering decision could be further improved in the LST policy. For example, instead of using a single threshold value, we could use another linear curve to determine the backordering decisions. However, this would make the LST policy more complicated in implementation because it requires the optimization of more control parameters.

In Table 9.2, the control parameters for production decisions appear to be more robust than the control parameters for raw material ordering decisions and demand backordering decisions. This reveals that the system performance is more sensitive

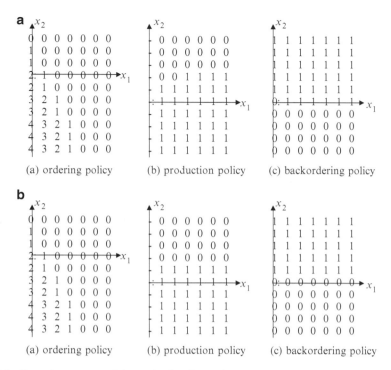

Fig. 9.1 Control structures of the optimal policy and the best LST in case 3. (**a**) The optimal policy. (**b**) The best LST policy

to the ordering and backordering decisions than to the production decisions. Such observation is useful in developing simpler threshold policies and optimizing the threshold parameters.

In all cases in Table 9.1, the customer demand rejecting cost is higher than the backlogging cost, which is reasonable in practice since lost sales may damage the company's reputation more severely than backordering. Next, we conduct another set of experiments to explore the sensitivity of system performance to the rejecting cost. Take case 5 in Table 9.1 as an example; Table 9.3 gives the optimal cost, the percentage of the cost under the best LST policy above the optimal cost, the best threshold parameters, for scenarios with different rejecting costs. The last column gives the optimal cost with completely backordering policy from Table 9.2. It can be seen from Table 9.3, as the rejecting cost increases, the optimal cost $J_{optimal}$ is increasing and converging to the optimal cost under the completely backordering policy J_{Opt_CB}. This is in agreement with the intuition. The results also show that the best LST policy is extremely close to the optimal policy, with less than 0.3% difference.

Table 9.3 The optimal costs with different rejecting costs

c_r	J_{optimal}	% above of J_{LST}	$(i_1, j_1)\ (i_2, j_2)\ (1, j_3)\ (i_4, j_4)\ (j_5)$	$J_{\text{Opt\_CB}}$
5	9.100	0.00%	$(0; 1)\ (3; -2)\ (1; 0)\ (1; 0)\ (1)$	15.747
10	14.729	0.22%	$(0; 3)\ (3; -3)\ (1; 1)\ (1; 1)\ (0)$	15.747
12	15.553	0.29%	$(0; 6)\ (3; -3)\ (1; 1)\ (1; 1)\ (-1)$	15.747
15	15.746	0.26%	$(0; 4)\ (3; -2)\ (1; 1)\ (1; 1)\ (-9)$	15.747
20	15.746	0.26%	$(0; 4)\ (3; -2)\ (1; 1)\ (1; 1)\ (-9)$	15.747

9.3 Threshold Control in a Failure-Prone Manufacturing Supply Chain with Backordering Decisions

In Chap. 4, we have shown that the optimal control policy in the failure-prone supply chain with backordering decisions (Fig. 4.5) with a discounted cost is a threshold control. This section extends the analytical results to the long-run average cost situation. The main purpose is to obtain the explicit form of the stationary distribution under the threshold policy, which can then be used to calculate various interesting steady-state performance measures and seek the optimal threshold values

9.3.1 System Stability and the Long-Run Average Cost

For the system in Fig. 4.5, the expected maximum production capacity is determined by $r\eta/(\eta + \xi)$ in long term. This is due to the fact that the machine produces nothing when it is down and its average up time fraction is $\eta/(\eta + \xi)$. As we know, the system is stable even under the completely backordering policy if the expected maximum production rate is greater than the demand arrival rate (Feng and Xiao 2002), that is,

$$r\eta/(\eta + \xi) > \mu. \tag{9.2}$$

Under the condition (9.2), the induced Markov chain is positive recurrent, and its stationary distribution exists. Define the long-run average cost as follows (independent of the initial state):

$$J^* = \min_{u,a} \lim_{T \to \infty} \frac{1}{T} E \int_0^T (g(x(t)) + c_r(1 - a(t)))\mathrm{d}t. \tag{9.3}$$

Under (9.2), the long-run average cost optimality equation holds (Sennott 1999)

$$
\begin{aligned}
w(1, x) + \frac{J^*}{\nu} = \frac{1}{\nu} \big[& c^+ x^+ + c^- x^- + \xi w(0, x) + \eta w(1, x) \\
& + r \cdot \min(w(1, x + 1), w(1, x)) \\
& + \min(\mu w(1, x - 1), c_r + \mu w(1, x)) \big]
\end{aligned} \tag{9.4}
$$

$$w\,(0,x) + \frac{J^*}{v} = \frac{1}{v}\left[c^+x^+ + c^-x^- + (\xi + r)\,w\,(0,x) + \eta w\,(1,x)\right.$$

$$\left. + \min\left(\mu w(0,x-1), c_r + \mu w\,(0,x)\right)\right] \tag{9.5}$$

where J^* is the optimal average cost and $w(\alpha, x)$ is a finite function.

Proposition 9.1. *(i) $w(\alpha, x+1) - w(\alpha, x)$ is increasing in x; (ii) $w(\alpha, x+1) - w(\alpha, x) \le 0$ for $x \le -1$; (iii) $w(1, x) - w(0, x)$ is increasing in x; (iv) $w(1, x) - w(0, x) \le 0$ for any x.*

Proof. Let $J^\beta(\alpha, x)$ denote the discounted cost defined in Sect. 4.3 with a discount factor β. Note that $w\,(\alpha, x) = \lim_{\beta \to 0}\left(J^\beta\,(\alpha, x) - J^\beta\,(1, 0)\right)$ by Sennott (1999) and recalling the results in Propositions 4.5 and 4.6, the assertions are true. This completes the proof. □

Proposition 9.2. *The optimal production and backordering policy $(u^*(\alpha, x), a^*(\alpha, x))$ for the average cost case is also a threshold-type control, that is,*

$$u^*\,(1,x) = \begin{cases} r & x < n^* \\ 0 & x \ge n^* \end{cases}, \quad u^*(0,x) \equiv 0; \tag{9.6}$$

$$a^*\,(1,x) = \begin{cases} 0 & x \le l^* \\ 1 & x > l^* \end{cases}, \quad a^*(0,x) = \begin{cases} 0 & x \le m^* \\ 1 & x > m^* \end{cases} \tag{9.7}$$

where

$n^* = \max\{x|w\,(1,x) - w\,(1,x-1) \le 0\};$
$l^* = \max\{x|w\,(1,x) - w\,(1,x-1) \le -c_r/\mu\};$
$m^* = \max\{x|w\,(0,x) - w(0,x-1) \le -c_r/\mu\}.$
In addition, we have $n^* \ge 0$ and $l^* \le m^* \le 0$.

Proposition 9.3. *(i) $\lim_{x \to +\infty} w(\alpha, x+1) - w(\alpha, x) = +\infty$; (ii) $\lim_{x \to -\infty} w(\alpha, x+1) - w(\alpha, x) = -\infty$.*

Proposition 9.4. *The optimal threshold values are finite integers and satisfy $-\infty < l^* \le m^* \le 0 \le n^* < +\infty$. This implies that the completely backordering policy is always not optimal for the long-run average cost case.*

It is interesting to compare Proposition 4.9 with Proposition 9.4. For the discounted cost case, suppose there are a very large number of backlogged demands. Backordering one more demand would incur an additional cost c^- over the time until it is satisfied. In the limit, the expected additional cost is $c^- \cdot \int_0^\infty e^{-\beta t}\,dt = c^-/\beta$. As long as this value is not greater than the one-off cost incurred by rejecting a demand (i.e., c_r/μ), the completely backordering policy is optimal. On the other hand, for the average cost case, the expected additional cost to satisfy a further backlogged demand when there are a large number of backlogged demands becomes infinity. Therefore, the completely backordering policy is always not optimal in the average cost situation.

If $r\eta/(\eta+\xi)\le\mu$, then the system is unstable under the completely backordering policy. Therefore, to ensure the stability of the system under a stationary policy, the system state space must be limited into a finite area by rejecting demands if the backlog length exceeds the limit. In this situation, the induced Markov chain has finite state space and is positive recurrent. The same results as stated in Propositions 9.1–9.2 hold. Thus, for the long-run average cost case, the optimal stationary policy is a threshold-type control with $-\infty < l^* \le m^* \le 0 \le n^* < +\infty$.

9.3.2 Stationary Distribution

Define a general threshold policy by replacing l^*, m^*, and n^* with l, m, and n in (9.6) and (9.7), respectively. This policy is denoted by $u_{l,m,n}$. In this section, we firstly derive the explicit form of the stationary distribution under $u_{l,m,n}$ subject to $-\infty < l \le m \le 0 \le n < +\infty$ based on the flow balance equations and the characteristic equation method, then compute the steady-state performance measures and find the optimal threshold values l^*, m^*, and n^* to implement the optimal policy.

Under a threshold control policy $u_{l,m,n}$, the induced system forms an irreducible Markov chain with finite states. Let $\{\pi_\alpha^{l,m,n}(x)\} = \text{Prob}\{\text{state }(\alpha, x)\}$ be the limiting state probability. It is well known that the limiting state probability in a finite irreducible Markov chain always exists and forms a stationary distribution, which is independent of the initial state. Clearly, the induced Markov chain under $u_{l,m,n}$ is equivalent to that of the Markov chain under a zero-inventory policy $u_{l-n,m-n,0}$. In fact, we have $\pi_\alpha^{l,m,n}(x) \equiv \pi_\alpha^{l-n,m-n,0}(x-n)$ for $l \le x \le n$ and $\alpha \in \{0, 1\}$. To simplify the narrative, denote $\pi_\alpha(x) := \pi_\alpha^{l,m,0}(x)$.

The stationary distribution $\{\pi_\alpha(x)\}$ can be found by solving a set of flow balance equations together with the normalization condition, which are given as follows:

$$\mu\pi_1(x+1) + r\pi_1(x-1) + \eta\pi_0(x) = (\mu+r+\xi)\pi_1(x) \quad \text{for } l < x < 0 \tag{9.8}$$

$$\mu\pi_0(x+1) + \xi\pi_1(x) = (\mu+h)\pi_0(x) \quad \text{for } m < x < 0 \tag{9.9}$$

$$\xi\pi_1(x) = \eta\pi_0(x) \quad \text{for } l \le x < m \tag{9.10}$$

$$\mu\pi_1(l+1) + \eta\pi_0(l) = (r+\xi)\pi_1(l) \tag{9.11}$$

$$\mu\pi_0(m+1) + \xi\pi_1(m) = \eta\pi_0(m) \tag{9.12}$$

$$r\pi_1(-1) + \eta\pi_0(0) = (\mu+\xi)\pi_1(0) \tag{9.13}$$

$$\xi \pi_1(0) = (\mu + \eta) \, \pi_0(0) \tag{9.14}$$

$$\sum_{x=l}^{0} \pi_1(x) + \sum_{x=l}^{0} \pi_0(x) = 1. \tag{9.15}$$

For $l \le x < m$, from (9.8), (9.10), and (9.11), we have

$$\pi_1(x) = (r/\mu)^{x-l} \pi_1(l) \quad \text{for } l \le x \le m \text{ and } \pi_0(x) = \xi \pi_1(x)/\eta \quad \text{for } l \le x < m. \tag{9.16}$$

The boundary conditions (9.12) and (9.8) at $x = m$ yield

$$\pi_0(m) = \left((r/\mu)^{m-l} \xi \pi_1(l) + \mu \pi_0 (m + 1) \right) / \eta \tag{9.17}$$

$$\pi_1(l) = (\mu/r)^{m-l+1} (\pi_1 (m + 1) + \pi_0 (m + 1)). \tag{9.18}$$

For $m < x < 0$, the characteristic equation approach can be used to derive the solution $\{\pi_i(x)\}$ (cf. Feng and Xiao 2002). We define a left-shift operator D: $D\pi_\alpha(x) := \pi_\alpha(x - 1)$. From (9.8) and (9.9),

$$\begin{pmatrix} rD^2 - (\mu + r + \xi)\, D + \mu & \eta D \\ \xi D & -(\mu + \eta)\, D + \mu \end{pmatrix} \begin{pmatrix} \pi_1 (x + 1) \\ \pi_0 (x + 1) \end{pmatrix} = 0. \tag{9.19}$$

Therefore, the characteristic equation for $m < x < 0$ is

$$(D - 1) \left[r (\mu + \eta)\, D^2 - \mu (\mu + r + \eta + \xi)\, D + \mu^2 \right] = 0. \tag{9.20}$$

Clearly, the characteristic equation (9.20) has three eigenvalues (counting the multiplicity), denoted by $y_1 < 1$, y_2 and $y_3 = 1$, where y_1 and y_2 are two distinct positive real roots of $r (\mu + \eta)\, y^2 - \mu (\mu + r + \eta + \xi)\, y + \mu^2 = 0$. A general solution to (9.8) and (9.9) for $m < \mu \le 0$ can be constructed based on the eigenvalues, that is,

$$\pi_1(x) = A_1 y_1^{-x} + B_1 y_2^{-x} + C_1 \quad \text{for } m < x \le 0 \tag{9.21}$$

$$\pi_0(x) = A_2 y_1^{-x} + B_2 y_2^{-x} + C_2 \quad \text{for } m < x \le 0. \tag{9.22}$$

The reason that $\pi_1(0)$ and $\pi_0(0)$ also have the forms in (9.21) and (9.22), respectively, follows from the Eqs. (9.8) and (9.9) with $x = -1$. If $y_2 = 1$, then $y_2^{-x} \equiv 1$, in which case the third term in RHS of (9.21) and (9.22) can be merged with the second term. Note that $y_2 = 1$ if and only if $r\eta = \mu(\eta + \xi)$. Therefore, we have either $r\eta \ne \mu (\eta + \xi)$ that guarantees $y_2 \ne 1$ or $r\eta = \mu (\eta + \xi)$ in which the last two terms are merged (i.e., $C_1 = C_2 = 0$). Substituting the above solution into the flow balance equation at $(x, 0)$, and comparing the coefficients of the terms with the same power, and then using $(r + \xi) y_k - \mu = (r + \xi + \eta + \mu) y_k - \mu - (\eta + \mu) y_k = r (\eta + \mu) y_k{}^2/\mu - (\eta + \mu) y_k$, we have

$$A_2 = A_1 (r y_1 - \mu)/\mu, B_2 = B_1 (r y_2 - \mu)/\mu, \text{ and } C_2 = C_1 \xi /\eta. \quad (9.23)$$

The boundary conditions (9.13) and (9.14) yield

$$r (A_1 y_1 + B_1 y_2 + C_1) + \eta (A_2 + B_2 + C_2) = (\mu + \xi)(A_1 + B_1 + C_1)$$

$$\xi (A_1 + B_1 + C_1) = (\mu + \eta)(A_2 + B_2 + C_2).$$

Using (9.23), it follows

$$[r (\mu + \eta) y_1 - \mu (\mu + \eta + \xi)] A_1 + [r (\mu + \eta) y_2 - \mu (\mu + \eta + \xi)] B_1$$
$$+ \mu (r - \mu) C_1 = 0 \quad (9.24)$$

$$[r (\mu + \eta) y_1 - \mu (\mu + \eta + \xi)] A_1 + [r (\mu + \eta) y_2 - \mu (\mu + \eta + \xi)] B_1$$
$$+ \xi \mu^2 C_1 /\eta = 0. \quad (9.25)$$

Comparing (9.24) and (9.25), we have $C_1 = 0$ and the relationship between A_1 and B_1 as follows:

$$C_1 = 0, B_1 = E_1 \cdot A_1 \text{ and } E_1 := \frac{-r (\mu + \eta) y_1 + \mu (\mu + \eta + \xi)}{r (\mu + \eta) y_2 - \mu (\mu + \eta + \xi)}. \quad (9.26)$$

Finally, substituting (9.18), (9.21), (9.22), and (9.26) into the normalization equation (9.15), we have

$$A_1 = 1/(F_1 + F_2) \quad (9.27)$$

where

$$F_1 = \frac{\eta + \xi}{\eta} \cdot \sum_{x=l}^{m} \left(\frac{r}{\mu}\right)^{x-m} y_1^{-m} + \frac{r y_1 - \mu}{\eta} y_1^{-m-1} + \frac{r}{\mu} \cdot \sum_{x=m+1}^{0} y_1^{-x+1} \quad (9.28)$$

$$F_2 = E_1 \cdot \left[\frac{\eta + \xi}{\eta} \cdot \sum_{x=l}^{m} \left(\frac{r}{\mu}\right)^{x-m} y_2^{-m} + \frac{r y_2 - \mu}{\eta} y_2^{-m-1} + \frac{r}{\mu} \cdot \sum_{x=m+1}^{0} y_2^{-x+1}\right]. \quad (9.29)$$

Proposition 9.5. *The stationary distribution* $\{\pi_\alpha(x)\}$ *under a threshold policy* $u_{l,m,0}$ *is given by (9.16), (9.17), and (9.18), (9.21) and (9.22), where the constant coefficients are given by (9.23) and (9.26), (9.27), (9.28), and (9.29).*

In (9.28) and (9.29), if we let m and n tends to $-\infty$, then the result is simplified into that in Feng and Xiao (2002).

9.3.3   Steady-State Performance Measures

Since we have obtained the explicit form of the stationary distribution under a threshold policy $u_{l,m,n}$, it is easy to calculate various steady-state performance measures such as the long-run average cost, the fraction of lost demands, and the stock-out probability (which is defined as the probability that the system has backordered demands).

Proposition 9.6. *The long-run average cost* $J(u_{l,m,n})$, *the fraction of lost demands (denoted by* P_{lost}), *and the stock-out probability (denoted by* $P_{stockout}$) *under* $u_{l,m,n}$ *can be explicitly expressed by*

$$J(u_{l,m,n}) = \sum_{x=l}^{0}\sum_{\alpha=0}^{1} -x\pi_\alpha^{l,m,n}(x)c^- + \sum_{x=1}^{n}\sum_{\alpha=0}^{1} x\pi_\alpha^{l,m,n}(x)c^+$$

$$+ \left(\pi_1^{l,m,n}(l) + \sum_{x=l}^{m}\pi_0^{l,m,n}(x)\right)c_r; \qquad (9.30)$$

$$P_{lost} = \pi_1^{l,m,n}(l) + \sum_{x=l}^{m}\pi_0^{l,m,n}(x); \qquad (9.31)$$

$$P_{stockout} = \sum_{x=l}^{-1}\sum_{\alpha=0}^{1}\pi_\alpha^{l,m,n}(x) \qquad (9.32)$$

where $\pi_\alpha^{l,m,n}(x) = \pi_\alpha^{l-n,m-n,0}(x-n)$ *for* $l \leq x \leq n$, $\alpha \in \{0, 1\}$ *and* $\pi_\alpha^{l-n,m-n,0}(x)$ *is given in Proposition 9.5 by replacing* l *with* $l-n$ *and* m *with* $m-n$.

Proposition 9.7. *The optimal average cost policy can be determined by* $(l^*, m^*, n^*) = \arg\min_{l \leq m \leq 0 \leq n} J(u_{l,m,n})$, *where* $J(u_{l,m,n})$ *is given in Proposition 9.6.*

Proposition 9.8. *The service level can be defined as* $SL := Ea(.) = 1 - P_{lost}$.

In practice, it is often interesting to optimize the average cost subject to service level constraint. This can be easily done as follows. Suppose there is a constraint on service level such as $SL \geq P_{SL}$. The optimization procedure includes two steps: (i) if $P_{lost}(l^*, m^*, n^*) \leq (1 - P_{SL})$ where (l^*, m^*, n^*) is given by Proposition 9.7, then the threshold policy u_{l^*,m^*,n^*} is optimal; (ii) otherwise, by gradually decreasing l^* and m^*, the service level constraint could be satisfied.

Table 9.4 The optimal discounted costs at (1, 0) in a failure-prone manufacturing supply chain with backordering decisions

c_r	μ	(l^*, m^*, n^*)	$J^\beta(1, 0)$
1.0	1.0	$(0, 0, 0)$	4.4444
5.0	1.0	$(-1, 0, 1)$	12.1534
10.0	1.0	$(-4, -2, 2)$	14.1556
20.0	1.0	$(-50, -50, 2)$	14.1712
40.0	1.0	$(-50, -50, 2)$	14.1712
80.0	1.0	$(-50, -50, 2)$	14.1712
20.0	2.0	$(-3, -1, 3)$	45.7243
20.0	3.0	$(-1, 0, 5)$	80.2012

Table 9.5 The optimal average costs, the lost demand, and stock-out probabilities in a failure-prone manufacturing supply chain with backordering decisions

c_r	μ	(l^*, m^*, n^*)	J^*_{anal}	J^*_{iter}	P_{lost}	$P_{stockout}$
1.0	1.0	$(0, 0, 0)$	1.0000	1.0000	1.0000	0.0000
5.0	1.0	$(0, 0, 2)$	2.2835	2.2835	0.1905	0.0000
10.0	1.0	$(-1, 0, 3)$	3.1324	3.1324	0.0701	0.0405
20.0	1.0	$(-1, 0, 3)$	3.8337	3.8337	0.0701	0.0405
40.0	1.0	$(-3, -1, 4)$	4.5225	4.5225	0.0213	0.0589
80.0	1.0	$(-6, -5, 4)$	4.9119	4.9119	0.0045	0.0782
20.0	2.0	$(0, 0, 6)$	6.7346	6.7346	0.2066	0.0000
20.0	3.0	$(0, 0, 9)$	9.6205	9.6205	0.4039	0.0000

9.3.4   Numerical Examples

This section gives a numerical example to illustrate the results. Consider a failure-prone manufacturing supply chain with parameters $r = 2$, $\beta = 0.9$, $\eta = 0.9$, $\xi = 0.1$, $c^+ = 1$, and $c^- = 10$. Let the demand rejecting cost c_r and the demand rate μ take different values. For both discounted cost and average cost cases, the value iteration algorithm (limiting the state space into $-50 \leq x \leq 50$) is performed to investigate the control structure of the optimal policies.

The analytical method in Sect. 9.3.3 is used to calculate the stationary distribution, the steady-state performance measures, and the optimal threshold values. The results verify that the optimal policies for both discounted cost and average cost cases are of threshold type and the analytical average cost obtained by Proposition 9.6 is the same as that obtained by the value iteration algorithm.

The optimal threshold values (l^*, m^*, n^*) and the optimal discounted cost (J^β) with initial state (1, 0) are given in Table 9.4. The optimal average costs from the analytical method (J^*_{anal}) and the optimal average costs from the numerical value iteration algorithm (J^*_{iter}) are given in Table 9.5. The fraction of lost demands (P_{lost}) and the probability of stock-out $(P_{stockout})$ are also given in Table 9.5.

It can be seen from Tables 9.4 and 9.5 that as c_r increases, l^*, m^*, and P_{lost} are decreasing while n^* is increasing. This is in agreement with the intuition that we tend to backorder more demands and maintain higher level of inventories in order to avoid higher penalty of losing customer demands.

For the cases with $\mu = 1$ and $c_r = 20$, 40, and 80, we have $c_r \geq \mu c^- / \beta$, and therefore, the completely backordering policy is optimal to minimize the discounted cost. However, if μ takes 2.0 or 3.0, we have $c_r < \mu c^- / \beta$, in which the completely backordering policy is no more optimal for the discounted cost. The results confirm that $-\infty < l^* \leq m^* \leq 0 \leq n^* < +\infty$ for the average cost and $l^* \leq m^* \leq 0 \leq n^* < +\infty$ for the discounted cost. Particularly, the completely backordering policy is optimal for the discounted cost if $c_r \geq \mu c^- / \beta$, but always not optimal for the average cost.

9.4   Notes

Section 9.3 is mainly based on Song (2006). In the literature of failure-prone manufacturing systems, completely backordering policy is often implied (e.g., Akella and Kumar 1986; Bielecki and Kumar 1988; Song and Sun 1998; Song and Sun 1999; Feng and Yan 2000; Feng and Xiao 2002; Khmelnitsky et al. 2011). In addition, no backordering (i.e., lost sales) is assumed in Hu (1995), and partially backordering or bounded backlogging is adopted in Martinelli and Valigi (2004). These types of order accepting policies can be regarded as prespecified special forms of the backordering decisions.

In the literature of supply chain systems with multiple classes of customers, customer demand accepting (admission) is sometimes jointly considered with inventory rationing. For example, when the supply chain is to meet two classes of customers, the high-priority classes often have a prespecified form of backordering decisions such as fully backordering or lost sales, whereas the low-priority classes are endogenously controllable, that is, backordered or rejected dynamically (e.g., Carr and Duenyas 2000; Isotupa 2006; Benjaafar et al. 2010; Iravani et al. 2012). Threshold policies similar to that in this chapter have been proposed to determine whether less important customer demands should be accepted or rejected (e.g., Mollering and Thonemann 2008; Benjaafar et al. 2010).

References

Akella, R., Kumar, P.R.: Optimal control of production rate in a failure prone manufacturing system. IEEE Trans. Autom. Control **31**(1), 116–126 (1986)

Benjaafar, S., ElHafsi, M., Huang, T.: Optimal control of a production-inventory system with both backorders and lost sales. Nav. Res. Logist. **57**(3), 252–265 (2010)

Bielecki, T., Kumar, P.R.: Optimality of zero-inventory policies for unreliable manufacturing systems. Oper. Res. **36**(4), 532–541 (1988)

Carr, S., Duenyas, I. Optimal admission control and sequencing in a make-to-stock/make-to-order production system. Oper. Res. **48**(5), 709–720 (2000)

Feng, Y.Y., Xiao, B.C.: Optimal threshold control in discrete failure-prone manufacturing systems. IEEE Trans. Autom. Control **47**(7), 1167–1174 (2002)

Feng, Y.Y., Yan, H.M.: Optimal production control in a discrete manufacturing system with unreliable machines and random demands. IEEE Trans. Autom. Control **45**(12), 2280–2296 (2000)

Hu, J.Q.: Production rate control for failure-prone production systems with no-backlog permitted. IEEE Trans. Autom. Control **40**(2), 291–295 (1995)

Iravani, S.M.R., Liu, T., Simchi-Levi, D.: Optimal production and admission policies in make-to-stock/make-to-order manufacturing systems. Prod. Oper. Manage. **21**(2), 224–235 (2012)

Isotupa, K.P.S.: An Markovian inventory system with lost sales and two demand classes. Math. Comput. Model. **43**(7–8), 687–694 (2006)

Khmelnitsky, E., Presman, E., Sethi, S.P.: Optimal production control of a failure-prone machine. Ann. Oper. Res. **182**(1), 67–86 (2011)

Martinelli, F., Valigi, P.: Hedging point policies remain optimal under limited backlog and inventory space. IEEE Trans. Autom. Control **49**(10), 1863–1869 (2004)

Mollering, K.T., Thonemann, U.W.: An optimal critical level policy for inventory systems with two demand classes. Nav. Res. Logist. **55**(7), 632–642 (2008)

Sennott, L.I.: Stochastic Dynamic Programming and the Control of Queueing Systems. Wiley, New York (1999)

Song, D.P.: Optimal production and backordering policy in failure-prone manufacturing systems. IEEE Trans. Autom. Control **51**(5), 906–911 (2006)

Song, D.P., Sun, Y.X.: Optimal service control of a serial production line with unreliable workstations and random demand. Automatica **34**(9), 1047–1060 (1998)

Song, D.P., Sun, Y.X.: Optimal control structure of an unreliable manufacturing system with random demands. IEEE Trans. Autom. Control **44**(3), 619–622 (1999)

Chapter 10
Threshold-Type Control of Supply Chain Systems with Preventive Maintenance Decisions

10.1 System Stability

In Chap. 5, the supply chain systems with preventive maintenance decision are discussed, in which the objective function is a discounted cost. In reality, the steady-state performance measures are often of interest. Therefore, it is necessary to ensure that the system is stable. A system is called stable if a stationary distribution of the system state exists for some admissible control policies. In other words, the system will not be increasingly backlogged in long term.

This can be done by estimating the supply chain output capacity in comparison with the customer demand rate. Clearly, the maximum raw material supply rate is $Q\lambda$. For the manufacturer, consider two extreme scenarios. First, if preventive maintenance is never performed, the long-run maximum production rate is $r_1\eta/(\eta + \xi_1)$; since the machine produces nothing during breakdown time, the average machine uptime is $\eta/(\eta + \xi_1)$, and the maximum machine production rate is r_1 during its uptime. Similarly, if preventive maintenance is always performed at the operational mode, the long-run maximum production rate is $r_0\eta/(\eta + \xi_0)$. Note that the warehouse capacity is assumed to be infinite. The system stability is guaranteed if both the supply capacity and the production capacity are greater than the demand rate.

Proposition 10.1. *The sufficient conditions to ensure the stability of the supply chain systems with preventive maintenance in Fig. 5.1 are $Q\lambda > \mu$ and* $\max\{r_0\eta/(\eta + \xi_0), r_1\eta/(\eta + \xi_1)\} > \mu.$

For the special failure-prone manufacturing supply chain without considering raw material supply activity, the system stability condition is simplified as follows:

Proposition 10.2. *The sufficient condition to ensure the stability of the failure-prone manufacturing supply chain with preventive maintenance in Fig. 5.4 is* $\max\{r_0\eta/(\eta + \xi_0), r_1\eta/(\eta + \xi_1)\} > \mu.$

D.-P. Song, *Optimal Control and Optimization of Stochastic Supply Chain Systems*, Advances in Industrial Control, DOI 10.1007/978-1-4471-4724-4_10, © Springer-Verlag London 2013

In the rest of this chapter, we always assume that the supply chain systems under consideration are stable, and our focus is on the development of threshold control policies and their effectiveness.

10.2 Threshold-Type Control for Ordering, Production, and Preventive Maintenance in a Supply Chain

For the supply chain system in Fig. 5.1, intuitively if the system is heavily backlogged, the maximum production capacity should be utilized in order to meet the backordered demands as soon as possible. Namely, if $r_1\eta/(\eta + \xi_1) \geq r_0\eta/(\eta + \xi_0)$, that is, the long-run maximum production rate without performing preventive maintenance is not less than the long-run maximum production rate with preventive maintenance being performed, we tend not to perform preventive maintenance operation when $x_2(t) \ll 0$. On the other hand, if $r_1\eta/(\eta + \xi_1) < r_0\eta/(\eta + \xi_0)$, we prefer to perform maintenance operation when $x_2(t) \ll 0$. Therefore, there are two alternatives to locate the control regions for preventive maintenance as shown in Fig. 10.1a, b.

As for the raw material ordering decision, it could be simplified by using local inventory level (Kanban) or the combination of the echelon inventory level and the local inventory level (base-stock and Kanban) to determine the ordering process. These two alternatives are illustrated in Fig. 10.1c, d.

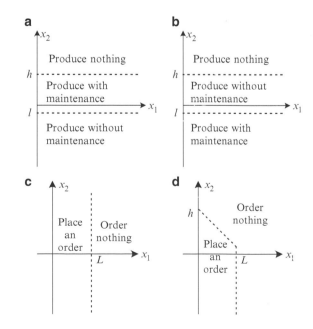

Fig. 10.1 Threshold control for RM ordering and preventive maintenance. (**a**) No maintenance when heavily backlogged. (**b**) Preventive maintenance when heavily backlogged. (**c**) Ordering based on RM inventory. (**d**) Ordering based on RM and FG inventory

Two alternatives of preventive maintenance, combined with two alternatives of raw material ordering, yield four types of threshold control policies. Mathematically, the four threshold control policies can be described as follows (here the order-up-to-point concept is applied when determining the raw material order sizes):

(I) Type-one threshold policy:

$$q(\alpha, \mathbf{x}) = \min\{L - x_1, Q\}, \quad \text{if } x_1 \leq L; q(\alpha, \mathbf{x}) = 0 \text{ otherwise.} \tag{10.1}$$

$$(\xi(1, \mathbf{x}), u(1, \mathbf{x})) = \begin{cases} (0, 0) & x_2 \geq h \\ (\xi_0, r_0) & l \leq x_2 < h \\ (\xi_1, r_1) & x_2 < l \end{cases}. \tag{10.2}$$

(II) Type-two threshold policy:

$$q(\alpha, \mathbf{x}) = \min\{L - x_1, Q\}, \quad \text{if } x_1 \leq L; q(\alpha, \mathbf{x}) = 0 \text{ otherwise.} \tag{10.3}$$

$$(\xi(1, x), u(1, x)) = \begin{cases} (0, 0) & x_2 \geq h \\ (\xi_1, r_1) & l \leq x_2 < h \\ (\xi_0, r_0) & x_2 < l \end{cases}. \tag{10.4}$$

(III) Type-three threshold policy:

$$q(\alpha, \mathbf{x}) = \min\{L - x_1, Q\} \cdot I\{x_2 + L \leq h\} + \min\{h - x_2 - x_1, Q\} \cdot$$
$$I\{x_2 + L > h\}, \quad \text{if } x_1 \leq L \text{ and } x_1 + x_2 \leq h; q(\alpha, \mathbf{x}) = 0 \text{ otherwise.} \tag{10.5}$$

$$(\xi(1, \mathbf{x}), u(1, \mathbf{x})) = \begin{cases} (0, 0) & x_2 \geq h \\ (\xi_0, r_0) & l \leq x_2 < h \\ (\xi_1, r_1) & x_2 < l \end{cases}. \tag{10.6}$$

(IV) Type-four threshold policy:

$$q(\alpha, \mathbf{x}) = \min\{L - x_1, Q\} \cdot I\{x_2 + L \leq h\} + \min\{h - x_2 - x_1, Q\} \cdot$$
$$I\{x_2 + L > h\}, \quad \text{if } x_1 \leq L \text{ and } x_1 + x_2 \leq h; q(\alpha, \mathbf{x}) = 0 \text{ otherwise.} \tag{10.7}$$

$$(\xi(1, \mathbf{x}), u(1, \mathbf{x})) = \begin{cases} (0, 0) & x_2 \geq h \\ (\xi_1, r_1) & l \leq x_2 < h \\ (\xi_0, r_0) & x_2 < l \end{cases}. \tag{10.8}$$

Each type of the above threshold control policies is characterized by three parameters, L, l, and h. Recalling the control structure of the optimal ordering,

Table 10.1 Optimal policy versus four threshold policies in supply chains with preventive maintenance

Policy	ξ_1	0.2	0.3	0.4
Optimal	J_{Opt}	128.5333	139.0710	139.6803
Type-one	J_{One}	134.6092	144.6021	146.4654
	% above J_{Opt}	4.7271%	3.9772%	4.8576%
	(L, l, h)	$(4, 3, 2)$	$(4, 3, -1)$	$(4, 3, -4)$
Type-two	J_{Two}	137.1958	145.9748	145.6126
	% above J_{Opt}	6.7395%	4.9642%	4.2471%
	(L, l, h)	$(4, 3, -4)$	$(4, 3, 2)$	$(4, 3, 2)$
Type-three	J_{Three}	130.1658	140.7926	142.7206
	% above J_{Opt}	**1.2701%**	**1.2379%**	2.1766%
	(L, l, h)	$(5, 4, 2)$	$(4, 4, -1)$	$(4, 4, -4)$
Type-four	J_{Four}	132.7821	142.1522	141.8054
	% above J_{Opt}	3.3056%	2.2156%	**1.5214%**
	(L, l, h)	$(5, 4, -4)$	$(4, 4, 2)$	$(4, 4, 2)$

production, and maintenance policy in Fig. 5.3, it can be seen that by appropriately designing these parameters, these threshold policies can closely approximate the optimal policy. This is illustrated by the example below:

Example 10.1. Consider a supply chain system with the following setting: the lead-time rate $\lambda = 0.8$, the maximum order capacity $Q = 3$, the demand rate $\mu = 0.7$, the machine repair rate $\eta = 0.9$, the machine failure rate $\xi_0 = 0.1$, the maximum production rates $r_0 = 1.0$ and $r_1 = 1.2$, the raw material inventory unit cost $c_1 = 1$, the finished goods inventory unit cost $c_2{}^+ = 2$, the backordering cost $c_2{}^- = 10$, and the preventive maintenance cost $c_m = 1.0$. The discount factor $\beta = 0.1$. The system state space is limited into a finite area with $x_1 \in [0, 10]$ and $x_2 \in [-30, 10]$. The iterative procedure will be terminated when the number of iterations exceeds 100. Consider three scenarios with different machine failure rates without preventive maintenance, for example, $\xi_1 = 0.2$, 0.3 and 0.4. Table 10.1 gives the optimal cost, the cost under four threshold-type policies, their percentage above the optimal policy, and the best threshold parameters of the corresponding threshold policies for three scenarios, respectively. The system initial state is assumed to be $(0, 0)$ for the discounted cost case.

It can be observed from Table 10.1 that (1) all four types of threshold policies are reasonably good if the threshold parameters (L, l, h) are optimized, for example, in the range of 1.2701 and 6.7395% above the optimal policy for the above three scenarios, and (2) type-three and type-four threshold policies are generally better than the first two types. This indicates that it is more beneficial to use both raw material inventory and finished goods inventory information to control the raw material ordering decisions than the raw material inventory information alone; (3) for the scenario with $\xi_1 = 0.4$, type-two is slight better than type-one, while type-four is slightly better than type-three. This reveals that performing preventive maintenance when the system is heavily backlogged is better, which reflects the intuition that if $r_1\eta/(\eta + \xi_1) < r_0\eta/(\eta + \xi_0)$, we should perform maintenance operation when $x_2(t) \ll 0$ so that the backlog can be reduce more quickly.

10.3 Threshold-Type Control for Production and Preventive Maintenance in a Manufacturing Supply Chain Without Raw Material Ordering Activity

For the failure-prone manufacturing supply chain in Fig. 5.4, this section presents two types of threshold policies to control the production and the preventive maintenance. Our main task is to derive the explicit form of the stationary distribution, evaluate the various performance measures, find the optimal threshold parameters, and demonstrate their effectiveness.

Depending on whether preventive maintenance should be performed when the system is heavily backlogged, two types of threshold control policies are given below:

(I) Type-one threshold policy:

$$(\xi(1, x), u(1, x)) = \begin{cases} (0, 0) & x \geq h \\ (\xi_0, r_0) & l \leq x < h \\ (\xi_1, r_1) & x < l \end{cases} \tag{10.9}$$

(II) Type-two threshold policy:

$$(\xi(1, \mathbf{x}), u(1, \mathbf{x})) = \begin{cases} (0, 0) & x \geq h \\ (\xi_1, r_1) & l \leq x < h \\ (\xi_0, r_0) & x < l \end{cases} \tag{10.10}$$

The above threshold policies are characterized by two integers, l and h. Let $\mathbf{u}_{l,h}$ denote the type-one threshold policy and $\mathbf{u}'_{l,h}$ denote the type-two threshold policy. The type-one threshold policy states that the machine should produce at full speed if the inventory level is less than h and produces nothing otherwise; preventive maintenance should not be performed if the inventory level is less than l and the maximum preventive maintenance is executed otherwise. It should be pointed out that the above policy is executed according to the system state, that is, machine operational mode and inventory-on-hand level. It does not depend on the age of the machine.

The ensuing arguments focus on the type-one threshold policy, and we assume the system is stable, that is, $r_1\eta/(\eta + \xi_1) > \mu$ by Proposition 10.2. For the type-two threshold policy, the arguments are very similar.

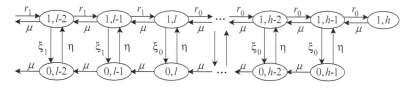

Fig. 10.2 State transition map under type-one threshold policy

10.3.1   State Transition Map Under Type-One Threshold Policy

It is clear that under the threshold policy in (10.9), the inventory position would never exceed h as soon as it enters a level not greater than h. When the inventory position is at h, the machine is idle and will not break down (i.e., no preventive maintenance is required). Therefore, $\mathbf{u}_{l,h} \equiv \mathbf{u}_{h,h}$ for $l > h$. Without any loss of the generality, we assume that $l \leq h$.

The system state transition map under the type-one threshold policy is partially shown in Fig. 10.2 (where the self-transition is omitted). Clearly, the induced system forms an irreducible, aperiodic and homogeneous Markov chain.

Let $\{\pi_i(k)\} = \text{Prob}\{\text{the system state } (\alpha, x) = (i, k)\}$ be the limiting state probability. From the probability theory, the limiting state probability in an irreducible aperiodic homogeneous Markov chain always exists, and it is independent of the initial state probability distribution. In addition, we have that either $\pi_i(k) \equiv 0$, in which case all states are transient or null recurrence, or $\pi_i(k) > 0$, in which case all states are positive recurrent and $\{\pi_i(k)\}$, forms a stationary distribution, that is, steady-state probability distribution.

10.3.2   Stationary Distribution Under Type-One
           Threshold Policy

This section aims to derive the stationary distribution $\{\pi_i(k)\}$ under the threshold control policy $\mathbf{u}_{l,h}$. Note that the stationary distribution of an irreducible aperiodic homogeneous Markov chain may be found by solving a set of flow balance equations and a normalization condition. The flow balance equations are given as follows

$$\mu\pi_1(k+1) + r_1\pi_1(k-1) + \eta\pi_0(k) = (\mu + r_1 + \xi_1)\pi_1(k), \quad \text{for } k < l, \tag{10.11}$$

$$\mu\pi_0(k+1) + \xi_1\pi_1(k) = (\mu + \eta)\pi_0(k), \quad \text{for } k < l, \tag{10.12}$$

$$\mu\pi_1(k+1) + r_0\pi_1(k-1) + \eta\pi_0(k) = (\mu + r_0 + \xi_0)\pi_1(k), \quad \text{for } l < k < h, \tag{10.13}$$

$$\mu\pi_0\,(k+1) + \xi_0\pi_1(k) = (\mu+\eta)\,\pi_0(k), \quad \text{for } l \le k < h-1, \tag{10.14}$$

$$\mu\pi_1\,(l+1) + r_1\pi_1\,(l-1) + \eta\pi_0(l) = (\mu+r_0+\xi_0)\,\pi_1(l), \tag{10.15}$$

$$r_0\pi_1\,(h-1) = \mu\pi_1(h), \tag{10.16}$$

$$\xi_0\pi_1\,(h-1) = (\mu+\eta)\,\pi_0\,(h-1)\,. \tag{10.17}$$

The normalization condition is

$$\sum_{-\infty}^{h-1} (\pi_1(k) + \pi_0(k)) + \pi_1(h) = 1. \tag{10.18}$$

The linear recursive equations (10.11), (10.12), (10.13), (10.14), (10.15), (10.16), and (10.17) contain infinite number of variables. Define a left-shifting operator $D{:}D\pi_i(k) := \pi_i\,(k-1)$, for $i = 0,1$.

For $k < l$, we have

$$\begin{pmatrix} r_1 D^2 - (\mu + r_1 + \xi_1)\,D + \mu & \eta D \\ \xi_1 D & -(\mu+\eta)\,D + \mu \end{pmatrix} \begin{pmatrix} \pi_1\,(k+1) \\ \pi_0\,(k+1) \end{pmatrix} = 0.$$

The characteristic equation for $k < l$ is

$$(D-1)\left[r_1\,(\mu+\eta)\,D^2 - \mu\,(\mu+r_1+\eta+\xi_1)\,D + \mu^2\right] = 0. \tag{10.19}$$

Similarly, the characteristic equation for $l < k < h-1$ is

$$(D-1)\left[r_0\,(\mu+\eta)\,D^2 - \mu\,(\mu+r_0+\eta+\xi_0)\,D + \mu^2\right] = 0. \tag{10.20}$$

From the characteristic equations (10.19) and (10.20), we can construct the solution $\{\pi_i(k)\}$ using the undetermined coefficient method and then find the undetermined coefficients based on the boundary conditions and the normalization condition. The applications of the characteristic equation approach can be referred to Sect. 9.3.2, Feng and Xiao (2002), and Song (2006).

Lemma 10.1. *The solution $\{\pi_i(k), k \le l,$ and $i = 0,1\}$ must satisfy*

$$\pi_1(k) = A_1 x_1^{-k} + B_1 x_2^{-k}, \tag{10.21}$$

$$\pi_0(k) = A_2 x_1^{-k} + B_2 x_2^{-k}, \tag{10.22}$$

$$A_2 = A_1\,(r_1 x_1 - \mu)\,/\mu, \quad \text{and} \quad B_2 = B_1\,(r_1 x_2 - \mu)\,/\mu, \tag{10.23}$$

*where x_1 and x_2 are the roots of equation $r_1(\mu+\eta)x^2-\mu(\mu+r_1+\eta+\xi_1)x+\mu^2=0$
and A_1 and B_1 are undetermined constants.*

Proof. Firstly we show that $\{\pi_i(k), k < l,$ and $i = 0, 1\}$ satisfies (11)–(13). Let $f(x) := r_1(\mu + \eta)x^2 - \mu(\mu + r_1 + \eta + \xi_1)x + \mu^2$. It is clear that $f(0) > 0$, $f(\mu/(\mu + \eta)) < 0$, and $0 < \mu/(\mu + \eta) < 1$. We also have $f(1) > 0$ due to $r_1\eta/(\eta + \xi_1) > \mu$. It follows that the equation $f(x) = 0$ has two different roots in the interval $(0, 1)$. Therefore, the characteristic equation (10.19) has three roots, 1, x_1, and x_2, where $x_1, x_2 \in (0, 1)$. Note that the state space is infinite with $k < l$; the root 1 cannot be used to construct the general solution for $\{\pi_i(k) \mid k < l\}$ due to the requirement of the normalization condition (10.18). That means, $\pi_i(k)$ for $k < l$ can be constructed in the forms of (10.21) and (10.22) with undetermined constants A_1, B_1, A_2, and B_2. Substitute two expressions (10.21) and (10.22) into (10.11) and (10.12), add them together, and compare the coefficients of x_1^{-k} and x_2^{-k} on both sides. We have $A_2 = A_1 (r_1x_1 - \mu)/\mu$ and $B_2 = B_1 (r_1x_2 - \mu)/\mu$.

Secondly, we show that $\{\pi_i(k), k = l,$ and $i = 0, 1\}$ also satisfies (10.21), (10.22), and (10.23). Let $k = l - 1$ in Eq. (10.11), together with the result that $\{\pi_i(k) \mid k < l\}$ satisfies (10.21), (10.22), and (10.23), we have

$$
\begin{aligned}
\mu\pi_1(l) &= (\mu + r_1 + \xi_1)\,\pi_1\,(l - 1) - r_1\pi_1\,(l - 2) - \eta\pi_0\,(l - 1) \\
&= (\mu + r_1 + \xi_1)\left[A_1x_1^{-l+1} + B_1x_2^{-l+1}\right] - r_1\left[A_1x_1^{-l+2} + B_1x_2^{-l+2}\right] \\
&\quad - \eta\left[A_2x_1^{-l+1} + B_2x_2^{-l+1}\right] \\
&= \left[(\mu + r_1 + \xi_1)x_1 - r_1x_1^2 - \eta(r_1x_1 - \mu)x_1/\mu\right]A_1x_1^{-l} \\
&\quad + \left[(\mu + r_1 + \xi_1)x_2 - r_1x_2^2 - \eta(r_1x_2 - \mu)x_2/\mu\right]B_1x_2^{-l} \\
&= \left[-r_1(\mu + \eta)x_1^2 + (\mu + r_1 + \xi_1 + \eta)x_1\right]A_1x_1^{-l}/\mu \\
&\quad + \left[-r_1(\mu + \eta)x_2^2 + (\mu + r_1 + \xi_1 + \eta)x_2\right]B_1x_2^{-l}/\mu \\
&= \mu A_1x_1^{-l} + \mu B_1x_2^{-l}.
\end{aligned}
$$

The last equation follows from the definitions of x_1 and x_2, that is, $f(x_1) = 0$ and $f(x_2) = 0$. Therefore, $\pi_1(l)$ takes the form of (10.21). Similarly, let $k = l - 1$ in Eq. (10.12); it can be shown that $\pi_0(l)$ has the same forms as (10.22) and (10.23). This completes the proof.   □

Lemma 10.2. *The solution $\{\pi_i(k), k \geq l,$ and $i = 0, 1\}$ must satisfy*

$$\pi_1(k) = C_1 y_1^{-k} + D_1 y_2^{-k}, \quad l \leq k \leq h, \tag{10.24}$$

$$\pi_0(k) = C_2 y_1^{-k} + D_2 y_2^{-k}, \quad l \leq k < h, \tag{10.25}$$

$$C_2 = C_1 (r_0 y_1 - \mu)/\mu, \quad \text{and} \quad D_2 = D_1 (r_0 y_2 - \mu)/\mu, \tag{10.26}$$

where y_1 and y_2 are the roots of equation $r_0(\mu+\eta)y^2-\mu(\mu+r_0+\eta+\xi_0)y+\mu^2=0$ and C_1 and D_1 are undetermined constants.

Proof. We consider the solution $\pi_i(k)$ in four parts: (i) $l<k<h-1$, (ii) $k=l$, (iii) $k=h-1$, and (iv) $k=h$ and $i=1$.

(i) For $l<k<h-1$, from the characteristic equation (10.20), a general solution to (10.13) and (10.14) can be expressed by

$$\pi_1(k) = C_1 y_1^{-k} + D_1 y_2^{-k} + E_1, \qquad (10.27)$$

$$\pi_0(k) = C_2 y_1^{-k} + D_2 y_2^{-k} + E_2. \qquad (10.28)$$

Let y_1 denote the smaller root of $r_0(\mu+\eta)y^2-\mu(\mu+r_0+\eta+\xi_0)y + \mu^2=0$. It is easy to show that $0<y_1<1$. Note that if $y_2=1$, then $y_2^{-k}\equiv 1$ and the third term in (10.27) and (10.28) is redundant. Therefore, we assume that $E_1=E_2=0$ in case $y_2=1$. Substitute (10.27) and (10.28) into (10.13) and (10.14), we have

$$C_2 = C_1(r_0 y_1-\mu)/\mu, \quad D_2 = D_1(r_0 y_2-\mu)/\mu, \quad E_2 = E_1\xi_0/\eta.$$
$$(10.29)$$

Later on, we will show that $E_1=0$.

(ii) For $k=l$. Firstly, let $k=l+1$ in (10.13), we have $\mu\pi_1(l+2) + r_0\pi_1(l) + \eta\pi_0(l+1) = (\mu+r_0+\xi_0)\pi_1(l+1)$. Substitute $\pi_1(l+2)$, $\pi_0(l+1)$ and $\pi_1(l+1)$ using (10.27) and (10.28), together with (10.29) and the definition of y_1 and y_2, it can be shown that $\pi_1(l)$ takes the form of (10.27). Secondly, let $k=l$ in (10.14). We have $\mu\pi_0(l+1)+\xi_0\pi_1(l) = (\mu+\eta)\pi_0(l)$. Plug $\pi_0(l+1)$ and $\pi_1(l)$ in the forms of (10.28) and (10.27); similarly, we can show that $\pi_0(l)$ takes the form of (10.28).

To show $E_1=0$. Plug $\pi_1(l+1)$ in the form of (10.27) and $\pi_1(l-1)$, $\pi_0(l)$, and $\pi_1(l)$ in the forms of (10.21) and (10.22) into equation (10.15); we have

$$\mu\left(C_1 y_1^{-l-1} + D_1 y_2^{-l-1} + E_1\right) + r_1\left(A_1 x_1^{-l+1} + B_1 x_2^{-l+1}\right)$$
$$+ \eta\left(A_2 x_1^{-l} + B_2 x_2^{-l}\right) = (\mu+r_0+\xi_0)\left(A_1 x_1^{-l} + B_1 x_2^{-l}\right). \quad (10.30)$$

In addition, let $k=l$ in (10.14) and substitute $\pi_0(l+1)$ using the form of (10.28) and $\pi_1(l)$ and $\pi_0(l)$ using the forms of (10.21) and (10.22); it follows

$$\mu\left(C_2 y_1^{-l-1} + D_2 y_2^{-l-1} + E_2\right) + \xi_0\left(A_1 x_1^{-l} + B_1 x_2^{-l}\right)$$
$$= (\mu+\eta)\left(A_2 x_1^{-l} + B_2 x_2^{-l}\right).$$

Plug (10.29) into the above equation, separate the right-hand side into two terms, and substitute A_2 and B_2 using (10.23) in one term; it yields

$$r_0 \left(C_1 y_1^{-l} + D_1 y_2^{-l} + E_1 \right) - \mu \left(C_1 y_1^{-l-1} + D_1 y_2^{-l-1} + E_1 \right)$$
$$+ \left(\mu \xi_0 + \mu \eta - r_0 \eta \right) E_1 / \eta + \xi_0 \left(A_1 x_1^{-l} + B_1 x_2^{-l} \right)$$
$$= \eta \left(A_2 x_1^{-l} + B_2 x_2^{-l} \right) + \left(r_1 x_1 - \mu \right) A_1 x_1^{-l} + \left(r_1 x_2 - \mu \right) B_1 x_2^{-l}. \tag{10.31}$$

Adding (10.30) and (10.31), it follows

$$r_0 \left(C_1 y_1^{-l} + D_1 y_2^{-l} + E_1 \right) + \left(\mu \xi_0 + \mu \eta - r_0 \eta \right) E_1 / \eta = r_0 \left(A_1 x_1^{-l} + B_1 x_2^{-l} \right).$$

On the other hand, from Lemma 10.1 and the above proof (ii), we know that $\pi_1(l)$ takes both forms of (10.21) and (10.27), namely,

$$\pi_1(l) = A_1 x_1^{-l} + B_1 x_2^{-l} = C_1 y_1^{-l} + D_1 y_2^{-l} + E_1. \tag{10.32}$$

The above two equations yield

$$\left(\mu \xi_0 + \mu \eta - r_0 \eta \right) E_1 / \eta = 0. \tag{10.33}$$

If $\mu \xi_0 + \mu \eta - r_0 \eta = 0$, it is easy to show that $y_2 = 1$. Thus, we have $E_1 = 0$.
(iii) For $k = h - 1$. Let $k = h - 2$ in (10.13) and (10.14); it is easy to show that $\pi_1(h-1)$ and $\pi_0(h-1)$ must have the same forms as (10.27) and (10.28).
(iv) For $k = h$ and $i = 1$. Similarly, let $k = h - 1$ in (10.13); it yields that $\pi_1(h)$ has the form of (10.27). This completes the proof. $\square$

Lemma 10.3. *The relationships between constants A_1, B_1, C_1, and D_1 are given by*

$$A_1 = \frac{y_1^{-l} \left(r_0 y_1 - r_1 x_2 \right) C_1 + y_2^{-l} \left(r_0 y_2 - r_1 x_2 \right) D_1}{r_1 x_1^{-l} \left(x_1 - x_2 \right)}, \tag{10.34}$$

$$B_1 = \frac{y_1^{-l} \left(r_0 y_1 - r_1 x_1 \right) C_1 + y_2^{-l} \left(r_0 y_2 - r_1 x_1 \right) D_1}{r_1 x_2^{-l} \left(x_2 - x_1 \right)}, \tag{10.35}$$

$$D_1 = \frac{\left(\mu - r_0 y_1 \right) y_2^h}{\left(r_0 y_2 - \mu \right) y_1^h} C_1. \tag{10.36}$$

Proof. Consider the boundary conditions at states with $k = l$. Note that both $\pi_1(l)$ and $\pi_0(l)$ have the forms in Lemma 10.1 and Lemma 10.2, that is,

$$\pi_0(l) = A_2 x_1^{-l} + B_2 x_2^{-l} = C_2 y_1^{-l} + D_2 y_2^{-l}.$$

Substitute (10.23) and (10.26) into the above equation; it follows

$$\left(r_1 x_1 - \mu \right) A_1 x_1^{-l} + \left(r_1 x_2 - \mu \right) B_1 x_2^{-l} = \left(r_0 y_1 - \mu \right) C_1 y_1^{-l} + \left(r_0 y_2 - \mu \right) D_1 y_2^{-l}. \tag{10.37}$$

Together with the Eq. (10.32), it leads to (10.34) and (10.35). Now consider the boundary condition (10.16); from Lemma 10.2, we have

$$r_0 \left(C_1 y_1^{-h+1} + D_1 y_2^{-h+1} \right) = \mu \left(C_1 y_1^{-h} + D_1 y_2^{-h} \right). \tag{10.38}$$

The above equation leads to (10.36). This completes the proof.   □

Lemma 10.4. *The constant C_1 is determined by*

$$C_1 = \frac{1}{F_1 + F_2} \tag{10.39}$$

where

$$F_1 = \frac{(r_1 + \eta + \xi_1) \, r_0 y_1 - \mu r_1}{r_1 \eta - \mu \, (\xi_1 + \eta)} \cdot y_1^{-l} + \frac{r_0}{\mu} \sum_{k=l}^{h-1} y_1^{-k+1} + \frac{1}{y_1^h} \tag{10.40}$$

$$F_2 = \left(\frac{(r_1 + \eta + \xi_1) \, r_0 y_2 - \mu r_1}{r_1 \eta - \mu \, (\xi_1 + \eta)} \cdot y_2^{-l} + \frac{r_0}{\mu} \sum_{k=l}^{h-1} y_2^{-k+1} + y_2^{-h} \right) \cdot \frac{(\mu - r_0 y_1) \, y_2^h}{(r_0 y_2 - \mu) \, y_1^h}. \tag{10.41}$$

Proof. From Lemma 10.1 and Lemma 10.3, we have

$$\sum_{k=-\infty}^{l-1} (\pi_1(k) + \pi_0(k)) = \frac{r_1}{\mu} \sum_{k=-\infty}^{l-1} \left(A_1 x_1^{-k+1} + B_1 x_2^{-k+1} \right) = \frac{r_1}{\mu} \left(\frac{x_1^{-l+2}}{1 - x_1} A_1 + \frac{x_2^{-l+2}}{1 - x_2} B_1 \right)$$

$$= \frac{r_1}{\mu} \cdot \frac{(r_0 y_1 \, (x_1 + x_2) - (r_0 y_1 + r_1) \, x_1 x_2) \, y_1^{-l} C_1 + (r_0 y_2 \, (x_1 + x_2) - (r_0 y_2 + r_1) \, x_1 x_2) \, y_2^{-l} D_1}{r_1 \, (1 - x_1) \, (1 - x_2)}.$$

Note that x_1 and x_2 are the roots of $r_1 (\mu + \eta) x^2 - \mu (\mu + r_1 + \eta + \xi_1) x + \mu^2 = 0$, which implies that $x_1 x_2 = \mu^2 / r_1 / (\mu + \eta)$ and $(x_1 + x_2) = \mu (\mu + r_1 + \xi_1 + \eta) / r_1 / (\mu + \eta)$. These lead to

$$\sum_{k=-\infty}^{l-1} (\pi_1(k) + \pi_0(k)) = \frac{(r_1 + \eta + \xi_1) \, r_0 y_1 - \mu r_1}{r_1 \eta - \mu \, (\xi_1 + \eta)} y_1^{-l} C_1$$

$$+ \frac{(r_1 + \eta + \xi_1) \, r_0 y_2 - \mu r_1}{r_1 \eta - \mu \, (\xi_1 + \eta)} y_2^{-l} D_1.$$

In addition, from Lemma 10.2, we have

$$\sum_{k=l}^{h-1} (\pi_1(k) + \pi_0(k)) + \pi_1(h) = \frac{r_0}{\mu} \sum_{k=l}^{h-1} C_1 y_1^{-k+1} + \frac{r_0}{\mu} \sum_{k=l}^{h-1} D_1 y_2^{-k+1} + C_1 y_1^{-h} + D_1 y_2^{-h}.$$

Using (10.36) and the normalization condition (10.18), we obtain (10.39), (10.40), and (10.41). This completes the proof. □

Proposition 10.3. *The stationary distribution $\{\pi_i(k)\}$ under the threshold policy $u_{l,h}$ can be explicitly given by (10.21), (10.22), (10.24), and (10.25), where the undetermined constants are given in Lemma 10.1–Lemma 10.4.*

In case $r_1\eta/(\eta + \xi_1) \geq r_0\eta/(\eta + \xi_0)$, the inequality $r_1\eta/(\eta + \xi_1) > \mu$ is actually a sufficient and necessary condition for the stability of the system under the threshold control policy $\mathbf{u}_{l,h}$. On the one hand, since we have obtained the explicit stationary distribution under the threshold policy with the assumption $r_1\eta/(\eta + \xi_1) > \mu$, it is sufficient. On the other hand, suppose $r_1\eta/(\eta + \xi_1) \leq \mu$. With the similar argument in Lemma 10.1, it is easy to show that there is only one root (say x_1) to the characteristic equation (10.19) that is less than 1. Therefore, we have $\pi_1(k) = A_1 x_1^{-k}$ and $\pi_0(k) = A_2 x_1^{-k}$ for $k \leq l$. In addition, Lemma 10.2 holds. From the boundary conditions at $k = l$, it is easy to show that $C_1 = 0$. Hence, the only solution to (10.11), (10.12), (10.13), (10.14), (10.15), and (10.16) is $\{\pi_i(k) \equiv 0\}$, which does not form a stationary distribution. Therefore, $r_1\eta/(\eta + \xi_1) > \mu$ is necessary for the stability of the system.

Proposition 10.4. *If $r_0\eta/(\eta + \xi_0) > \mu$, then the stationary distribution $\{\pi_i(k)\}$ under the type-two threshold policy $\mathbf{u}'_{l,h}$ exists and can be explicitly given by (10.21), (10.22), (10.24), and (10.25) by swapping r_1 with r_0, ξ_1 with ξ_0, and x_i with y_i.*

10.3.3   Steady-State Performance Measures

This section produces formula to compute system steady-state performance measures including long-run average cost, stock-out probability, average backlog level, average inventory level, and machine utilization. The optimal threshold parameters can then be obtained by minimizing a specific performance measure.

10.3.3.1   Long-Run Average Cost

The long-run average cost under a control policy $\mathbf{u}$ is defined as

$$J(\mathbf{u}) := \sum_k [G(k)(\pi_1(k) + \pi_0(k)) + H(k)\pi_1(k)] \tag{10.42}$$

where $G(k)$ represents the inventory and backlog cost and $H(k)$ represents the preventive maintenance cost, for example,

$$G(k) = c^+ \max(k,0) + c^- \max(-k,0), \tag{10.43}$$

$$H(k) = c_m \cdot (\xi_1 - \xi(k)) \cdot I\{u(k) > 0\}. \tag{10.44}$$

Under the threshold policy $\mathbf{u}_{l,h}$, for the case $l \geq 0$, the average cost can be explicitly given by (for the case $l < 0$, similar results can be obtained)

$$J\left(\mathbf{u}_{l,h}\right) = \frac{r_1 c^-}{\mu}\left(\frac{A_1 x_1^2}{(1-x_1)^2} + \frac{B_1 x_2^2}{(1-x_2)^2}\right) + \sum_{k=1}^{l-1} \frac{r_1 c^+ k}{\mu}\left(A_1 x_1^{-k+1} + B_1 x_2^{-k+1}\right)$$

$$+ \sum_{k=l}^{h-1} \frac{r_0 c^+ k}{\mu}\left(C_1 y_1^{-k+1} + D_1 y_2^{-k+1}\right) + c^+ h\left(C_1 y_1^{-h} + D_1 y_2^{-h}\right)$$

$$+ \sum_{k=l}^{h-1} \frac{c_m\left(\xi_1 - \xi_0\right)}{\mu}\left(C_1 y_1^{-k} + D_1 y_2^{-k}\right). \tag{10.45}$$

10.3.3.2 Stock-Out Probability

The stock-out probability is defined as the probability that the system has backordered demands, that is, $\mathrm{SP}\left(\mathbf{u}\right) := \sum_{k<0}\left(\pi_1(k) + \pi_0(k)\right)$. As an example, for the case $l \geq 0$, we have

$$\mathrm{SP}\left(\mathbf{u}_{l,h}\right) = \frac{r_1 A_1 x_1^2}{\mu\left(1 - x_1\right)} + \frac{r_1 B_1 x_2^2}{\mu\left(1 - x_2\right)}. \tag{10.46}$$

A related measure is the service level, which may be defined as the percentage of customer demands that are satisfied from the inventory directly, that is, $\sum_{k>0}\left(\pi_1(k) + \pi_0(k)\right)$.

10.3.3.3 Average Backlog Level and Average Inventory Level

The average backlog level is defined as $\mathrm{BL}\left(\mathbf{u}\right) := \sum_{k<0}(-k)\left(\pi_1(k) + \pi_0(k)\right)$ and the average inventory level is defines as $\mathrm{IL}\left(\mathbf{u}\right) := \sum_{k>0} k \cdot \left(\pi_1(k) + \pi_0(k)\right)$. For the case $l \geq 0$, we have

$$\mathrm{BL}\left(\mathbf{u}_{l,h}\right) = \frac{r_1 A_1 x_1^2}{\mu(1-x_1)^2} + \frac{r_1 B_1 x_2^2}{\mu(1-x_2)^2}, \tag{10.47}$$

$$\mathrm{IL}\left(\mathbf{u}_{l,h}\right) = \sum_{k=1}^{l-1} \frac{r_1 k}{\mu}\left(A_1 x_1^{-k+1} + B_1 x_2^{-k+1}\right) + \sum_{k=l}^{h-1} \frac{r_0 k}{\mu}\left(C_1 y_1^{-k+1} + D_1 y_2^{-k+1}\right)$$

$$+ h\left(C_1 y_1^{-h} + D_1 y_2^{-h}\right). \tag{10.48}$$

The average backlog level indicates how many customer demands are backordered on average at any time. The average inventory level is useful to calculate the buffer or warehouse utilization.

10.3.3.4   Machine Utilization

The machine utilization is defined as the fraction of time that the machine is in operation, that is, MU $(\mathbf{u}) := \sum_{\lambda(k)>0} \pi_1(k)$. For the case $l \ge 0$, we have

$$
\text{MU}\,(\mathbf{u}_{l,h}) = \frac{x_1^{-l+1}}{1-x_1}A_1 + \frac{x_2^{-l+1}}{1-x_2}B_1 + \frac{1-y_1^{h-l}}{1-y_1}y_1^{-h+1}C_1 + \frac{1-y_2^{h-l}}{1-y_2}y_2^{-h+1}D_1.
$$
(10.49)

Since the system is stable, all the demands must be satisfied eventually. Under the threshold policy, the machine's effective production rate is the same as the demand rate μ.

In the remainder of this section, we present some results about the optimal threshold values by minimizing the long-run average cost.

Proposition 10.5. *To minimize the average cost, the optimal threshold value h^* must be nonnegative, that is, $h^* \ge 0$.*

Proof. It suffices to show that $J(\mathbf{u}_{l,-1}) \ge J(\mathbf{u}_{l+1,0})$. Let $\{\pi_i(k)\}$ and $\{\pi_i'(k)\}$ be the stationary distributions under the threshold policies $\mathbf{u}_{l,-1}$ and $\mathbf{u}_{l+1,0}$, respectively. Clearly, $\pi_i'(k+1) = \pi_i(k)$ for any i and k. To simplify the notation, let $\pi_0(h) = \pi_0(-1) := 0$ (because the machine never breaks down at the state h). From the definition of the average cost, we have

$$
J\,(\mathbf{u}_{l,-1}) = \sum_{-\infty}^{-1} G(k)\,(\pi_1(k) + \pi_0(k)) + \sum_{l}^{-1} H(k)\pi_1(k)
$$

$$
J\,(\mathbf{u}_{l+1,0}) = \sum_{-\infty}^{0} G(k)\,(\pi_1\,(k-1) + \pi_0\,(k-1)) + \sum_{l+1}^{0} H(k)\pi_1\,(k-1).
$$

Note that $H(k) = c_m(\xi_0 - \xi_1)$ for $k < h$, it follows

$$
J\,(\mathbf{u}_{l,-1}) - J\,(\mathbf{u}_{l+1,0}) = \sum_{-\infty}^{0} (G\,(k-1) - G(k))\,(\pi_1\,(k-1) + \pi_0\,(k-1)).
$$

The right-hand side of the above equation is apparently nonnegative since more backordered demands incur more penalties in reality. This completes the proof. $\square$

Since we have obtained the explicit form of the long-run average cost $J(\mathbf{u}_{l,h})$ in (10.45), the optimal threshold values (l^*, h^*) can be obtained through numerical optimization approaches, that is, $(l^*, h^*) = \arg\min_{l \le h, h \ge 0} J\,(\mathbf{u}_{l,h})$. Physically, the optimal threshold value h^* is helpful to design the warehouse size for finished goods since the inventory level will never exceed h^* under such threshold policy. On the other hand, together with the average inventory level (IL), it yields the utilization of the buffer.

In some circumstances, we may aim to find the optimal threshold policy to minimize the average cost subject to stock-out probability (or service level) requirements, for example, $SP(\mathbf{u}_{l,h}) < \alpha$, where α is a prespecified target. This can be easily done by imposing constraints to the numerical optimization procedure in search for the optimal threshold parameters.

If the production process is operating in just-in-time mode (also called make-to-order), no inventory should be maintained. In such case, we only need to optimize the preventive maintenance threshold value. The optimal preventive maintenance threshold value l^* is determined by $l^* = \arg \min_{l \leq 0} J(\mathbf{u}_{l,0})$, which is a straightforward single parameter optimization problem.

10.3.4 Numerical Examples

This section provides numerical examples to demonstrate the results. The purpose is threefold: (1) to show the effectiveness of the proposed threshold policies in a wide range of scenarios using a full factorial experiment, (2) to investigate the sensitivity of the steady-state performance measures in response to some specific system parameters, and (3) to illustrate the structural properties of the optimal policy in comparison with the proposed threshold policies.

We compare the proposed threshold policies (type-one and type-two) with three other policies: the optimal integrated production and preventive maintenance policy, the optimal always-maintenance production policy, and the optimal non-maintenance production policy. Under the always-maintenance policy, the machine has the maximum production rate r_0 and a constant failure rate ξ_0, while under the non-maintenance policy, the machine has the maximum production rate r_1 and a constant failure rate ξ_1. Therefore, we only need to optimize the production rate for these two policies since the maintenance action is implied. The average costs for the above three policies (denoted as J^*, J^a, and J^n, respectively) are evaluated using the value iteration algorithm with 1,500 iterations and a truncated state space $-50 < x(t) < 50$, while the performance measures for the type-one and type-two threshold policies are calculated analytically based on the propositions in Sect. 10.3.3. It should be pointed out that the value iteration approach can yield the long-run average cost but cannot provide other steady-state performance measures.

From the assumptions of exponential distributions, we know that $1/\eta$ represents the average machine downtime and $1/\xi$ represents the average machine uptime; it is therefore reasonable to assume that η is greater than ξ (with or without preventive maintenance). The machine production rate should exceed the demand rate. The backlog cost is greater than the inventory cost and the maintenance cost. More specifically, system parameters should generally follow $\eta > \xi_1 > \xi_0$, $r_1 \geq r_0 > \mu$, $c^- > c^+$, and $c^- \geq c_m$.

Example 10.2. With the above guidelines, a full factorial experiment is designed to compare the cost performance under different policies in a range of parameter values. Six factors are considered. Each factor takes three or two levels, that

Table 10.2 Relative difference of the costs between specific policies and the optimal policy

	$J(\mathbf{u}_{l,h})$	$J(\mathbf{u}'_{l,h})$	$\min(J(\mathbf{u}),J(\mathbf{u}'))$	J^a	J^n
Mean	0.21%	0.71%	0.05%	48.76%	15.85%
SD	0.68	1.80	0.22	66.35	30.09
Min	0.00%	0.00%	0.00%	0.00%	0.00%
Max	5.21%	12.66%	2.21%	312.48%	183.21%

is, η (0.7, 0.8, 0.9); ξ_1 (0.3, 0.4); ξ_0 (0.1, 0.2); r_1 (2.0, 2.5, 3.0); r_0 (1.5, 1.8, 2.0); and c_m (1, 5, 10), which gives total $3*2*2*3*3*3 = 324$ different scenarios. Other parameters are fixed, for example, $\mu = 1.0$, $c^+ = 1$, and $c^- = 10$. The statistics corresponding to the relative differences of the costs between a specific policy (e.g., type-one threshold policy, type-two threshold policy, the minimum of two threshold policies, the optimal always-maintenance policy, the optimal non-maintenance policy) and the optimal policy are given in Table 10.2. Take the type-one threshold policy ($\mathbf{u}_{l,h}$) as an example, the relative difference is defined as $(J(\mathbf{u}_{l,h}) - J^*)/J^* \cdot 100\%$. These statistics include the average value (i.e., mean), the standard deviation (SD), the minimum value, and the maximum value over total 324 scenarios for each policy.

It can be observed from Table 10.2 that both types of threshold policies perform very well for the experimented 324 scenarios (less than 1% away from the optimal policy on average). More importantly, the minimum of these two threshold policies is extremely close to the optimal policy and much better than each individual type. Even in the worst scenario, it is only 2.21% above the optimal cost. This demonstrates the effectiveness of the proposed threshold policies and also reflects the fact that both types have their relative merit. As for the optimal always-maintenance policy and the optimal non-maintenance policy (in the last two columns of Table 10.2), they perform significantly worse than the optimal policy with large standard deviations. Although in some scenarios they could be as good as the optimal policy (indicated by "Min" row of Table 10.2), in other scenarios, they are up to 312 and 183% worse than the optimal policy. This full factorial experiment shows that the proposed threshold policies are very effective in a reasonably wide range of scenarios.

Example 10.3. Next, we want to examine the sensitivity of the steady-state performance measures in response to some specific system parameters (e.g., demand rate μ, production capacity r_0, maintenance cost c_m) and make a detailed comparison between these control policies. Consider a system with parameter settings as follows: $r_1 = 2.0$, $\eta = 0.9$, $\xi_1 = 0.3$, $\xi_0 = 0.1$, $c^+ = 1$, and $c^- = 10$. Four specific groups of cases are investigated:

(A) $r_0 = 1.6$, $c_m = 1.0$ with different demand rate μ
(B) $\mu = 1.0$, $c_m = 1.0$ with different production capacity r_0 during performing maintenance

Table 10.3 Steady-state performance measures in group A with different demand rate μ under threshold policy $\mathbf{u}_{l,h}$

μ	0.80	0.9	1.0	1.1	1.2
(l^*,h^*)	$(-1, 4)$	$(-2, 5)$	$(-1, 7)$	$(-1, 9)$	$(0, 13)$
$J(\mathbf{u}_{l,h})$	4.682	5.901	7.611	10.147	14.464
SP	0.066	0.077	0.070	0.084	0.089
IL	2.830	3.426	4.795	5.897	8.261
BL	0.176	0.237	0.270	0.412	0.607
MU	0.493	0.557	0.616	0.676	0.734
J^*	4.682	5.901	7.611	10.147	14.464
J^a	4.736	6.004	7.793	10.585	15.637
J^n	5.139	6.513	8.306	10.971	15.341

Table 10.4 Steady-state performance measures in group B with different r_0 under threshold policy $\mathbf{u}_{l,h}$

r_0	1.4	1.5	1.6	1.7	1.8	1.9
(l^*,h^*)	$(4, 7)$	$(2, 7)$	$(-1, 7)$	$(-50, 6)$	$(-50, 5)$	$(-50, 4)$
$J(\mathbf{u}_{l,h})$	8.204	8.017	7.611	6.734	5.964	5.437
SP	0.086	0.081	0.070	0.072	0.077	0.091
IL	4.558	4.599	4.795	4.150	3.448	2.698
BL	0.356	0.331	0.270	0.247	0.24	0.263
MU	0.629	0.632	0.616	0.588	0.556	0.526
J^*	8.204	8.017	7.611	6.734	5.964	5.437
J^a	12.056	9.413	7.793	6.734	5.964	5.437
J^n	8.306	8.306	8.306	8.306	8.306	8.306

(C) $r_0 = 1.6$, $\mu = 1.0$ with different maintenance cost c_m

(D) $r_0 = 1.9$, $\mu = 1.0$ with different maintenance cost c_m

More specifically, in group A, the demand rate μ varies to represent different traffic scenarios. When μ is approaching to the production rate r_0, the system tends to be unstable. In group B, the production rate r_0 varies to represent different degrees of maintenance disruption to the production. In group C, the preventive maintenance cost c_m is varying. In the last group, the type-two threshold policy is examined.

Tables 10.3, 10.4, 10.5, and 10.6 give the optimal threshold values (l^*,h^*) and the average cost $J(\mathbf{u}_{l,h})$ using the analytical results in Propositions 10.3 and 10.4 associated with a numerical optimization procedure. Other steady-state performance measures such as stock-out probability (SP), average inventory level (IL), average backlog level (BL), and machine utilization (MU) are also provided. In addition, the costs corresponding to the optimal policy (J^*) and the optimal always-maintenance policy (J^a) and the optimal non-maintenance policy (J^n) are given in the last three rows.

It can be seen from Table 10.3 that (1) the optimal type-one threshold policy is indeed the same as the optimal policy in this group. In terms of the average cost, the optimal always-maintenance policy is 1.2–8.1% worse than the optimal threshold policy, whereas the optimal non-maintenance policy is about 10% worse than the optimal threshold policy. As the demand rate increases, the non-maintenance policy

Table 10.5 Steady-state performance measures in group C with $r_0 = 1.6$ and different maintenance cost c_m under threshold policy $\mathbf{u}_{l,h}$

c_m	1	2	4	6	8	10
(l^*,h^*)	$(-1, 7)$	$(-1, 7)$	$(-1, 7)$	$(-1, 7)$	$(7, 7)$	$(7, 7)$
$J(\mathbf{u}_{l,h})$	7.611	7.727	7.959	8.192	8.306	8.306
SP	0.070	0.070	0.070	0.070	0.085	0.085
IL	4.795	4.795	4.795	4.795	4.795	4.795
BL	0.270	0.270	0.270	0.270	0.351	0.351
MU	0.616	0.616	0.616	0.616	0.500	0.500
J^*	7.611	7.727	7.946	8.085	8.180	8.240
J^a	7.793	7.918	8.168	8.418	8.668	8.918
J^n	8.306	8.306	8.306	8.306	8.306	8.306

Table 10.6 Steady-state performance measures in group D with $r_0 = 1.9$ and different c_m under threshold policy $\mathbf{u}'_{l,h}$

c_m	1	2	4	6	8	10
(l^*,h^*)	$(4, 4)$	$(4, 4)$	$(4, 4)$	$(4, 5)$	$(4, 5)$	$(4, 5)$
$J(\mathbf{u}'_{l,h})$	5.437	5.542	5.753	5.960	6.090	6.220
SP	0.091	0.091	0.091	0.072	0.072	0.072
IL	2.698	2.698	2.698	3.459	3.459	3.459
BL	0.263	0.263	0.263	0.211	0.211	0.211
MU	0.526	0.526	0.526	0.516	0.516	0.516
J^*	5.437	5.542	5.753	5.960	6.090	6.220
J^a	5.437	5.542	5.753	5.963	6.174	6.384
J^n	8.306	8.306	8.306	8.306	8.306	8.306

is getting better than the always-maintenance policy because in this group, the effective production capacity under the non-maintenance policy is greater than that under the always-maintenance policy. (2) The threshold value h^*, the average inventory level, the average backlog level, and the machine utilization are increasing as the demand rate μ increases. This is in agreement with the intuition that higher demand rates require higher inventory levels and higher machine utilization.

Table 10.4 also shows that the type-one threshold policy is the same as the optimal policy in terms of minimizing the average cost. The always-maintenance policy performs poorly when r_0 is small (e.g., $r_0 < 1.6$) but is approaching to the optimal policy when r_0 becomes adequately large (e.g., $r_0 \geq 1.7$). This indicates that the benefit of machine reliability improvement cannot offset the capacity reduction due to the disruption of performing maintenance if r_0 is too small. The non-maintenance policy does not change for different r_0. As r_0 increases, the benefit of performing preventive maintenance is increasing substantially (up to 52%). This reflects the fact that the less the production capacity is sacrificed for a fixed degree of improvement of machine reliability, the more benefit can be achieved. If r_0 is greater than 1.70, we have $l^* = -50$, which means always performing preventive maintenance is a better choice. In fact, when $r_0 \geq 1.70$, we have $r_1\eta/(\eta + \xi_1) < r_0\eta/(\eta + \xi_0)$. Namely, the average production rate with maintenance is larger than the average production rate without maintenance. As expected, in this situation, performing maintenance is preferred when $x(t) \ll 0$ and the type-two threshold policy is recommended.

More numerical experiments have been performed, which confirms that the type-one policy is preferable if $r_1\eta/(\eta+\xi_1) \geq r_0\eta/(\eta+\xi_0)$, while the type-two policy is preferable if $r_1\eta/(\eta+\xi_1) < r_0\eta/(\eta+\xi_0)$. For example, type-one policy is actually the same as the optimal policy in the cases reported in Tables 10.3 and 10.4, whereas type-two policy becomes optimal in the cases reported in Table 10.6.

Table 10.5 shows that the type-one threshold policy is optimal when the maintenance cost c_m is small (e.g., $c_m < 4.0$), but it is gradually deviating from the optimal policy and converges to the optimal non-maintenance policy as c_m increases. This reflects the intuition that if the preventive maintenance incurs too much cost, it is preferable not to perform it. However, the results reveal that the threshold policy is still very close to the optimal policy (e.g., only 0–1.5% worse than the optimal policy in Table 10.5). The always-maintenance policy outperforms the non-maintenance policy when c_m is small (e.g., $c_m \leq 4.0$), while the non-maintenance policy performs better when c_m is large (e.g., $c_m \geq 6.0$).

As mentioned in the discussion of group B, when $r_0 = 1.90$, we have $r_1\eta/(\eta+\xi_1) < r_0\eta/(\eta+\xi_0)$, and the type-two threshold policy $\mathbf{u}'_{l,h}$ is preferred. Table 10.6 gives the optimal threshold values, the average costs, and the corresponding steady-state performance measures under $\mathbf{u}'_{l,h}$. In addition, the average costs under the optimal policy (J^*), the always-maintenance policy (J^a), and the non-maintenance policy (J^n) are also provided. It can be observed that in group D, the type-two threshold policy $\mathbf{u}'_{l,h}$ is indeed the same as the optimal policy, which is 33–52% better than the optimal non-maintenance policy. The always-maintenance policy performs fairly well. It is the same as the optimal policy when c_m is small (e.g., $c_m \leq 4.0$), whereas it is deviating from the optimal policy as c_m increases. The threshold value h^* in Table 10.6 is lower than that in Table 10.5, which reflects the fact that the production rate r_0 is larger in group D.

Tables 10.3, 10.4, 10.5, and 10.6 show that the average inventory level and the average backlog level are much more sensitive to μ than to r_0 or c_m. In all cases, the stock-out probabilities (SP) are low between 6 and 9%. This is due to the parameter setting in which the backlog unit cost is ten times of the inventory unit cost.

From the above specific numerical examples, it can be seen that in most cases, the optimal policy takes either the type-one threshold policy (group A, group B, and the first two cases in group C) or the type-two threshold policy (group D). For some cases in group C (with $c_m = 4.0, 6.0, 8.0$, and 10.0), the optimal policy is different from both types of threshold policies. It is worthwhile to explore the detailed structural properties of the optimal policy when it is different from the threshold policy. As an example, the optimal policies for the fourth and fifth cases ($c_m = 6.0$ and 8.0) in group C, denoted as case C(4) and case C(5), are shown in Figs. 10.3 and 10.4, respectively.

It can be seen that in case C(4), the optimal policy and the type-one threshold policy only differ at states $k = 5$ and 6, while in case C(5), they differ at states $k = 0 \sim 3$. Physically, $\xi(k) = \xi_1$ indicates that no preventive maintenance is

Fig. 10.3 Optimal policy versus type-one threshold policy in case C(4). (**a**) Optimal policy. (**b**) Optimal type-one threshold policy

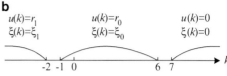

Fig. 10.4 Optimal policy versus type-one threshold policy in case C(5). (**a**) Optimal policy. (**b**) Optimal type-one threshold policy

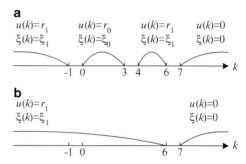

performed, $\xi(k) = \xi_0$ indicates that a preventive maintenance action is taken, and $u(k) = 0$ means that the machine is not producing and therefore will not break down. It appears that the optimal policy can be characterized by three threshold parameters instead of two. However, the derivation of the stationary distribution for a three-parameter threshold policy is more demanding. In addition, it is unknown whether such structural property is true for more general cases. Nevertheless, it can be observed that the proposed threshold policies are fairly close to the optimal policy and easy to implement.

10.4   Notes

This chapter focuses on threshold-type control policies for the supply chains with preventive maintenance decisions presented in Chap. 5. Section 10.3 is mainly based on Song (2009).

In the context of failure-prone manufacturing systems/supply chains producing discrete items with maintenance decisions, a number of studies have presented and evaluated threshold-type policies to control production and maintenance jointly. For example, Srinivasan and Lee (1996) considered a production facility to meet Poisson demand process in which the failure rate of the facility is increasing in operations. They presented a specific threshold control policy for operating the facility: Whenever the inventory level is raised to a prespecified value, S, a preventive maintenance operation is triggered; whenever the inventory level drops below another prespecified value, s, the facility should continue to produce items and try to bring the inventory level back to S. However, if the facility breaks down during operation, it is minimally repaired and put back into operation. Van der Duyn Schouten and Vanneste (1995) proposed a suboptimal preventive maintenance policy, which is based on the age of the machine and the inventory of the subsequent buffer. Das and Sarkar (1999) considered a production–inventory system that produces a single product type with Poisson demand process. They used a single threshold parameter S to control production and two parameters to determine the preventive maintenance. More specifically, whenever the inventory level reaches a certain level and the production count (i.e., the number of products produced since the last repair/maintenance) reaches another prespecified amount, the machine will be maintained.

Iravani and Duenyas (2002) considered a single machine system with multiple operational states and introduced a double-threshold policy to characterize the decisions of production and maintenance. Yao et al. (2005) studied joint preventive maintenance and production policies for an unreliable production–inventory system in which maintenance/repair times are non-negligible and stochastic. They demonstrated that the optimal production and maintenance policies have the control-limit structure and the optimal actions on the entire state space are divided into regions. Threshold-type policies can be constructed based on their switching structure of the optimal policy. Rezga et al. (2004) considered a production line consisting of multiple machines without intermediate buffers. They evaluated various maintenance and production strategies using simulation and genetic algorithms. Rezga et al. (2008) presented a production and maintenance policy characterized by three threshold parameters, in which a maintenance action is triggered as soon as the machine reaches a certain age or failure, whichever occurs first. A numerical procedure is developed to optimize the control parameters.

References

Das, T.K., Sarkar, S.: Optimal preventive maintenance in a production inventory system. IIE Trans. **31**, 537–551 (1999)

Feng, Y.Y., Xiao, B.C.: Optimal threshold control in discrete failure-prone manufacturing systems. IEEE Trans. Autom. Control **47**(7), 1167–1174 (2002)

Iravani, S., Duenyas, I.: Integrated maintenance and production control of a deteriorating production system. IIE Trans. **34**, 423–435 (2002)

Rezga, N., Xie, X., Mati, Y.: Joint optimization of preventive maintenance and inventory control in a production line using simulation. Int. J. Prod. Res. **42**(10), 2029–2046 (2004)

Rezga, N., Dellagia, S., Chelbib, A.: Joint optimal inventory control and preventive maintenance policy. Int. J. Prod. Res. **46**(19), 5349–5365 (2008)

Song, D.P.: Optimal production and backordering policy in failure-prone manufacturing systems. IEEE Trans. Autom. Control **51**(5), 906–911 (2006)

Song, D.P.: Production and preventive maintenance control in a stochastic manufacturing system. Int. J. Prod. Econ. **119**, 101–111 (2009)

Srinivasan, M.M., Lee, H.S.: Production/inventory system with preventive maintenance. IIE Trans. **28**, 879–890 (1996)

Van der Duyn Schouten, F.A., Vanneste, S.G.: Maintenance optimization of a production system with buffer capacity. Eur. J. Oper. Res. **82**, 323–338 (1995)

Yao, X.D., Xie, X.L., Fu, M.C., Marcus, S.I.: Optimal joint preventive maintenance and production policies. Nav. Res. Logist. **52**, 668–681 (2005)

Chapter 11
Threshold-Type Control of Supply Chain Systems with Assembly Operations

11.1 System Stability

For the supply chain systems with assembly operations described in Chap. 6, the system stability is subject to the raw material supply rates, the assembly production rate, and the customer demand rate. Note that the warehouse capacities for raw materials and for finished goods are assumed to be sufficiently large, and the assembly operation depends on the availability of all types of raw materials. Therefore, the minimum raw material supply rate (among all raw material supply rates) constrains the raw material availability, and the maximum assembly rate constrains the production rate.

In addition, the supply rate for raw material i is constrained by the product of the replenishment lead-time rate λ_i and the maximum order quantity Q_i. To avoid customer demands being increasingly backlogged, it requires that both the assembly production rate and the minimum raw material supply rate should be greater than the demand arrival rate μ. We therefore have the following result.

Proposition 11.1. *The sufficient condition to ensure the stability of the supply chain systems in Fig. 6.1 with reliable assembly machines is $min\{Q_i\lambda_i, r \mid i \in \{1, 2, \ldots, n\}\} > \mu$.*

When the assembly machine is failure-prone, the expected maximum production rate is further subject to the machine's failure rate and repair rate, which is determined by $r\eta/(\eta + \xi)$. The system stability conditions for the supply chains with failure-prone assembly machine are given in the following proposition.

Proposition 11.2. *The sufficient condition to ensure the stability of the supply chain systems in Fig. 6.1 with failure-prone assembly machines is $min\{Q_i\lambda_i, r\eta/(\eta + \xi) \mid i \in \{1, 2, \ldots, n\}\} > \mu$.*

Under the conditions in Propositions 11.1 and 11.2, the assembly supply chain is stable and the optimal control problem in the long-run average cost case can be

D.-P. Song, *Optimal Control and Optimization of Stochastic Supply Chain Systems*, Advances in Industrial Control, DOI 10.1007/978-1-4471-4724-4_11, © Springer-Verlag London 2013

examined (similar to Chap. 8). However, this chapter concentrates on the discount cost case with an emphasis on the development of simple threshold control policies and the evaluation of their effectiveness.

11.2　Threshold Control Policies for Reliable Assembly Supply Chains

This section first illustrates the control structure of the optimal raw material ordering and finished goods assembly policy for the supply chain systems with reliable assembly machines. Then, we construct simple threshold control policies and use numerical examples to demonstrate their effectiveness.

11.2.1　Illustration of the Switching Structure of the Optimal Policy

To simplify the narrative and facilitate the visualization of the results, we consider an assembly supply chain with two types of raw materials, that is, $n = 2$. The system parameters are set as follows: the lead-time rate for replenishing raw materials $\lambda_1 = \lambda_2 = 0.8$, the maximum ordering quantity $Q_1 = Q_2 = 3$, the maximum assembly production rate $r = 1.0$, the demand rate $\mu = 0.7$, the inventory unit cost for raw materials $c_1 = c_2 = 1$, the finished goods inventory unit cost $c_2^+ = 5$, and the backordering cost $c_2^- = 20$. The discount factor $\beta = 0.5$.

To conduct the value iteration algorithm and reduce the computation complexity, the system state space is limited into a finite area with $x_1 \leq 20, x_2 \leq 20, x_{n+1} \in [-20, 20]$. The iterative procedure will be terminated when the number of iterations exceeds 100. Part of the optimal ordering and assembly policy is illustrated in Fig. 11.1. Figure 11.1a–c describes the ordering decisions for raw material 1 in the $x_1 - x_3$ plane when x_2 takes 0, 1, and 2, respectively. Figure 11.1d–f describes the ordering decisions for raw material 2 in the $x_1 - x_3$ plane when x_2 takes 0, 1, and 2, respectively. The numbers in Fig. 11.1a–f indicate the order size of the decisions. Figure 11.1g–i describes whether an assembly operation should be performed in the $x_1 - x_3$ plane when x_2 takes 0, 1, and 2, respectively, in which 1 means "yes" and 0 means "no." The solid lines in Fig. 11.1a–i show the boundary of two actions, for example, placing an order and ordering nothing in (a)–(f), assembly or no assembly in (g)–(i).

When there are two types of raw materials ($n = 2$), the switching manifolds defined in Chap. 6 are reduced to be switching surfaces. Tables 11.1, 11.2, and 11.3 show the significant parts of the switching surfaces $S_1(x_2, x_3)$, $S_2(x_1, x_3)$, and $S_3(x_1, x_2)$. It can be seen that the switching surface $S_1(x_2, x_3)$ is increasing in x_2 and decreasing in x_3; $S_2(x_1, x_3)$ is increasing in x_1 and decreasing in x_3; and $S_3(x_1, x_2)$

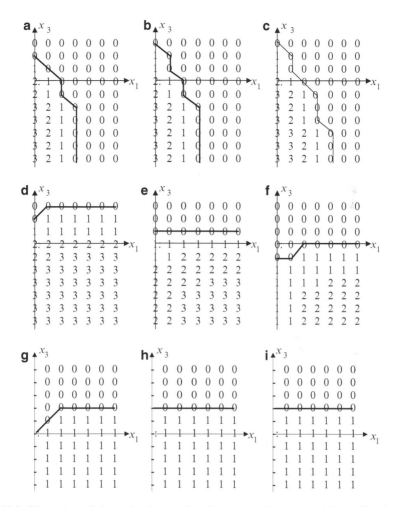

Fig. 11.1 Illustration of the optimal control policy for supply chains with reliable assembly machines. (**a**) Ordering policy for RM 1 when $x_2 = 0$. (**b**) Ordering policy for RM 1 when $x_2 = 1$. (**c**) Ordering policy for RM 1 when $x_2 = 2$. (**d**) Ordering policy for RM 2 when $x_2 = 0$. (**e**) Ordering policy for RM 2 when $x_2 = 1$. (**f**) Ordering policy for RM 2 when $x_2 = 2$. (**g**) Assembly policy when $x_2 = 1$. (**h**) Assembly policy when $x_2 = 2$. (**i**) Assembly policy when $x_2 = 3$

is increasing in both x_1 and x_2. This verifies the analytical results in Chap. 6. Note that $\mathbf{x} = (x_1, x_2, x_3)$ in this example. These switching surfaces can be visualized in the three-dimension state space, as shown in Fig. 11.2, in which RM 1 refers to raw material 1.

The above results verify the monotonicity and asymptotic behaviors of the switching manifolds established in Chap. 6.

Table 11.1 Switching surface $S_1(x_2, x_3)$ in the optimal policy

	$x_3 \geq 3$	$x_3 = 2$	$x_3 = 1$	$x_3 = 0$	$x_3 = -1$	$x_3 = -2$	$x_3 = -3$	$x_3 \leq -4$
$x_2 = 0$	0	0	1	2	2	3	3	3
$=1$	0	1	1	2	2	3	3	3
$=2$	0	1	1	2	3	3	4	4
$=3$	0	1	1	2	3	3	4	4
$=4$	0	1	1	2	3	4	4	4
≥ 5	0	1	1	2	3	4	4	4

Table 11.2 Switching surface $S_2(x_1, x_3)$ in the optimal policy

	$x_3 \geq 3$	$x_3 = 2$	$x_3 = 1$	$x_3 = 0$	$x_3 = -1$	$x_3 = -2$	$x_3 = -3$	$x_3 \leq -4$
$x_1 = 0$	0	0	1	2	2	3	3	3
$=1$	0	1	1	2	2	3	3	3
$=2$	0	1	1	2	3	3	4	4
$=3$	0	1	1	2	3	3	4	4
$=4$	0	1	1	2	3	4	4	4
≥ 5	0	1	1	2	3	4	4	4

Table 11.3 Switching surface $S_3(x_1, x_2)$ in the optimal policy

	$x_2 = 0$	$=1$	$=2$	$=3$	$=4$	≥ 5
$x_1 = 0$	–	–	–	–	–	–
$=1$	–	1	2	2	2	2
$=2$	–	2	2	2	2	2
$=3$	–	2	2	2	2	2
$=4$	–	2	2	2	2	2
≥ 5	–	2	2	2	2	2

11.2.2   Threshold Control Policies

It can be seen that the switching manifolds that characterize the optimal raw materials ordering and finished goods assembly policy could be quite complicated, although they demonstrate good structural properties such as monotonicity and asymptotic convergence. Especially when there are more than two types of raw materials, it is difficult to visualize the control structure and implement the optimal policy. Developing simple but good quality control policies is therefore necessary.

By utilizing the monotonicity and asymptotic behaviors of the switching manifolds established in Chap. 6, we present two simple threshold control policies below.

The first policy is based on the idea of Kanban systems and the concept of order-up-to-point. It is termed as Kanban threshold control (KTC) policy, which is characterized by a set of threshold parameters serving as target levels of local inventories for raw materials and finished goods. Mathematically, the ordering and assembly decisions, $(u(\mathbf{x}), q_i(\mathbf{x})$, for $i = 1, 2, \ldots, n)$, are given by

$$u(\mathbf{x}) = r, \quad \text{if } x_{n+1} \leq H \text{ and } x_i > 0 \quad \text{for } 1 \leq i \leq n; u(\mathbf{x}) = 0, \text{ otherwise} \tag{11.1}$$

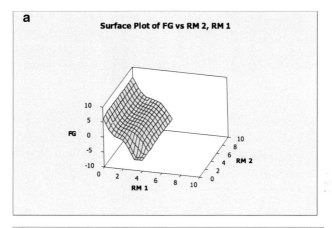

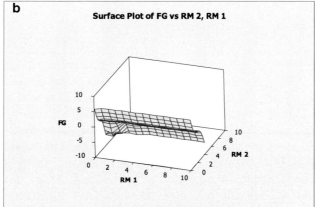

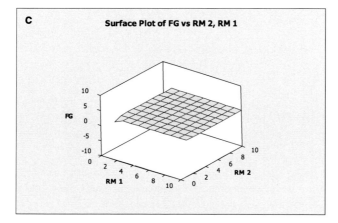

Fig. 11.2 Switching surfaces in the state space (x_1, x_2, x_3). (**a**) The switching surface $S_1(x_2, x_3)$ in the state space (x_1, x_2, x_3). (**b**) The switching surface $S_2(x_1, x_3)$ in the state space (x_1, x_2, x_3). (**c**) The switching surface $S_3(x_1, x_2)$ in the state space (x_1, x_2, x_3)

Fig. 11.3 KTC policy for $n = 2$. (**a**) Ordering policy for raw material 1. (**b**) Ordering policy for raw material 2. (**c**) Assembly operation policy

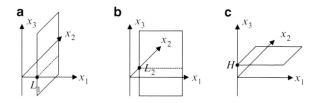

Fig. 11.4 BKC policy for $n = 2$. (**a**) Ordering policy for raw material 1. (**b**) Ordering policy for raw material 2. (**c**) Assembly operation policy

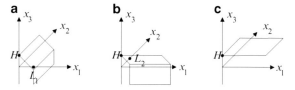

$$q_i \left(\mathbf{x} \right) = \min \left\{ L_i - x_i, Q_i \right\}, \quad \text{if } x_i \leq L_i; q_i \left(\mathbf{x} \right) = 0 \text{ otherwise.} \qquad (11.2)$$

Physically, KTC states that the assembly operation should be performed if the finished goods inventory level drops below H and there are raw materials available for all types; an order should be placed to supplier i if the inventory level of raw materials i is below L_i, and the order size is subject to the order-up-to-point L_i and the maximum order size Q_i. Therefore, this policy is determined by $n + 1$ threshold parameters. When $n = 2$, the KTC policy can be illustrated in Fig. 11.3, in which the switching surfaces become the switching planes.

The second policy is based on the concepts of base-stock, Kanban, and order-up-to-point. It is termed as base-stock Kanban control (BKC) policy. Its ordering and assembly decisions, $\left(u \left(\mathbf{x} \right), q_i \left(\mathbf{x} \right), \text{ for } i = 1, 2, \ldots, n \right)$, are given by

$$u \left(\mathbf{x} \right) = r, \quad \text{if } x_{n+1} \leq H \text{ and } x_i > 0 \quad \text{for } 1 \leq i \leq n; u \left(\mathbf{x} \right) = 0, \text{ otherwise} \qquad (11.3)$$

$$q_i \left(\mathbf{x} \right) = \min \left\{ L_i - x_i, Q_i \right\} \cdot I \left\{ x_{n+1} + L_i \leq H \right\}$$
$$+ \min \left\{ H - x_{n+1} - x_i, Q_i \right\} \cdot I \left\{ x_{n+1} + L_i > H \right\}, \quad \text{if } x_i \leq L_i, x_i$$
$$+ x_{n+1} \leq H; q_i \left(\mathbf{x} \right) = 0, \text{ otherwise.} \qquad (11.4)$$

The BKC policy has the same control mechanism as KTC in terms of the assembly operation. However, the raw material ordering decisions are different. Under BKC policy, the ordering decision for raw material i is not only subject to its inventory-on-hand but also the echelon base-stock level, which is defined as the sum of inventory-on-hand of raw material i and the inventory level of finished goods. When $n = 2$, the BKC policy can be illustrated in Fig. 11.4. Intuitively, BKC can better approximate the optimal control regions shown in Fig. 11.2.

Table 11.4 Parameter
setting for different cases

Case	Q_i	λ_i	r	c_i
1	3	0.8	1.0	1
2	1	0.8	1.0	1
3	5	0.8	1.0	1
4	3	0.4	1.0	1
5	3	1.0	1.0	1
6	3	0.8	0.8	1
7	3	0.8	1.2	1
8	3	0.8	1.0	2
9	3	0.8	1.0	4

11.2.3   Effectiveness of Threshold Control Policies

To show the effectiveness of the threshold control policies, we compare them with the optimal policies in a range of scenarios. The common system settings are the demand rate $\mu = 0.7$, the inventory unit cost for finished goods $c_{n+1}^+ = 5$, and the backordering cost $c_{n+1}^- = 20$. The discount factor $\beta = 0.5$. The value iteration algorithm is applied to compute the optimal cost and is used iteratively to search for the best threshold parameters. To reduce the computational complexity, the system state space is limited into a finite area with $x_i \leq 10$, $x_{n+1} \in [-20, 20]$, and the value iteration algorithm is terminated when the number of iterations exceeds 100.

The scenarios are classified into four groups. Each group has nine different cases with the parameter settings in Table 11.4:

- Group A: The assembly supply chain has two types of raw materials, that is, $n = 2$, and they have the same raw material supply parameters, that is, $Q_1 = Q_2$; $\lambda_1 = \lambda_2$; $c_1 = c_2$.
- Group B: The assembly supply chain has two types of raw materials, that is, $n = 2$, and they may have different raw material supply parameters, that is, (Q_1, λ_1, c_1) are specified in Table 11.4 for nine cases, while (Q_2, λ_2, c_2) take the values in case 1.
- Group C: The assembly supply chain has three types of raw materials, that is, $n = 3$, and they have the same raw material supply parameters, that is, $Q_1 = Q_2 = Q_3$; $\lambda_1 = \lambda_2 = \lambda_3$; $c_1 = c_2 = c_3$.
- Group D: The assembly supply chain has three types of raw materials, that is, $n = 3$, and they may have different raw material supply parameters, that is, (Q_1, λ_1, c_1) are specified in Table 11.4 for nine cases, while (Q_2, λ_2, c_2) and (Q_3, λ_3, c_3) take the values in case 1.

The results of group A and group B are given in Table 11.5; the results of group C and group D are given in Table 11.6. These include the optimal cost under the optimal policy, the percentage of the cost under the best threshold control policy above the optimal cost, and the best threshold values for both KTC and BKC policies. It can be observed from Tables 11.5 and 11.6 that the best KTC policy is 0.56–3.34% above the optimal policy, while the best BKC policy is 0.00–1.97%

Table 11.5 Results of group A and group B

Case	J_{Opt}	% of J_{KTC} above	(L_1, L_2, H) of KTC	% of J_{BKC} above	(L_1, L_2, H) of BKC
Group A
1	39.07	1.45	(2; 2; 0)	0.28	(3; 3; 2)
2	41.07	0.56	(2; 2; 0)	0.32	(3; 3; 2)
3	39.07	1.46	(2; 2; 0)	0.28	(3; 3; 2)
4	46.71	0.95	(3; 3; 0)	0.17	(3; 3; 2)
5	36.70	1.48	(2; 2; 0)	0.47	(3; 3; 2)
6	41.70	0.81	(2; 2; 1)	0.51	(3; 3; 2)
7	36.88	2.34	(2; 2; 0)	0.48	(3; 3; 2)
8	42.56	2.36	(2; 2; 1)	0.43	(2; 2; 1)
9	47.54	1.28	(1; 1; 2)	1.01	(1; 1; 1)
Group B
1	39.07	1.45	(2; 2; 0)	0.28	(3; 3; 2)
2	40.28	1.06	(2; 2; 0)	0.73	(3; 2; 2)
3	39.07	1.45	(2; 2; 0)	0.28	(3; 3; 2)
4	43.51	1.28	(3; 2; 0)	0.68	(4; 3; 2)
5	37.92	1.56	(2; 2; 0)	0.46	(3; 3; 2)
6	41.70	0.81	(2; 2; 1)	0.51	(3; 3; 2)
7	36.88	2.34	(2; 2; 0)	0.48	(3; 3; 2)
8	40.83	1.95	(2; 2; 0)	0.96	(3; 3; 1)
9	43.46	3.34	(1; 2; 1)	1.08	(2; 3; 1)

above the optimal policy out of total 36 scenarios. This indicates that if the threshold parameters in the KTC and BKC policies have been optimized, their performance is fairly close to the optimal costs.

11.3   Threshold Control Policies for Failure-Prone Assembly Supply Chains

This section illustrates the control structure of the optimal ordering and assembly policy for the supply chain systems with failure-prone assembly machines, presents simple threshold control policies, and then uses numerical examples to demonstrate their effectiveness.

11.3.1   Illustration of the Switching Structure of the Optimal Policy

Consider the same example as that in Sect. 11.2.1. However, now the assembly machine is failure-prone with failure rate ξ and repair rate η. Assume $\xi = 0.1$ and

Table 11.6 Results of group C and group D

Case	J_{Opt}	% of J_{KTC} above	(L_1, L_2, L_3, H) of KTC	% of J_{BKC} above	(L_1, L_2, L_3, H) of BKC
Group C					
1	44.47	1.33	(2; 2; 2; 0)	0.50	(3; 3; 3; 1)
2	45.80	0.80	(2; 2; 2; 1)	0.51	(2; 2; 2; 1)
3	44.47	1.33	(2; 2; 2; 0)	0.50	(3; 3; 3; 1)
4	51.97	0.58	(2; 2; 2; 1)	0.24	(2; 2; 2; 1)
5	41.74	1.81	(2; 2; 2; 0)	0.58	(3; 3; 3; 1)
6	46.57	1.17	(2; 2; 2; 1)	0.33	(2; 2; 2; 1)
7	42.72	1.64	(2; 2; 2; 0)	0.66	(3; 3; 3; 1)
8	49.04	1.46	(1; 1; 1; 1)	0.49	(2; 2; 2; 1)
9	54.64	1.64	(1; 1; 1; 3)	0.00	(1; 1; 1; 0)
Group D					
1	44.47	1.33	(2; 2; 2; 0)	0.50	(3; 3; 3; 1)
2	45.07	1.38	(2; 2; 2; 1)	0.76	(2; 2; 2; 1)
3	44.47	1.33	(2; 2; 2; 0)	0.50	(3; 3; 3; 1)
4	47.77	1.74	(2; 2; 2; 0)	1.05	(3; 2; 2; 1)
5	43.64	1.53	(2; 2; 2; 0)	0.59	(3; 3; 3; 1)
6	46.57	1.17	(2; 2; 2; 1)	0.33	(2; 2; 2; 1)
7	42.72	1.64	(2; 2; 2; 0)	0.66	(3; 3; 3; 1)
8	46.05	2.30	(2; 2; 2; 1)	0.64	(2; 2; 2; 1)
9	48.32	2.96	(1; 1; 1; 1)	1.97	(2; 2; 2; 1)

$\eta = 0.9$. The optimal control for the failure-prone assembly supply chain with two types of raw materials (i.e., $n = 2$) is shown in Fig. 11.5, where α represents the assembly machine state (up or down).

As two types of raw materials are symmetrical in this example, the ordering policy for raw material 2 is the same as that for raw material 1 and therefore omitted. The dotted lines in Fig. 11.5 are the switching boundaries for the reliable machine situation. From Fig. 11.5a–c, it shows that the switching boundaries are moving downward slightly for the failure-prone machine situation. Physically, it means that when the assembly machine is failure-prone, we should order raw materials less frequently or in smaller sizes. Comparing the raw material ordering decisions when the machine is up (Fig. 11.5a–c) to those when the machine is down (Fig. 11.5d–f), the differences are more notable, which represents the intuition that we should order fewer raw materials when the machine is down. As for the assembly actions, Fig. 11.5g shows a slight difference from the reliable machine situation, that is, the switching boundary should move upward. Again, this is in agreement with the intuition that we should assemble more finished goods to buffer against the unreliability of the assembly machine.

When there are two types of raw materials ($n = 2$), the switching manifolds defined in Chap. 6 are reduced to be switching surfaces. Tables 11.7, 11.8, and 11.9 show the significant parts of the switching surfaces $S_1(1, x_2, x_3)$, $S_1(0, x_2, x_3)$, and

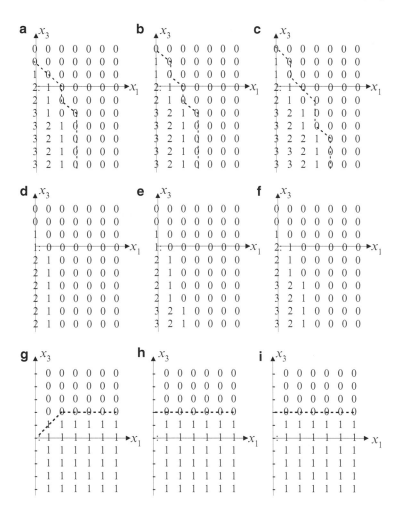

Fig. 11.5 Illustration of the optimal control policy for supply chains with unreliable assembly machine. (**a**) Ordering RM 1 when $x_2 = 0$, $\alpha = 1$. (**b**) Ordering RM 1 when $x_2 = 1$, $\alpha = 1$. (**c**) Ordering RM 1 when $x_2 = 2$, $\alpha = 1$. (**d**) Ordering RM 1 when $x_2 = 0$, $\alpha = 0$. (**e**) Ordering RM 1 when $x_2 = 1$, $\alpha = 0$. (**f**) Ordering RM 1 when $x_2 = 2$, $\alpha = 0$. (**g**) Assembly policy when $x_2 = 1$, $\alpha = 1$. (**h**) Assembly policy when $x_2 = 2$, $\alpha = 1$. (**i**) Assembly policy when $x_2 = 3$, $\alpha = 1$

$S_3(1, x_1, x_2)$. Due to the symmetry of two raw materials, the switching surface $S_2(\alpha, x_1, x_3)$ is the same as $S_1(\alpha, x_2, x_3)$ and therefore omitted.

It can be seen that the switching surface $S_1(\alpha, x_2, x_3)$ is increasing in x_2 and decreasing in x_3, and $S_3(1, x_1, x_2)$ is increasing in both x_1 and x_2. The above results verify the monotonicity and asymptotic behaviors of the switching manifolds established in Sect. 6.5 in Chap. 6.

Table 11.7 Switching surface $S_1(1, x_2, x_3)$ in the optimal policy $\alpha = 1$

	$x_3 \geq 3$	$x_3 = 2$	$x_3 = 1$	$x_3 = 0$	$x_3 = -1$	$x_3 = -2$	$x_3 = -3$	$x_3 \leq -4$
$x_2 = 0$	0	0	1	2	2	2	3	3
$= 1$	0	1	1	2	2	3	3	3
$= 2$	0	1	1	2	2	3	3	3
$= 3$	0	1	1	2	3	3	3	4
$= 4$	0	1	1	2	3	3	4	4
≥ 5	0	1	1	2	3	3	4	4

Table 11.8 Switching surface $S_1(0, x_2, x_3)$ in the optimal policy $\alpha = 0$

	$x_3 \geq 3$	$x_3 = 2$	$x_3 = 1$	$x_3 = 0$	$x_3 = -1$	$x_3 = -2$	$x_3 = -3$	$x_3 \leq -4$
$x_2 = 0$	0	0	1	1	2	2	2	2
$= 1$	0	0	1	1	2	2	2	3
$= 2$	0	0	1	2	2	2	3	3
$= 3$	0	0	1	2	3	3	3	3
$= 4$	0	0	1	2	3	3	3	3
≥ 5	0	0	1	2	3	3	3	3

Table 11.9 Switching surface $S_3(1, x_1, x_2)$ in the optimal policy $\alpha = 1$

	$x_2 = 0$	$= 1$	$= 2$	$= 3$	$= 4$	≥ 5
$x_1 = 0$	–	–	–	–	–	–
$= 1$	–	2	2	2	2	2
$= 2$	–	2	2	2	2	2
$= 3$	–	2	2	2	2	2
$= 4$	–	2	2	2	2	2
≥ 5	–	2	2	2	2	2

11.3.2   Threshold Control Policies

Similar to Sect. 11.2.2, we present two simple threshold control policies: Kanban threshold control (KTC) policy and base-stock Kanban control (BKC) policy below.

Under the KTC policy, the ordering and assembly decisions, $(u(\alpha, \mathbf{x}), q_i(\alpha, \mathbf{x})$, for $i = 1, 2, \ldots, n)$, are given by (note that the assembly operation can only be performed when the machine is up, while the raw material orders can be placed when the machine is up or down)

$$u(\alpha, \mathbf{x}) = r, \quad \text{if } \alpha = 1, x_{n+1} \leq H \text{ and } x_i > 0 \text{ for } 1 \leq i \leq n; u(\alpha, \mathbf{x}) = 0, \text{ otherwise}$$
(11.5)

$$q_i(\alpha, \mathbf{x}) = \min\{L_i - x_i, Q_i\}, \quad \text{if } x_i \leq L_i; q_i(\alpha, \mathbf{x}) = 0, \text{ otherwise.} \quad (11.6)$$

Under the BKC policy, its ordering and assembly decisions, $(u(\alpha, \mathbf{x}), q_i(\alpha, \mathbf{x}),$ for $i = 1, 2, \ldots, n)$, are given by (note that the assembly production can only be performed when the machine is up)

$$u(\alpha, \mathbf{x}) = r, \quad \text{if } \alpha = 1, x_{n+1} \le H \text{ and } x_i > 0 \text{ for } 1 \le i \le n; u(\alpha, \mathbf{x}) = 0, \text{ otherwise} \tag{11.7}$$

$$
\begin{aligned}
q_i(\alpha, \mathbf{x}) = {} & \min\{L_i - x_i, Q_i\} \cdot I\{x_{n+1} + L_i \le H\} \\
& + \min\{H - x_{n+1} - x_i, Q_i\} \cdot I\{x_{n+1} + L_i > H\}, \quad \text{if } x_i \le L_i, x_i \\
& + x_{n+1} \le H; q_i(\alpha, \mathbf{x}) = 0, \text{ otherwise.}
\end{aligned}
\tag{11.8}
$$

In the above KTC and BKC policies, the numbers of threshold parameters are the same as those in the reliable machine situation. In other words, whether the assembly machine is up or down, we adopt the same raw material ordering policy.

11.3.3   Effectiveness of Threshold Control Policies

We compare the threshold control policies with the optimal policies in a range of scenarios. The machine failure rate $\xi = 0.1$ and repair rate $\eta = 0.9$. Other system parameters are the same as those in Sect. 11.2.3. The scenarios are classified into four groups. Each group has nine cases with the parameter settings in Table 11.4:

- Group E: The assembly supply chain has two types of raw materials with the same supply parameters, that is, $n = 2$, $Q_1 = Q_2$; $\lambda_1 = \lambda_2$; $c_1 = c_2$.
- Group F: The assembly supply chain has two types of raw materials with different supply parameters, that is, $n = 2$; (Q_1, λ_1, c_1) are specified in Table 11.4 for nine cases, while (Q_2, λ_2, c_2) take the values in case 1.
- Group G: The assembly supply chain has three types of raw materials with the same supply parameters, that is, $n = 3$, $Q_1 = Q_2 = Q_3$; $\lambda_1 = \lambda_2 = \lambda_3$; $c_1 = c_2 = c_3$.
- Group H: The assembly supply chain has three types of raw materials with different supply parameters, that is, $n = 3$, (Q_1, λ_1, c_1) are specified in Table 11.4 for nine cases, while (Q_2, λ_2, c_2) and (Q_3, λ_3, c_3) take the values in case 1.

The results, including the optimal cost (under the optimal policy), the percentage of the cost under the best threshold control policy above the optimal cost, and the best threshold values for both KTC and BKC policies, are given in Table 11.10 for groups E and F and in Table 11.11 for groups G and H.

It can be observed that the best KTC policy is 0.53–2.83% above the optimal policy, while the best BKC policy is 0.16–1.79% above the optimal policy out of total 36 scenarios. This indicates that after the threshold parameters in the KTC and BKC policies are optimized, their performance is fairly close to the optimal cost.

Table 11.10 Results of group E and group F

Case	J_{Opt}	% of J_{KTC} above	(L_1, L_2, H) of KTC	% of J_{BKC} above	(L_1, L_2, H) of BKC
Group E					
1	40.52	1.21	(2; 2; 0)	0.35	(3; 3; 2)
2	42.21	0.53	(2; 2; 1)	0.37	(3; 3; 2)
3	40.52	1.21	(2; 2; 0)	0.36	(3; 3; 2)
4	47.62	0.88	(2; 2; 0)	0.17	(3; 3; 2)
5	38.27	1.32	(2; 2; 0)	0.60	(3; 3; 2)
6	43.03	0.67	(2; 2; 1)	0.42	(2; 2; 2)
7	38.42	1.89	(2; 2; 0)	0.36	(3; 3; 2)
8	43.94	2.36	(2; 2; 1)	0.32	(2; 2; 1)
9	48.76	0.95	(1; 1; 2)	0.68	(1; 1; 1)
Group F					
1	40.52	1.21	(2; 2; 0)	0.35	(3; 3; 2)
2	41.54	0.97	(2; 2; 1)	0.68	(3; 2; 2)
3	40.52	1.21	(2; 2; 0)	0.35	(3; 3; 2)
4	44.64	1.26	(3; 2; 0)	0.74	(4; 3; 2)
5	39.42	1.37	(2; 2; 0)	0.58	(3; 3; 2)
6	43.03	0.67	(2; 2; 1)	0.42	(2; 2; 2)
7	38.42	1.89	(2; 2; 0)	0.36	(3; 3; 2)
8	42.25	1.77	(2; 2; 1)	0.88	(3; 3; 1)
9	44.82	2.83	(1; 2; 1)	1.19	(2; 3; 1)

In comparison to the results in the reliable machine scenarios, the optimal costs are increasing; the best threshold values are also increasing but only marginally. These reflect the impact of unreliability of the assembly machine on the supply chain performance and its management. In addition, the sensitivity of the supply chain performance under the optimal policy or the threshold policies to the system parameters is similar to what we observed in the reliable machine situation.

11.4 Notes

This chapter numerically verifies the structural characteristics of the optimal ordering and assembly policy for the assembly supply chains described in Chap. 6. We consider both reliable assembly and unreliable assembly situations. Threshold control policies based on the concepts of Kanban, base-stock, and order-up-to-point are presented and evaluated in a range of scenarios in comparison with the optimal policies.

A number of studies in the literature have developed and investigated the effectiveness of specific control policies in the assembly context with uncertainty. Similar to the case of multistage serial supply chain systems, those specific control policies can be classified into base-stock policy, Kanban policy, and other type of push/pull policy.

Table 11.11 Results of group G and group H

Case	J_{Opt}	% of J_{KTC} above	(L_1, L_2, L_3, H) of KTC	% of J_{BKC} above	(L_1, L_2, L_3, H) of BKC
Group G					
1	45.67	1.26	(2; 2; 2; 1)	0.50	(3; 3; 3; 1)
2	46.74	0.82	(2; 2; 2; 1)	0.44	(2; 2; 2; 1)
3	45.67	1.26	(2; 2; 2; 1)	0.50	(3; 3; 3; 1)
4	52.53	0.62	(2; 2; 2; 1)	0.16	(2; 2; 2; 1)
5	43.14	1.73	(2; 2; 2; 1)	0.58	(3; 3; 3; 1)
6	47.66	1.21	(2; 2; 2; 1)	0.19	(2; 2; 2; 1)
7	44.03	1.44	(2; 2; 2; 0)	0.51	(3; 3; 3; 1)
8	50.10	0.93	(1; 1; 1; 1)	0.72	(2; 2; 2; 1)
9	55.20	1.35	(0; 0; 0; 0)	0.12	(1; 1; 1; 0)
Group H					
1	45.67	1.26	(2; 2; 2; 1)	0.50	(3; 3; 3; 1)
2	46.15	1.38	(2; 2; 2; 1)	0.60	(2; 2; 2; 1)
3	45.67	1.26	(2; 2; 2; 1)	0.50	(3; 3; 3; 1)
4	48.73	1.60	(2; 2; 2; 1)	0.84	(3; 2; 2; 1)
5	44.91	1.45	(2; 2; 2; 1)	0.58	(3; 3; 3; 1)
6	47.66	1.21	(2; 2; 2; 1)	0.19	(2; 2; 2; 1)
7	44.03	1.44	(2; 2; 2; 0)	0.51	(3; 3; 3; 1)
8	47.23	2.32	(2; 2; 2; 1)	0.50	(2; 2; 2; 1)
9	49.41	2.36	(1; 1; 1; 1)	1.79	(1; 2; 2; 1)

In the aspect of base-stock policies, Rosling (1989) showed the optimality of the echelon order-up-to-point policy in multistage assembly system under certain conditions. Langenhof and Zijm (1990) decomposed the multidimensional control problems in multi-echelon systems (e.g., assembly, serial, and distribution) into one-dimensional problems and established the optimality of base-stock control policies. De Kok and Visschers (1999) considered more general assembly systems in which components may be used in a number of different subassemblies and end products. Under the base-stock control policies, they decomposed the assembly network into purely divergent multi-echelon systems, which enables to calculate near-optimal order-up-to-levels subject to service level constraints. The above three papers all assumed random demands, fixed lead times, and periodic-review scheme.

Sbiti et al. (2003) presented a queuing network model for base-stock-controlled assembly systems and evaluated the system steady-state performance based on a decomposition method. Zhao (2008) applied a base-stock policy to a general class of supply chains including assembly, distribution, tree, and two-level networks as special cases. Avsar et al. (2009) considered base-stock-controlled assembly systems with Poisson demand arrivals and exponential single facility for manufacturing and assembly operations. An approximate model is developed to obtain the steady-state probability distribution. Huh and Janakiraman (2010) studied an assembly system with capacity constraints under an echelon base-stock policy. Benjaafar and Elhafsi (2006) and Benjaafar et al. (2011) showed that state-dependent base-stock policies are optimal in a multistage assembly system with a unit or variable batch sizes.

In the aspect of Kanban control policies, Hazra et al. (1999) considered assembly systems with tree structure and random processing times. Under a constant WIP control, the system performance is evaluated through an aggregation–disaggregation approach. Matta et al. (2005) evaluated the performance of assembly systems under Kanban control using queuing network techniques. Topan and Avsar (2011) considered the same type of Kanban-controlled assembly systems as the ones in Matta et al. (2005) and proposed a different approximation to evaluate the steady-state performance measures. Ramakrishnan and Krishnamurthy (2012) presented an approximate analysis of a kitting station (feeding into the assembly line) with uncertain supply under a Kanban inventory control policy.

In the aspect of other types of push/pull policies, CONWIP is a typical example, which has been applied to various assembly systems, for example, Duenyas and Hopp (1992, 1993). Ip et al. (2007) applied and compared two types of CONWIP control mechanisms to a lamp assembly production line. Genetic algorithms are used to optimize control parameters. Dellaert and De Kok (2000) compared base-stock-based push and pull strategies in multistage assembly systems with a gamma distribution of periodic demands subject to service level constraints. Chaouiya et al. (2000) presented and compared pull-type production control policies, which are the combination of base-stock and Kanban control, for the production coordination of assembly manufacturing systems with random processing times and inter-arrival times of demands. Khojasteh-Ghamari (2009) compared the performance between Kanban and CONWIP policies in an assembly system with stochastic processing times and unlimited demand for finished goods. Ghrayeb et al. (2009) proposed a hybrid push/pull policy in an assembly-to-order system. Simulation is used to compare the hybrid policy with pure push and pull policies.

References

Avsar, Z.M., Zijm, W.H., Rodoplu, U.: An approximate model for base-stock-controlled assembly systems. IIE Trans. **41**(3), 260–274 (2009)

Benjaafar, S., Elhafsi, M.: Production and inventory control of a single product assemble-to-order system with multiple customer classes. Manage. Sci. **52**(12), 1896–1912 (2006)

Benjaafar, S., ElHafsi, M., Lee, C.Y., Zhou, W.H.: Optimal control of an assembly system with multiple stages and multiple demand classes. Oper. Res. **59**(2), 522–529 (2011)

Chaouiya, C., Liberopoulos, G., Dallery, Y.: The extended Kanban control system for production coordination of assembly manufacturing systems. IIE Trans. **32**(10), 999–1012 (2000)

De Kok, A.G., Visschers, J.W.C.H.: Analysis of assembly systems with service level constraints. Int. J. Prod. Econ. **59**(1–3), 313–326 (1999)

Dellaert, N.P., De Kok, A.D.: Push and pull strategies in multi-stage assembly systems. Stat. Neerl. **54**(2), 175–189 (2000)

Duenyas, I., Hopp, W.J.: CONWIP assembly with deterministic process and random outages. IIE Trans. **24**(4), 97–109 (1992)

Duenyas, I., Hopp, W.J.: Estimating the throughput of an exponential CONWIP assembly system. Queueing Syst. **14**(1–2), 135–157 (1993)

Ghrayeb, O., Phojanamongkolkij, N., Tan, B.A.: A hybrid push/pull system in assemble-to-order manufacturing environment. J. Intell. Manuf. **20**(4), 379–387 (2009)

Hazra, J., Schweitzer, P.J., Seidmann, A.: Analyzing closed Kanban-controlled assembly systems by iterative aggregation disaggregation. Comput. Oper. Res. **26**(10–11), 1015–1039 (1999)

Huh, W.T., Janakiraman, G.: Base-stock policies in capacitated assembly systems: convexity properties. Nav. Res. Logist. **57**(2), 109–118 (2010)

Ip, W.H., Huang, M., Yung, K.L., Wang, D.W., Wang, X.W.: CONWIP based control of a lamp assembly production line. J. Intell. Manuf. **18**(2), 261–271 (2007)

Khojasteh-Ghamari, Y.: A performance comparison between Kanban and CONWIP controlled assembly systems. J. Intell. Manuf. **20**(6), 751–760 (2009)

Langenhof, L.J.G., Zijm, W.H.M.: An analytical theory of multi-echelon production/distribution systems. Stat. Neerl. **44**(3), 149–174 (1990)

Matta, A., Dallery, Y., Di Mascolo, M.: Analysis of assembly systems controlled with Kanbans. Eur. J. Oper. Res. **166**(2), 310–336 (2005)

Ramakrishnan, R., Krishnamurthy, A.: Performance evaluation of a synchronization station with multiple inputs and population constraints. Comput. Oper. Res. **39**(3), 560–570 (2012)

Rosling, K.: Optimal inventory policies for assembly systems under random demands. Oper. Res. **37**(4), 565–579 (1989)

Sbiti, N., Di Mascolo, M., Bennani, T., Amghar, M.: Modeling and performance evaluation of base-stock controlled assembly systems. In: Gershwin, S.B., Dallery, Y., Papadopoulos, C.T., Smith, J.M. (eds.) Analysis and Modeling of Manufacturing Systems. Kluwer's International Series in Operations Research and management Science, pp. 307–341. Kluwer, Norwell (2003)

Topan, E., Avsar, Z.M.: An approximation for Kanban controlled assembly systems. Ann. Oper. Res. **182**(1), 133–162 (2011)

Zhao, Y.: Evaluation and optimization of installation base-stock policies in supply chains with compound Poisson demand. Oper. Res. **56**(2), 437–452 (2008)

Chapter 12
Threshold-Type Control of Supply Chain Systems with Multiple Products

12.1 System Stability

Chapter 7 discussed the stochastic supply chains systems producing multiple types of products. The output capacity of the supply chain system is determined by the raw material replenishment rate and the overall production capacity. On the other hand, note that the demand processes for each type of products are independent; the overall demand rate is equal to the sum of the demand rates for all product types. To avoid increasingly backlogging any type of products, the supply chain output capacity must be greater than the overall demand rate. Similar to Proposition 11.1 and Proposition 11.2, we have:

Proposition 12.1. *The sufficient condition to ensure the stability of the supply chain systems with multiple products in Fig. 7.1 is* $\min\{Q\lambda, r\} > \sum_i \mu_i$.

Proposition 12.2. *The sufficient condition to ensure the stability of the supply chain systems producing multiple products in Fig. 7.1 with failure-prone machines is* $\min\{Q\lambda, r\eta/(\eta + \xi)\} > \sum_i \mu_i$.

Clearly, for the failure-prone manufacturing supply chain producing multiple products without considering raw material ordering activity, its sufficient stability condition is simplified to be $r\eta/(\eta + \xi) > \sum_i \mu_i$.

This chapter will consider three types of supply chain systems and develop their corresponding threshold-type control policies. The first type is the supply chain in Fig. 7.1. The second type is the one in Sect. 7.3. The third type is a manufacturing supply chain producing two part-types with externally specified priority.

For the third type of supply chains, we assume that the production times for part-type i follow an exponential distribution with rate r_i (i.e., r_i is the production rate for part-type i). Similar to the model in Sect. 7.2, the system is assumed to be completely flexible, namely, the machine can switch between two part-types without significant setup but can only produce one product at a time. However, different from the model in Sect. 7.2, the maximum production rates may be different

D.-P. Song, *Optimal Control and Optimization of Stochastic Supply Chain Systems*, Advances in Industrial Control, DOI 10.1007/978-1-4471-4724-4_12,
© Springer-Verlag London 2013

(i.e., r_1 may not be equal to r_2). The customer demands for two products arrive as independent Poisson processes with rate μ_1 and μ_2, respectively. They have different priorities. Unmet demands for part-type one (i.e., high-priority product) may be partially backlogged. Unmet demands for part-type two (i.e., low-priority product) are completely backlogged. If both types of demands are completely backlogged, the system is equivalent to an $M/M/1$ queuing system. The stability condition is $\mu_1/r_1 + \mu_2/r_2 < 1$. However, when the first type of demands is partially backordered, the system stability is more complicated. The matrix analytic method will be used to derive a sufficient and necessary condition for the stability in Sect. 12.4.

12.2   Threshold-Type Control for Ordering and Production in a Supply Chain with Multiple Products

This section first numerically explores the structure of the optimal joint raw material ordering and production capacity allocation policy in the supply chain system with multiple products shown in Fig. 7.1. Then, a threshold-type control policy is proposed.

Example 12.1. To simplify the narrative and facilitate the visualization of the results, we consider the supply chain with two types of products, that is, $n = 2$. The system parameters are set as follows: the lead-time rate for replenishing raw materials $\lambda = 0.8$, the maximum ordering quantity $Q = 3$, the maximum production rate $r = 2.0$, demand rate $\mu_1 = \mu_2 = 0.5$, the inventory unit cost for raw materials $c_0 = 1$, the finished goods inventory unit cost $c_1^+ = c_2^+ = 5$, and the backlog costs $c_1^- = 30$ and $c_2^- = 20$. The discount factor $\beta = 0.5$.

To conduct the value iteration algorithm and reduce the computation complexity, the system state space is limited into a finite area with $x_0 \leq 20$, $x_1, x_2 \in [-30, 10]$. The iterative procedure will be terminated when the number of iterations exceeds 100. Part of the optimal ordering and production policy is illustrated in Fig. 12.1.

Figure 12.1a–c describes the ordering decisions for the raw material in the (x_0, x_2) plane when x_1 takes 2, 1, and 0, respectively. The numbers in the Fig. 12.1a–c indicate the order size of the decisions at the corresponding system state. Figure 12.1d, e describes which product should be produced in the (x_1, x_2) plane when x_0 takes 1 and 2, respectively, in which 1 means the maximum production capacity should be allocated to produce type-one product, 2 means to allocate the maximum production capacity to type two, and 0 means producing nothing.

The following structural properties of the optimal policy can be observed. From Fig. 12.1a–c, it appears that the raw material ordering decisions follow the order-up-to-point rule subject to the maximum order size when the inventory levels of two types of finished goods are fixed. However, the target inventory level of the raw material varies when the inventory levels of two types of products vary. Interestingly, as the sum of the inventories of two types of products decreases, the target inventory level of the raw material is increasing.

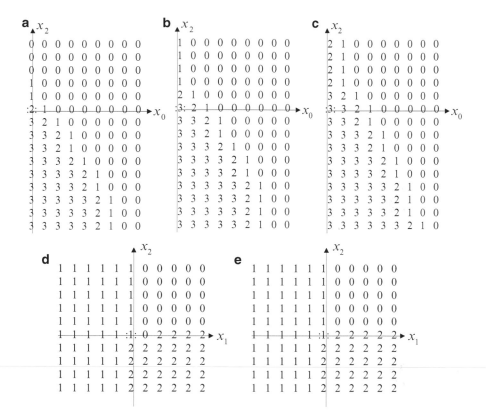

Fig. 12.1 Illustration of the optimal policy for supply chain with two products. (**a**) Ordering RM when $x_1 = 2$. (**b**) Ordering RM when $x_1 = 1$. (**c**) Ordering RM when $x_1 = 0$. (**d**) Production allocation when $x_0 = 1$. (**e**) Production allocation when $x_0 = 2$

From Fig. 12.1d, e, it can be seen that there is no need to produce any type of products if their inventories-on-hand are above certain levels. If one type of products has nonnegative inventory-on-hand whereas the other type has backlogged demands, then the maximum production capacity should be allocated to produce the backlogged type of products. When both types of products are backlogged, the maximum production capacity is allocated to produce part-type one. This is in agreement with the intuition that part-type one has a higher backlog unit cost than part-type two.

Based on the control structure in Fig. 12.1 and the above discussions, we present the following threshold control policy. For the raw material ordering decision, it is characterized by a threshold parameter serving as the total target inventory level of the raw material and all finished goods subject to the maximum order capacity Q. For the production capacity allocation, it is characterized by a set of threshold parameters to act as the target inventory levels for individual types of products and

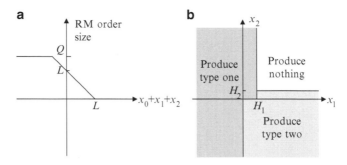

Fig. 12.2 A threshold control policy for a supply chain producing two part-types. (**a**) RM ordering decisions. (**b**) Production allocation decisions

a priority rule to break the tie when more than two types of products are below the target levels. An example of priority rule is the part-type with higher backlog cost should be produced first if both are backlogged.

In the case of producing two part-types (i.e., $n = 2$), the threshold control policy can be illustrated in Fig. 12.2, in which the priority is given to part-type one if both types are backlogged or both types have positive inventories that are below their target levels. Figure 12.2a shows the raw material ordering decisions when $x_0 + x_1 + x_2$ takes different values. Figure 12.2b shows the production capacity allocation decisions over two part-types when there are raw materials available, in which the switching curves divide the state space into three regions to determine three different control actions.

Mathematically, the raw material ordering and production capacity allocation decisions for the supply chain producing two part-types, $(q\,(\mathbf{x}),\, u_1\,(\mathbf{x}), u_2\,(\mathbf{x}))$, are given by (assuming part-type one is given the priority to break the tie)

$$q\,(\mathbf{x}) = \min\{L - x_0 - x_1 - x_2, Q\}, \quad \text{if } x_0 + x_1 + x_2 \le L; q\,(\mathbf{x}) = 0 \text{ otherwise.} \tag{12.1}$$

$$(u_1\,(\mathbf{x}),\, u_2\,(\mathbf{x})) = \begin{cases} (0,0) & \{\mathbf{x}|x_0 = 0\} \cup \{\mathbf{x}|x_1 > H_1, x_2 > H_2\} \\ (r,0) & \{\mathbf{x}|x_1 \le H_1, x_2 \ge 0\} \cup \{\mathbf{x}|x_1 < 0\}. \\ (0,r) & \text{otherwise} \end{cases} \tag{12.2}$$

Physically, the threshold control policy in (12.1) and (12.2) states as follows: An order for raw materials should be placed to the supplier if the inventory level of raw materials plus the inventory levels of all finished goods is below L, and the order size is order-up-to-point L subject to the maximum order size Q. In other words, the ordering policy is a kind of echelon base-stock policy. No production is possible (when there is no raw material available) or needed if both types of finished goods exceed their target inventory threshold levels; the maximum production capacity should be allocated to the high-priority product if its inventory drops below the target inventory level (H_1), but the other type has nonnegative inventories; the maximum production capacity should also allocated to the high-priority product if it has backorders regardless the state of the other type, whereas in other situations, the maximum production capacity should be allocated to the low-priority product type. Clearly, this policy is determined by three threshold parameters, L, H_1, and H_2, together with the endogenously determined priority rules.

12.3   Threshold-Type Control for Production Rate Allocation in a Failure-Prone Manufacturing Supply Chain with Two Part-Types

Since the characteristics of the optimal production rate allocation policy have been established in Sect. 7.3, we are able to construct near-optimal threshold-type control policies. Consider the failure-prone manufacturing supply chain producing two part-types as described in Sect. 7.3. From Proposition 7.11, we know that the optimal production rate allocation for two part-types case when the machine is up is of region switching structure and can be described by three monotone curves, $S_1(x_2)$, $S_2(x_1)$, and $S_3(x_2)$. Moreover, they converge to nonnegative finite asymptotes. By replacing three switching curves with four straight lines, we can obtain a linear switching threshold (LST) control policy as shown in Fig. 12.3.

Mathematically, the linear switching threshold (LST) control policy in Fig. 12.3 when the machine is up, $(u_1 (1, \mathbf{x}), u_2 (1, \mathbf{x}))$, is given by

$$(u_1 (1, \mathbf{x}), u_2 (1, \mathbf{x})) = \begin{cases} (r, 0) & \mathbf{x} \in R_1 \\ (0, r) & \mathbf{x} \in R_2 \\ (0, 0) & \mathbf{x} \in R_3 \end{cases} \tag{12.3}$$

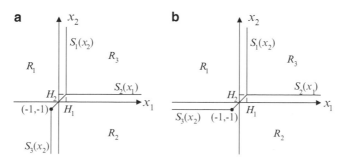

Fig. 12.3 Linear switching threshold control policy in a failure-prone manufacturing supply chain producing two part-types. (**a**) When $c_1^- > c_2^-$. (**b**) When $c_1^- < c_2^-$

Let k be the slope of the linear segment between two points (H_1, H_2) and $(-1, -1)$ in the switching curve $S_3(x_2)$, which is defined as $k := (H_2 + 1)/(H_1 + 1)$. When $c_1^- \geq c_2^-$, the control regions R_1, R_2, and R_3 can be defined as follows (corresponding to Fig. 12.3a):

$$R_1 := \{\mathbf{x}|x_1 \leq H_1, x_2 \geq k \cdot x_1 + (H_2 - k \cdot H_1)\} \cup \{\mathbf{x}|x_1 \leq -1\};$$

$$R_2 := \{\mathbf{x}|x_2 \leq H_2, x_2 < k \cdot x_1 + (H_2 - k \cdot H_1), x_1 > -1\};$$

$$R_3 := \{\mathbf{x}|x_2 > H_2, x_1 > H_1\}. \tag{12.4}$$

When $c_1^- < c_2^-$, the control regions R_1, R_2, and R_3 can be defined as (corresponding to Fig. 12.3b)

$$R_1 := \{\mathbf{x}|x_1 \leq H_1, x_2 > k \cdot x_1 + (H_2 - k \cdot H_1), x_2 > -1\};$$

$$R_2 := \{\mathbf{x}|x_2 \leq H_2, x_2 \leq k \cdot x_1 + (H_2 - k \cdot H_1)\} \cup \{\mathbf{x}|x_2 \leq -1\};$$

$$R_3 := \{\mathbf{x}|x_2 > H_2, x_1 > H_1\}. \tag{12.5}$$

The linear switching threshold policy can be further simplified by redefining the control regions. For example, we call the following policy the naïve threshold control (NTC) policy, which is essentially characterized by two straight lines ($x_1 = H_1$ and $x_2 = H_2$):

$$R_1 := \{\mathbf{x}|x_1 \leq H_1\} \quad \text{if } c_1^- \geq c_2^-; R_1 := \{\mathbf{x}|x_1 \leq H_1, x_2 > H_2\} \quad \text{if } c_1^- < c_2^-;$$

$$R_2 := \{\mathbf{x}|x_2 \leq H_2, x_1 > H_1\} \quad \text{if } c_1^- \geq c_2^-; R_2 := \{\mathbf{x}|x_2 \leq H_2\} \quad \text{if } c_1^- < c_2^-;$$

$$R_3 := \{\mathbf{x}|x_2 > H_2, x_1 > H_1\}. \tag{12.6}$$

The advantage of the threshold policies in (12.3) is apparent. It is determined by two parameters H_1 and H_2. These two control parameters may be estimated analytically or numerically since they are closely related to the asymptotes of the switching curves $S_1(x_2)$ and $S_2(x_1)$. They may also be optimized using off-line simulation to avoid solving the Bellman optimality equation, which is extremely difficult when the types of products are more than three.

By appropriately designing the endogenous priority rules, the above threshold policies can be extended to the failure-prone manufacturing supply chains producing more than two part-types.

In the remainder of this section, numerical examples are given to verify the analytical results in Chap. 7 and to demonstrate the effectiveness of the proposed threshold policies given in (12.3).

Example 12.2. Consider a failure-prone manufacturing supply chain producing two types of products, that is, $n = 2$. The system parameters are set as follows: the maximum production rate $r = 2.0$, the machine failure rate $\xi = 0.1$, the repair

Fig. 12.4 The optimal
control regions for a
failure-prone manufacturing
supply chain with two
products

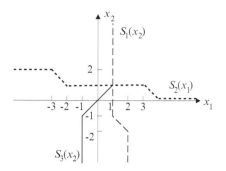

rate $\eta = 0.9$, the demand rate $\mu_1 = \mu_2 = 0.5$, the finished goods inventory unit cost
$c_1^+ = c_2^+ = 1$, and the backlog unit costs $c_1^- = 30$ and $c_2^- = 20$. The discount
factor $\beta = 0.5$.

The system state space is limited into a finite area with x_1, $x_2 \in [-30, 10]$. The
value iteration procedure will be terminated when the number of iterations exceeds
100. Part of the optimal production policy is illustrated in Fig. 12.4, in which the
dashed line represents $S_1(x_2)$, the dotted line represents and $S_2(x_1)$, and the solid line
represents $S_3(x_2)$.

More specifically, the three switching curves are given as follows:

$$S_1(x_2) = \begin{cases} 3 & x_2 \leq -8 \\ 2 & -7 \leq x_2 \leq -2 . \\ 1 & x_2 \geq -1 \end{cases} \tag{12.7}$$

$$S_2(x_1) = \begin{cases} 3 & x_1 \leq -10 \\ 2 & -9 \leq x_1 \leq -3 \\ 1 & -2 \leq x_1 \leq 3 \\ 0 & x_1 \geq 4 \end{cases} . \tag{12.8}$$

$$S_3(x_2) = \begin{cases} -1 & x_2 \leq -1 \\ 0 & x_2 = 0 \\ i & x_2 = i \geq 1 \end{cases} . \tag{12.9}$$

From Fig. 12.4 and (12.7), (12.8), and (12.9), it can be seen that the optimal
production rate allocation in the failure-prone manufacturing supply chain is indeed
characterized by three switching curves. The switching curves $S_1(x_2)$ and $S_2(x_1)$
are monotonic decreasing, and the switching curve $S_3(x_2)$ is monotonic increasing.
This verifies the analytical results in Chap. 7. In the next example, we explore the
effectiveness of the proposed threshold control policies in a range of scenarios.

Example 12.3. Consider a failure-prone manufacturing supply chain with two types
of products, that is, $n = 2$. Nine scenarios with different combinations of the system
parameters are investigated (see Table 12.1). Other system parameters are the same
for all scenarios such as $c_2^+ = 1$, $\beta = 0.5$.

Table 12.1 Parameter setting for different cases

Case	r	ξ	η	μ_1	μ_2	$c_1{}^+$	$c_1{}^-$	$c_2{}^-$
1	2.0	0.1	0.9	0.5	0.5	1	30	20
2	1.2	0.1	0.9	0.5	0.5	1	30	20
3	2.0	0.3	0.9	0.5	0.5	1	30	20
4	2.0	0.1	0.7	0.5	0.5	1	30	20
5	2.0	0.1	0.9	0.3	0.5	1	30	20
6	2.0	0.1	0.9	0.7	0.5	1	30	20
7	2.0	0.1	0.9	0.5	0.5	5	30	20
8	2.0	0.1	0.9	0.5	0.5	1	10	20
9	2.0	0.1	0.9	0.5	0.5	5	10	20

Table 12.2 Optimal policy, LST policy, and NTC policy for different cases

Case	J_{Opt}	% of J_{LST} above	(H_1, H_2)	% of J_{NYC} above	(H_1, H_2)
1	17.4856	0.0001%	(1, 1)	12.8779%	(0, 1)
2	32.2016	0.0000%	(1, 1)	9.3502%	(0, 2)
3	22.1879	0.0000%	(1, 1)	11.2823%	(0, 1)
4	18.1010	0.0000%	(1, 1)	12.5917%	(0, 1)
5	12.7921	0.0000%	(0, 1)	15.5436%	(0, 1)
6	23.4165	0.0000%	(1, 1)	12.1418%	(0, 1)
7	21.7356	0.0000%	(0, 1)	10.8243%	(−1, 1)
8	11.1202	0.9628%	(0, 0)	9.2749%	(0, 0)
9	12.4493	0.0000%	(−1, 1)	12.3555%	(−1, 0)

The value iteration algorithm is used to evaluate the performance. The results under the linear switching threshold (LST) policy and the naïve threshold control (NTC) policy are shown in Table 12.2, which are compared with the optimal control policy. The columns of "% above" represent the percentage of the cost under the threshold policy above the optimal cost. The columns of "(H_1, H_2)" in Table 12.2 give the best threshold parameters for the corresponding threshold policy.

The results of those nine scenarios show that the LST policy is extremely close to the optimal policy. In eight out of nine scenarios, the difference is less than 0.0001%. In the worst scenario, it is still within 1% of the optimal policy. This indicates that the LST policy, after optimizing its threshold parameters, is near to optimal.

The naïve threshold control policy, after optimizing its threshold parameters, performs 9–16% worse than the optimal policy. This indicates that the NTC policy may be oversimplified. Note that both LST and NTC policies need to find two appropriate threshold parameters, which means the computational effort to seek the best control parameters is similar. Therefore, it is recommended that the LST policy should be used in practice.

12.4 Prioritized Base-Stock Threshold Control for a Manufacturing Supply Chain Producing Two Part-Types with Given Priority

Consider a reliable manufacturing supply chain producing two types of products with externally given priority, in which the raw material ordering activities are not considered. Unmet demands for part-type one (i.e., high-priority product) are partially backlogged. Namely, the newly arriving demands will be lost if the current backlog length of part-type one exceeds a nonnegative number l. This assumption represents the situations in which the high-priority customers can observe the current queue length before deciding to join. Unmet demands for part-type two (i.e., low-priority product) are completely backlogged. Practically, the first type of customer has a limit of patience and will go for other suppliers if the expected waiting time would exceed a certain level. On the other hand, the second type of customer has sufficient patience or is loyal to the manufacturer.

We use a prioritized base-stock policy to control the production, which is characterized by two base-stock levels m and n with a static priority assigned to part-type-one subject to $-l \leq m$. More specifically, the policy, denoted by $u_{m,n} := \left(u^1_{m,n}(\mathbf{x}), u^2_{m,n}(\mathbf{x}) \right)$, is defined by

$$u^1_{m,n}(\mathbf{x}) = \begin{cases} r_1 & \text{if} \quad -l \leq x_1 < m \\ 0 & \text{otherwise} \end{cases}; \quad u^2_{m,n}(\mathbf{x}) = \begin{cases} r_2 & \text{if} \quad x_1 = m, x_2 < n \\ 0 & \text{otherwise} \end{cases}.$$

$$(12.10)$$

Physically, this policy states that the machine will always produce part-type one unless its inventory level reaches m. When the inventory of part-type one is maintained at m, the machine will produce part-type two whenever its inventory falls below n. Under the prioritized base-stock policy $u_{m,n}$, apart from the initial transient period, the inventory position of part-type one will never go beyond m, and the inventory position of part-type two will never go beyond n. We are interested in the steady-state performance and therefore assume the initial inventory positions for type one and two are not greater than m and n, respectively. Under the control of $u_{m,n}$, the system state space can be represented by $X = \{(x_1, x_2) \mid -l \leq x_1 \leq m \text{ and } x_2 \leq n\}$. The state transition map of the induced Markov chain is shown in Fig. 12.5.

12.4.1 System Stability

We apply the matrix analytic method (Latouche and Ramaswami 1999) to establish the system stability condition. Sequence the system states as follows: (m, n), $(m-1, n)$, $\ldots$, $(-l, n)$, $(m, n-1)$, $(m-1, n-1)$, $\ldots$, $(-l, n-1)$, $\ldots$, in which the

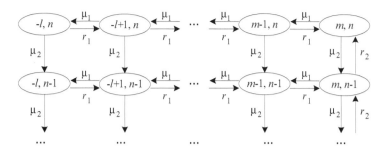

Fig. 12.5 State transition map under a prioritized base-stock policy in a manufacturing supply chain with two products

part-type-one state (i.e., x_1) is treated as different phases from m to $-l$ and part-type-two state (i.e., x_2) is treated as different levels from n to $-\infty$. This is a quasi-birth–death (QBD) Markov chain. The infinitesimal generator Q is given by

$$
Q = \begin{pmatrix}
B_0 & A_0 & 0 & 0 & \cdots \\
A_2 & A_1 & A_0 & 0 & \cdots \\
0 & A_2 & A_1 & A_0 & \cdots \\
0 & 0 & A_2 & A_1 & \cdots \\
\cdots & \cdots & \cdots & \cdots & \cdots
\end{pmatrix}
\tag{12.11}
$$

where B_0, A_1, A_2, and A_0 are $(m+l+1) \times (m+l+1)$ matrices given by

$$
B_0 = \begin{pmatrix}
-\mu_1 - \mu_2 & \mu_1 & & & \\
r_1 & -r_1 - \mu_1 - \mu_2 & \mu_1 & & \\
& \cdots & & \cdots & \\
& & r_1 & -r_1 - \mu_1 - \mu_2 & \mu_1 \\
& & & r_1 & -r_1 - \mu_2
\end{pmatrix};
$$

$$
A_0 = \begin{pmatrix}
\mu_2 & & & & \\
& \mu_2 & & & \\
& & \cdots & & \\
& & & \mu_2 & \\
& & & & \mu_2
\end{pmatrix};
\quad
A_2 = \begin{pmatrix}
r_2 & & & \\
& 0 & & \\
& & \cdots & \\
& & & 0 \\
& & & & 0
\end{pmatrix};
$$

$$
A_1 = \begin{pmatrix}
-r_2 - \mu_1 - \mu_2 & \mu_1 & & & \\
r_1 & -r_1 - \mu_1 - \mu_2 & \mu_1 & & \\
& \cdots & & \cdots & \cdots \\
& & r_1 & -r_1 - \mu_1 - \mu_2 & \mu_1 \\
& & & r_1 & -r_1 - \mu_2
\end{pmatrix}.
$$

Proposition 12.3. *The necessary and sufficient condition for the stability of the system is* $\rho := \rho_2 \cdot \left(1 + \rho_1 + \rho_1^2 + \cdots + \rho_1^{m+l}\right) < 1$, *where* $\rho_i := \mu_i / r_i$. *This stability condition can be easily rewritten as a more compact geometrical form when* $\rho_1 \neq 1$.

Proof. Let $A = A_1 + A_2 + A_0$. It is clear that in the directed graph associated with the matrix A, each pair of neighboring vertices is connected. Therefore, there exists a path from each vertex to every other vertex. Namely, the associated directed graph is strongly connected. It follows that the matrix A is irreducible (Varga 1962).

Solving the equations $\pi A = 0$ and $\pi e = 1$, where e is a column vector of dimension $m + l + 1$ consisting entirely of ones, it gives the unique solution

$$\pi_0 = \frac{1}{(1 + \rho_1 + \cdots + \rho_1^{m+1})} \quad \text{and} \quad \pi_i = \rho_1^{\,i} \pi_0 \quad \text{for } i = 1, \ldots, m + l.$$

From Lemma 8.1, the induced Markov chain under $u_{m,n}$ is positive recurrent if and only if $\pi(A_2 - A_0)e > 0$. That is, $r_2 \cdot \pi_0 - \mu_2 > 0$. This completes the proof. $\square$

From Proposition 12.3, it is clear that (i) the stability condition does not depend on n but depends on $m + l$. In fact, if $m + l$ is fixed, then the induced Markov chains for different n are equivalent from the state transition map in Fig. 12.5. (ii) If $m + l = 0$, the stability condition is simplified as $\rho_2 < 1$, in which case all demands for part-type one are lost and the machine is dedicated to meet the demands for part-type two. (iii) As $m + l$ increases to infinity, the system tends to be completely backordering the part-type one as well. The system stability condition becomes $\rho_1 + \rho_2 < 1$, which is intuitively true. It should be pointed out that if $\rho_1 + \rho_2 = 1$ and $m + l$ is finite, then our system is stable, while the $M/M/1$ system is not. (iv) The above stability condition can provide useful management insights. For example, for any given ρ_1 and $m + l$, it can be used to compute the upper bound of ρ_2 in order to ensure the system stability. (v) If the system is stable under $u_{m,n}$, it is also stable under $u_{k,n}$ with $k < m$. This can be physically interpreted as follows. For the fixed maximum backlog length for part-type one l, when the base-stock level m decreases, the system stores less stock to meet the demands for part-type one and equivalently the chance to lose the demands for part-type one increases. In other words, by sacrificing losing more demands for part-type one, the system gains more capacity to meet the demands for part-type two and ensure the stability.

Corollary 12.1. *The necessary and sufficient condition for the stability of the system is* $r_1 r_2 > \mu_2 \cdot (r_1 + \mu_1)$ *for the following two special cases: (i) The maximum backlog length for part-type one is 1, and the manufacturer adopts the zero base-stock level policy for part-type one; (ii) the type-one customers are completely impatient (i.e., must be met by inventory), and the manufacturer adopts the base-stock policy with inventory level for part-type one equaling 1.*

The Markov chain in Fig. 12.5 is apparently irreducible. Proposition 12.3 provides the condition to ensure that the induced Markov chain is positive recurrent. Therefore, it is ergodic. Let $\left\{ p_i^{l,m,n}(j) | i = -l, -l+1, \ldots, m \text{ and } j = n, n-1, \ldots \right\}$

be the stationary distribution under the control policy $u_{m,n}$ with the maximum backlog length l for part-type one. Here to simplify the narrative, (i, j) is used to denote the system state (x_1, x_2).

Proposition 12.4. *The stationary distributions under different parameters l, m, and n have the following relationship:* $p_i^{l,m,n}(j) \equiv p_{i+l}^{0,m+l,0}(j-n)$ *for any* $i = -l, -l+1,$ *..., m and $j = n, n-1, \ldots, -\infty$, where $p_i^{0,m+l,0}(j)$ is the stationary distribution under $u_{m+l,0}$ with the maximum backlog length for part-type one being zero.*

Proposition 12.4 justifies the reduction of the problem to an $(l + m, 0)$ policy, that is, under the control of $u_{m+l,0}$ with the maximum backlog length for part-type one being zero. To simplify the narrative, in the remainder of this section, we omit the superscript of $p_i^{0,m+l,0}(j)$, that is, $p_i(j) := p_i^{0,m+l,0}(j)$, and introduce a new notation $L := m + l$.

12.4.2   Stationary Distribution

To find the stationary distribution of the recurrent QBD process $u_{L,0}$, either the spectral expansion method or matrix analytic method can be utilized. Mitrani and Chakka (1995) compared the spectral expansion with the matrix-geometric solution and stated that spectral expansion offers considerable advantages in efficiency. Moreover, the spectral expansion method provides a good basis for obtaining asymptotic and approximate result. For example, the tail of the steady-state distribution is dominated by the term in the expansion containing the eigenvalue with the largest modulus less than 1 (Mitrani and Chakka 1995). We therefore use the spectral expansion approach.

The spectral expansion approach involves the following steps: (1) from the infinitesimal generator Q or the flow balance equations, obtain the characteristic equation; (2) solve the characteristic equation to obtain eigenvalues; (3) based on eigenvalues, construct a general solution (i.e., stationary distribution) to the flow balance equations with undetermined constants; (4) using boundary flow balance conditions and the probability normalization condition, determine the unknown constants

For the quasi-birth–death Markov process in (12.11), the associated characteristic matrix polynomial, $Q(y)$, is defined as

$$Q(y) = A_0 + A_1 y + A_2 y^2. \tag{12.12}$$

The characteristic equation of the flow balance equations is given by the scalar polynomial $\det(Q(y)) = 0$, where $\det(Q(y))$ is the determinant of matrix $Q(y)$.

Next, we first consider the special case with $L = 1$ and then derive the stationary distribution for the general case.

12.4.2.1   Stationary Distribution for the Special Case with $L = 1$

In the case $L = 1$, we have

$$Q(y) = \begin{bmatrix} \mu_2 - (\mu_1 + \mu_2 + r_2) \, y + r_2 y^2 & \mu_1 y \\ r_1 y & \mu_2 - (r_1 + \mu_2) \, y \end{bmatrix}. \tag{12.13}$$

This follows: $\det (Q(y)) = (1 - y)[r_2 \, (r_1 + \mu_2) \, y^2 - \mu_2 \, (r_1 + r_2 + \mu_1 + \mu_2) \, y + \mu_2{}^2] = 0$. Let $h(y) := r_2 \, (r_1 + \mu_2) \, y^2 - \mu_2 \, (r_1 + r_2 + \mu_1 + \mu_2) \, y + \mu_2{}^2$. Clearly, $h(0) > 0, h \, (\mu_2 / (r_1 + \mu_2)) < 0$ and $h(1) > 0$. The last inequality is due to Corollary 12.1. It follows that the equation $h(y) = 0$ has two real roots in the intervals $(0, \mu_2 / (r_1 + \mu_2))$ and $(\mu_2 / (r_1 + \mu_2), 1)$, respectively. Therefore, the characteristic equation $\det(Q(y)) = 0$ has three eigenvalues, y_0, y_1, and y_2, satisfying $0 < y_0 < y_1 < y_2 = 1$.

The flow balance equations for the case $L = 1$ are

$$\begin{cases} (\mu_1 + \mu_2 + r_2) \, p_1(j) = \mu_2 p_1 \, (j + 1) + r_2 p_1 \, (j - 1) + r_1 p_0(j) \\ (\mu_2 + r_1) \, p_0(j) = \mu_2 p_0 \, (j + 1) + \mu_1 p_1(j) \end{cases}. \tag{12.14}$$

The solution to (12.14) can be constructed from the eigenvalues. Due to the infinite number of states for part-type two, the eigenvalue $y_2 = 1$ cannot be used to construct the solution. A general solution to (12.14) can be expressed as

$$p_i(j) = a_{0i} \, y_0{}^{-j} + a_{1i} \, y_1{}^{-j} \quad \text{for } i = 0, 1 \text{ and } j \leq 0. \tag{12.15}$$

Substituting (12.15) into (12.14), we have

$$a_{i0} = \frac{-r_2 y_i^2 + (\mu_1 + \mu_2 + r_2) \, y_i - \mu_2}{r_1 y_i} a_{i1} \quad \text{for } i = 0, 1. \tag{12.16}$$

The flow balance equation at the state $\mathbf{x} = (L, 0)$ gives the boundary conditions, which yields

$$a_{11} = -\frac{(\mu_2 + r_1) \, y_1 - \mu_2}{(\mu_2 + r_1) \, y_0 - \mu_2} a_{01}. \tag{12.17}$$

Moreover, the normalization condition $\sum_{j=0}^{-\infty} (p_1(j) + p_0(j)) = 1$ and Eqs. (12.15), (12.16), and (12.17) yield

$$a_{01} = \frac{(1 - y_0) \, (1 - y_1) \, ((\mu_2 + r_1) \, y_0 - \mu_2)}{(\mu_1 + r_1) \, (y_0 - y_1)}. \tag{12.18}$$

Proposition 12.5. *In the case $L = 1$, the stationary distribution $\{p_i(j), i = 0 \text{ and } 1\}$ is given by*

$$p_1(j) = \frac{(1 - y_0)(1 - y_1)}{(\mu_1 + r_1)(y_0 - y_1)} \left[((\mu_2 + r_1) y_0 - \mu_2) y_0^{-j} - ((\mu_2 + r_1) y_1 - \mu_2) y_1^{-j} \right],$$

$$p_0(j) = \frac{\mu_1 (1 - y_0)(1 - y_1)}{(\mu_1 + r_1)(y_0 - y_1)} \left(y_0^{1-j} - y_1^{1-j} \right),$$

where y_0 and y_1 are the roots of $r_2 (r_1 + \mu_2) y^2 - \mu_2 (r_1 + r_2 + \mu_1 + \mu_2) y + \mu_2^2 = 0$.

12.4.2.2   Stationary Distribution for the General Case

In the general case, the associated characteristic matrix polynomial $Q(y)$, defined in (12.12), is a $(L + 1) \times (L + 1)$ matrix. From the structure of the matrix $Q(y)$, if we define

$$g_0 = 1,$$

$$g_1 = \mu_2 - (\mu_1 + \mu_2 + r_2) y + r_2 y^2,$$

$$g_i = (\mu_2 - (\mu_1 + \mu_2 + r_1) y) g_{i-1} - r_1 \mu_1 y^2 g_{i-2} \quad \text{for } 1 < i \le L, \text{ and}$$

$$g_{L+1} = (\mu_2 - (\mu_2 + r_1) y) g_L - r_1 \mu_1 y^2 g_{L-1}.$$

Then, we have $\det(Q(y)) = g_{L+1}$. It is therefore clear that $\det(Q(y)) = 0$ is a polynomial with degree $L + 2$.

In addition, $y = 1$ is a root of $\det(Q(y)) = 0$. From Mitrani and Chakka (1995), the necessary and sufficient condition for the ergodicity of the induced QBD process is that the number of eigenvalues of $Q(y)$ strictly inside the unit disk is equal to $L + 1$ (each counted according to its multiplicity). We have shown that the QBD process is ergodic under the condition in Proposition 12.3. Thus, $\det(Q(y)) = 0$ has $L + 1$ roots inside the unit disk. Moreover, from the sequence of $\{g_0, g_1, \ldots, g_{L+1}\}$, it is clear that $g_i > 0$ if $y < 0$, which implies that all eigenvalues of $Q(y)$ are positive.

To obtain the further structural properties of the eigenvalues of $Q(y)$, we introduce the following result.

Lemma 12.1. (*Grassmann 2002*). *Consider a quasi-birth–death Markov chain given in (12.11). Let $q_{i,j}^{(k)}$ be the rate of going from phase i to phase j while increasing the level by $1 - k$. If for each i, either $q_{i,i-1}^{(0)} = 0$ or $q_{i-1,i}^{(0)} = 0$, then all eigenvalues of $Q(y)$ are real. If in addition to this, $q_{i,i}^{(0)} > 0$, then all eigenvalues are distinct.*

Proposition 12.6. *The characteristic equation $\det(Q(y)) = 0$ has $L + 2$ distinct real roots: $y_0, y_1, \ldots, y_{L+1}$, such that $0 < y_0 < y_1 < \ldots < y_L < y_{L+1} = 1$.*

Proof. Recall the infinitesimal generator Q, where the part-type-one state is treated as a phase and the part-type-two state is treated as a level. The transition rate from phase i to phase $i - 1$ while increasing the level 1 is zero, that is, $q_{i,i-1}^{(0)} = 0$ for

$i = 0, 1, \ldots, L$. In addition, the transition rate from phase i to phase i by increasing the level 1 is μ_2, that is, $q_{i,i}^{(0)} = \mu_2$ for $i = 0, 1, \ldots, L$. From Lemma 12.1, it follows that all eigenvalues of $Q(y)$ are distinct and real. Note that we have stated that $\det(Q(y)) = 0$ has $L + 1$ positive roots inside the unit disk and another root $y = 1$. Therefore, the assertion is true. $\qquad\square$

The flow balance equations for the states $\mathbf{x} = (i, j)$ with $i = 1, \ldots, L$ are

$$\begin{cases} (\mu_1 + \mu_2 + r_2)\, p_L(j)) = \mu_2 p_L\, (j + 1) + r_2 p_L\, (j - 1) + r_1 p_{L-1}(j) \\ (\mu_1 + \mu_2 + r_2)\, p_i(j) = \mu_2 p_i\, (j + 1) + \mu_1 p_{i+1}(j) + r_1 p_{i-1}(j), \quad for\, i = 1, 2, \ldots, L-1 \end{cases}$$
$$(12.19)$$

A general solution to (12.19) can be expressed as

$$p_i(j) = a_{0i}\, y_0^{-j} + a_{1i}\, y_1^{-j} + \cdots + a_{Li}\, y_L^{-j} \quad \text{for } i = 0, 1, \ldots, L \text{ and } j \le 0,$$
$$(12.20)$$

where a_{ik} for $i, k = 0, 1, \ldots, L$ are undetermined constants. Substituting (12.20) into (12.19), we have

$$a_{kL-1} = \frac{-r_2 y_k^2 + (\mu_1 + \mu_2 + r_2)\, y_k - \mu_2}{r_1 y_k} a_{kL} \quad \text{for } 0 \le k \le L \qquad (12.21)$$

$$a_{ki-1} = \frac{((\mu_1 + \mu_2 + r_1)\, y_k - \mu_2)\, a_{ki} - \mu_1 y_k a_{ki+1}}{r_1 y_k} \quad \text{for } 0 \le k \le L \text{ and } 0 < i < L.$$
$$(12.22)$$

Introducing the following notation, let $f_0\, (y_k) = 1$, $f_1\, (y_k) = (-r_2 y_k^2 + (\mu_1 + \mu_2 + r_2) y_k - \mu_2)/(r_1 y_k)$, and $f_{i+1}\, (y_k) = (((\mu_1 + \mu_2 + r_1)\, y_k - \mu_2)\, f_i\, (y_k) - \mu_1 y_k f_{i-1}\, (y_k))/(r_1 y_k)$ for $1 < i < L$. Then, Eqs. (12.21) and (12.22) can be rewritten as

$$a_{ki} = f_{L-i}\, (y_k)\, a_{kL} \quad \text{for } 0 \le k \le L \text{ and } 0 \le i \le L. \qquad (12.23)$$

The boundary conditions for the states $\mathbf{x} = (i, 0)$ with $i = 0, \ldots, L-1$ are

$$(\mu_2 + r_1)\, p_0(0) = \mu_1 p_1(0)$$
$$(\mu_1 + \mu_2 + r_1)\, p_i(0) = \mu_1 p_{i+1}(0) + r_1 p_{i-1}(0) \quad \text{for } i = 1, 2, \ldots, L-1.$$

Using (12.20) and (12.23), it follows

$$\sum_{k=0}^{L} [(\mu_2 + r_1)\, f_L\, (y_k) - \mu_1 f_{L-1}\, (y_k)]\, a_{kL} = 0 \qquad (12.24)$$

$$\sum_{k=0}^{L} [(\mu_1+\mu_2+r_1)\, f_{L-i}\,(y_k) - \mu_1 f_{L-i-1}\,(y_k) - r_1 f_{L-i+1}\,(y_k)]\, a_{kL} = 0 \text{ for } 0<i<L.$$

$$(12.25)$$

The normalization condition yields

$$\sum_{k=0}^{L} \frac{f_L\,(y_k) + f_{L-1}\,(y_k) + \cdots + f_0\,(y_k)}{1-y_k} a_{kL} = 1. \qquad (12.26)$$

The boundary condition and the normalization condition provide total $(L+1)$ linear equations in (12.24), (12.25), and (12.26), which can determine $L+1$ unknown constants $\{a_{0L}, a_{1L}, \ldots, a_{LL}\}$. This gives the explicit form of the stationary distribution of the induced Markov chain.

Proposition 12.7. *The stationary distribution under the prioritized base-stock policy $u_{m,n}$ with the maximum backlog length l for part-type one is given by $p_i^{l,m,n}(j) = p_{i+l}(j-n)$ for $i = -l,\ldots,m$ and $j \leq n$, where $p_{i+l}(j-n)$ is given in (12.20) and (12.23), (12.24), (12.25), and (12.26).*

12.4.3   Steady-State Performance Measures

Based on the stationary distribution, it is straightforward to compute relevant steady-state performance measures.

The most common steady-state performance measure is the long-run average cost. Let $g(i, j)$ denote the cost per unit of time at the state $\mathbf{x} = (i, j)$. For example, a linear form of the cost function $g(i, j)$ may be defined as

$$g\,(i, j) = c_r \cdot I\,\{i = -l\} + c_1^{+} \cdot i^{+} + c_1^{-} \cdot i^{-} + c_2^{+} \cdot j^{+} + c_2^{-} \cdot j^{-}, \quad (12.27)$$

where c_r is the lost-sale penalty for part-type one. From the ergodicity of the system, the long-run average cost function under the prioritized base-stock policy $u_{m,n}$ is given by

$$J\,(u_{m,n}) = \sum_{j=n}^{-\infty} \sum_{i=-l}^{m} p_i^{l,m,n}(j) g\,(i, j) = \sum_{j=0}^{-\infty} \sum_{i=0}^{m+l} p_i(j) g\,(i-l, j+n). \quad (12.28)$$

Equation (12.28) gives the explicit form of the long-run average cost since $p_i(j)$ has been derived in Proposition 12.7. The result is applicable to any form of $g(i, j)$.

Other interesting steady-state performance measures include the stock-out probability for each part-type; the fraction of lost demands for part-type one; the service level, which may be defined as the average percentage of demands for each part-type that are satisfied from inventory (i.e., the fraction of orders that are filled without any waiting time); the machine utilization (i.e., the fraction of time that the machine is on operation); the expected backlog length; and the expected waiting time to fill an order for part-type one.

These performance measures can be easily calculated from the stationary distribution. As an example, the stock-out probability and the fraction of lost demands are explicitly given as follows. Let SP_k denote the stock-out probability for part-type k, which is defined as the fraction of time that there exist unmet demands for part-type k. Let F_{loss} denote the fraction of the lost demands for part-type one. Then,

$$SP_1 = \sum_{j=0}^{-\infty} \sum_{i=0}^{l-1} p_i(j) \quad \text{and} \quad SP_2 = \sum_{j=-n-1}^{-\infty} \sum_{i=0}^{m+l} p_i(j) \tag{12.29}$$

$$F_{\text{loss}} = \sum_{j=0}^{-\infty} p_0(j). \tag{12.30}$$

12.4.4   Optimal Base-Stock Levels

Although we have obtained the explicit performance measures, it is still not trivial to find the optimal base-stock levels, which is a two-variable optimization problem. This section analyzes the properties of the cost function and present a two-step optimization procedure.

Proposition 12.8. *If $g(i, j)$ takes the linear form of (12.27), then the expected cost $J(u_{m,n})$ is convex in n and $J(u_{m,n}) \geq J(u_{m,0})$ for $n < 0$.*

Proof. The cost function (12.27) can be divided into two parts and rewritten as $g(i, j) = g_1(i) + g_2(j)$. Note that for the fixed base-stock level m, the stationary distribution $\{p_i(j)\}$ is the same for any n. We have

$$J(u_{m,n}) - J(u_{m,0}) = \sum_{j=0}^{-\infty} \sum_{i=0}^{m+l} p_i(j) [g_2(j+n) - g_2(j)] = -nc_2^-, \quad \text{for any } n < 0;$$

$$\tag{12.31}$$

$$J(u_{m,n+1}) - J(u_{m,n}) = \sum_{j=0}^{-\infty} \sum_{i=0}^{m+l} p_i(j) [g_2(j+n+1) - g_2(j+n)]$$

$$= \sum_{j=-n}^{0} \sum_{i=0}^{m+l} p_i(j)(j+n+1-j-n)c_2^+ + \sum_{j=-\infty}^{-n-1} \sum_{i=0}^{m+l} p_i(j)(-j-n-1+j+n)c_2^-$$

$$= (c_2^+ + c_2^-) \sum_{j=-n}^{0} \sum_{i=0}^{m+l} p_i(j) - c_2^-, \quad \text{for any } n \geq 0.$$

$$\tag{12.32}$$

Firstly, consider the case with $n < 0$. From (12.31), it is clear that $J(u_{m,n}) - J(u_{m,0}) \geq 0$ and $J(u_{m,n})$ is monotone decreasing in n for $n < 0$. Secondly, consider the case with $n \geq 0$. From (12.32), it is clear that as n increases, $J(u_{m,n+1}) - J(u_{m,n})$ is monotone increasing and converging to a positive number $c_2{}^+$. On the other hand, from (12.31), we have $J(u_{m,n+1}) - J(u_{m,n}) \equiv -c_2{}^-$ for $n < 0$. From (12.32), we have $J(u_{m,1}) - J(u_{m,0}) = (c_2^+ + c_2^-) \sum_{i=0}^{m+l} p_i(0) - c_2^- \geq -c_2^-$. Thus, $J(u_{m,n+1}) - J(u_{m,n})$ is increasing in n for any n. It follows that $J(u_{m,n})$ is convex in n. This completes the proof. □

Proposition 12.9. *If $g(i, j)$ takes the linear form of (12.27), then the optimal base-stock levels m^* and n^* to minimize the long-run average cost can be determined in two steps:*

(i) For any fixed m, the optimal base-stock level for part-type two $n^*(m)$ is the minimum nonnegative integer n such that

$$\sum_{j=0}^{-n} \sum_{i=0}^{m+l} p_i(j) \geq \frac{c_2^-}{c_2^+ + c_2^-}.$$

(ii) The optimal base-stock level for part-type one m^* is given by

$$m^* = \arg\min_m \left\{ J\left(u_{m,n^*(m)}\right) \mid -l \leq m \text{ and } \rho < 1 \right\}.$$

Proof. From Proposition 12.8, the optimal base-stock level $n^*(m)$ is the minimum nonnegative number n such that $J(u_{m,n+1}) - J(u_{m,n}) \geq 0$. It follows from (12.32) that step (i) is true. The second step varies m subject to the constraints of the maximum backlogged length and the stability index so that we can find the actual optimal control parameters m^* and $n^*(m^*)$, where $n^*(m)$ is determined by step (i). This completes the proof. □

As we know, for the fixed base-stock level m, the stationary distribution $\{p_i(j)\}$ is independent of parameter n. Therefore, Proposition 12.9(i) is actually a simple one-variable optimization problem for a convex function $J(u_{m,n})$. From the application point of view, if the base-stock level for part-type one has been predetermined, then Proposition 12.9(i) can be used straightforward to determine the optimal base-stock level for part-type two. For example, some company may adopt make-to-order (i.e., zero inventory) policy for producing part-type one, in which the base-stock level for part-type one is pre-specified to be zero and the optimal base-stock level for part-type two $n^*(0)$ can be easily obtained by step (i).

Proposition 12.9(ii) is another one-variable optimization problem but involves much more numerical computation because different m leads to different stationary distribution and Proposition 12.7 must be reapplied whenever m changes. It appears to be difficult to establish the analytical properties of the expected cost for different control parameter m. However, from intuition, it is reasonable to assume that m, the inventory of part-type one, should not be too large. Hence, Proposition 12.9(ii) may be applied by starting m from $-l$ upward until $J\left(u_{m,n^*(m)}\right)$ is no longer decreasing.

Corollary 12.2. *If the cost function $g(i, j)$ takes the linear form of (12.27) and the base-stock level for part-type one m is fixed, then the optimal base-stock level for part-type two $n^* = 0$ if and only if*

$$\sum_{i=0}^{m+l} \sum_{k=0}^{m+l} f_k(x_i) a_{im+l} \geq \frac{c_2^-}{c_2^+ + c_2^-}.$$

Proof. From Proposition 12.9, it suffices to show that $\sum_{i=0}^{m+l} p_i(0) \geq c_2^- / (c_2^- + c_2^+)$. From Eqs. (12.20) and (12.23), we have

$$\sum_{i=0}^{m+l} p_i(0) = \sum_{i=0}^{m+l} \sum_{k=0}^{m+l} a_{ik} = \sum_{i=0}^{m+l} \sum_{k=0}^{m+l} f_k(x_i) a_{im+l}.$$

This completes the proof. □

The above corollary provides a necessary and sufficient condition for the optimality of the make-to-order (zero inventory) policy for part-type two provided that the base-stock level for part-type one is pre-specified.

Apart from minimizing the long-run average cost, sometimes we are interested in design the base-stock levels in terms of other steady-state performances. Particularly, the lost-sale fraction and the stock-out probability are highly related to service level, which a company wants to achieve at a certain level. For example:

Proposition 12.10. *The minimum base-stock levels (m^*, n^*) such that the fraction of the lost sales of part-type one is less than α and the stock-out probability of part-type k is less than β_k, that is, $F_{loss} < \alpha$, $SP_1 < \beta_1$ and $SP_2 < \beta_2$, can be determined by (subject to $-l \leq m$ and $\rho < 1$),*

$$\min \left\{ (m, n) \mid \sum_{j=0}^{-\infty} p_0^{0,m+l,0}(j) < \alpha, \sum_{j=0}^{-\infty} \sum_{i=0}^{l-1} p_i^{0,m+l,0}(j) < \beta_1, \sum_{j=n-1}^{-\infty} \sum_{i=0}^{m+l} p_i^{0,m+l,0}(j) < \beta_2 \right\}.$$

This is nontrivial two-variable optimization problem. Suppose the system is stable for sufficiently large m. Note that as the base-stock level m increases, all the demands for part-type one will be satisfied eventually. That means the lost sales F_{loss} tend to be zero. Similarly, the stock-out probability of part-type one is decreasing as the base-stock level for part-type one increases. We present the following observations:

Observation 12.1. (i) $p_i^{0,l+m,0}(j) \geq p_i^{0,l+m+1,0}(j)$ for $i = 0, 1, \ldots,$ $l+m$ and $j \leq 0$; (ii) $p_i^{0,l+m,0}(j) \geq p_{i+1}^{0,l+m+1,0}(j)$ for $i = 0, 1, \ldots, l+m$ and $j \leq 0$; (iii) F_{loss} and SP_1 are decreasing in m and converging to 0 as m tends to infinity; (iv) SP_2 is decreasing in n and converging to 0 as n tends to infinity.

Although we cannot prove all the above observations, they appear to conform to the intuitions. Consider two Markov chains under control of $u_{m,n}$ and $u_{m+1,n}$, denoted by $X_{m,n}$ and $X_{m+1,n}$, respectively. From the transition map (Fig. 12.5), these two Markov chains have the exactly same structure with the same transition rates except that $X_{m+1,n}$ has one more column than $X_{m,n}$. Intuitively, $X_{m+1,n}$ splits its steady-state probability more widely than $X_{m,n}$, which may be represented by the above observation (i) and (ii). The observation (iii) follows from (i) and (ii). The observation (iv) can be easily derived by the definition of SP_2. These observations have been demonstrated using numerical examples in Song (2009).

Based upon the Observation 12.1, the following two-step procedure can solve the optimization problem in Proposition 12.10:

Step 1: Increasing m starting from $-l$ until it satisfies $F_{loss} < \alpha$ and $SP_1 < \beta_1$
Step 2: After finding m^*, increasing n starting from 0 until it satisfies $SP_2 < \beta_2$

The first step involves the repeatedly application of Proposition 12.7 to calculate stationary distributions for different m. The second step is simple since the stationary distribution is independent of n. However, it should be pointed out that such m^* and n^* may not exist if the demand rates are too high (i.e., intensive traffic situations).

Up to now, we have treated the maximum backlog length for part-type one l as an exogenously determined parameter (e.g., by customers). It is also interesting to regard it as one of the manufacturer's decision variables, as we did in Chaps. 4 and 9. Let $u_{l,m,n}$ and $J(u_{l,m,n})$ denote the corresponding control policy and the expected cost. This is a three-variable optimization problem and can be solved by extending Proposition 12.9.

Proposition 12.11. *If $g(i, j)$ takes the linear form of (12.27) and l is controllable, then the optimal control parameters l^*, m^*, and n^* to minimize the long-run average cost can be determined by the following three-step procedure:*

(i) For any fixed $L = l + m$, the optimal $l^*(L)$ is independent of the base-stock level n, and it is the minimum integer l satisfying $0 \leq l \leq L$ and

$$c_1^- \sum_{i=0}^{l} \sum_{j=0}^{-\infty} p_i(j) - c_1^+ \sum_{i=l+1}^{L} \sum_{j=0}^{-\infty} p_i(j) \geq 0.$$

(ii) For any fixed $L = l + m$ and $l^*(L)$, the optimal base-stock level for part-type two $n^*(L)$ is the minimum nonnegative integer n satisfying

$$\sum_{j=0}^{-n} \sum_{i=0}^{L} p_i(j) \geq \frac{c_2^-}{c_2^+ + c_2^-}.$$

(iii) The optimal parameter L^* is given by

$$L^* = \arg\min_{L} \left\{ J\left(u_{l^*(L),L-l^*(L),n^*(L)}\right) \mid L \geq 0 \text{ and } \rho < 1 \right\}.$$

That yields $l^* = l^*(L^*)$, $m^* = L^* - l^*(L^*)$ and $n^* = n^*(L^*)$.

Table 12.3 System parameters for different cases

Case	(μ_1, μ_2)	(r_1, r_2)	$(c_1{}^+, c_2{}^+)$	$(c_1{}^-, c_2{}^-)$	c_r
A	(0.2, 0.4)	(2, 1)	(1, 1)	(40, 5)	80
B	(0.4, 0.4)	(2, 1)	(1, 1)	(40, 5)	80
C	(0.8, 0.4)	(2, 1)	(1, 1)	(40, 5)	80
D	(0.2, 0.4)	(2, 1)	(1, 2)	(40, 5)	80
E	(0.2, 0.4)	(2, 1)	(2, 1)	(40, 5)	80

Proof. For any fixed $L = l + m$, from (12.28) we have

$$J\left(u_{l+1, m-1, n}\right) - J\left(u_{l, m, n}\right) = \sum_{j=0}^{-\infty} \sum_{i=0}^{L} p_i(j) \left(-c_1^+ I\{i \geq l+1\} + c_1^- I\{i \leq l\} \right)$$

$$= c_1^- \sum_{i=0}^{l} \sum_{j=0}^{-\infty} p_i(j) - c_1^+ \sum_{i=l+1}^{L} \sum_{j=0}^{-\infty} p_i(j).$$

It is clear that the right-hand side of the above equation is monotonically increasing in l and takes the value c_1^- when $l = L$. In other words, $J(u_{l, m, n})$ is convex in l for fixed L. It reaches its minimum at the minimum integer $l^*(L)$ satisfying $J(u_{l+1, m-1, n}) - J(u_{l, m, n}) \geq 0$. Since $J(u_{l+1, m-1, n}) - J(u_{l, m, n})$ does not depend on n, it follows that $l^*(L)$ is independent of the control parameter n. Step (ii) and Step (iii) are similar to the two steps in Proposition 12.9. This completes the proof. ☐

In Proposition 12.11, Steps (i) and (ii) are simple since both of them are a single variable optimization problem for a convex function with explicit form. These two steps yield the optimal $l^*(L)$ and $n^*(L)$ for fixed L. Particularly, here $l^*(L)$ is consistently optimal for any n with fixed L.

12.4.5 Numerical Examples

Consider five cases with system parameters given in Table 12.3 and the cost function takes the form of (12.27). The maximum backlog length for part-type one is fixed at $l = 3$. From case A to case C, the demand rate μ_1 is increasing. Case D and case E have different combinations of inventory holding costs.

The optimal base-stock levels (m^*, n^*) by minimizing the expected cost can be obtained by Proposition 12.9. Table 12.4 gives the system stability index ρ, the optimal base-stock levels, the expected cost under u_{m^*, n^*}, the stock-out probabilities, and the fraction of lost sales under u_{m^*, n^*}. The last column in Table 12.4 gives the optimal long-run average cost among all feedback control policies, which is obtained numerically by the stochastic dynamic programming and the value iteration algorithm, in which the iteration number is 8,000 and the state space is limited by $x_1 \leq 300$ and $-300 < x_2 \leq 300$.

Table 12.4 Optimal base-stock levels and steady-state performance measures

Case	ρ	(m^*, n^*)	$J(u_{m,n})$	PS_1	PS_2	F_{loss}	J^*
A	0.44	(1, 2)	3.58	0.0100	0.0966	0.0001	3.58
B	0.50	(2, 2)	4.99	0.0079	0.1492	0.0003	4.81
C	0.67	(3, 5)	9.71	0.0240	0.1483	0.0025	9.28
D	0.44	(1, 1)	4.39	0.0100	0.2093	0.0001	4.39
E	0.44	(1, 2)	4.48	0.0100	0.0966	0.0001	4.47

As the demand rate μ_1 increases from case A to case C, the stability index ρ is increasing and the system could be unstable if μ_1 becomes too large. The optimal base-stock levels (m^*, n^*) are increasing in μ_1 (in cases A, B, and C). Comparing case A with case D, the base-stock level for part-type two decreases due to the increase of penalty on the inventory of part-type two. The stock-out probability and the lost-sale fraction for part-type one are very low: this is in agreement with the parameter setting, that is, high penalties for lost sales and part-type-one stock-out. Comparing the forth column with the last column, it shows that the prioritized base-stock policy is very close to the optimal state-feedback control policy in some cases (A, D, and E). This may be explained by the fact that when ρ is small, the machine has sufficient capacity to meet both types of customers and the optimal policy tends to be the static prioritized policy.

12.5   Discussion and Notes

Section 12.4 is based on Song (2009). When a supply chain involves producing multiple products, it has to deal with two types of decisions: (i) on/off decision of the production activity and (ii) which type of products to produce if it is on and when the production should be switched from one product to another.

In the literature, base-stock policy is often used to determine the on/off decision, for example, Zheng and Zipkin (1990), Wein (1992), and Pena-Perez and Zipkin (1997). Priority rules based on cost coefficient, production rates, and inventory positions are presented to determine which product to produce and when to switch between products. For example, Zheng and Zipkin (1990) stated that whenever an allocation decision is required, the manufacturer should produce the product with the lower net inventory. Wein (1992) used cost and service rate-based priority rules. Pena-Perez and Zipkin (1997) presented a static priority rule based on average workload and cost rate. Ha (1997) established a simple index rule to determine which product should be produced when the corresponding product is backlogged. A similar index rule is developed in a two-stage make-to-order system in Zhao and Lian (2011) to determine which class of customers should be served.

When the priority of products/customers is exogenously specified, the second type of decisions focuses on when the production should be switched from one product to another or when the lower priority demand should be rejected or backordered. Such decision may be characterized by an inventory threshold parameter. For example, Shiue and Altiok (1993) considered a two-stage multiproduct inventory system with Poisson demands. Products have a priority structure and a switching rule. The on/off production decisions at two stages follow continuous-review threshold control policies. They evaluated the steady-state performance measures using an approximation procedure. Melchiors et al. (2000) considered a continuous-review (s, Q) model with two prioritized demand classes. A single threshold level policy is applied to ration the inventory among the two demand classes. Isotupa (2006) considered a lost-sale Markovian inventory system with two types of customers (a priority type and an ordinary type). Under the (s, Q) ordering policy, they sought the optimal values for reorder level and reorder quantity.

References

Grassmann, W.: Real eigenvalues of certain tridiagonal matrix polynomials, with queueing applications. Linear Algebra Appl. **342**, 93–106 (2002)

Ha, A.: Optimal dynamic scheduling policy for a make-to-stock production system. Oper. Res. **45**(1), 42–53 (1997)

Isotupa, K.P.S.: An Markovian inventory system with lost sales and two demand classes. Math. Comput. Model. **43**(7–8), 687–694 (2006)

Latouche, G., Ramaswami, V.: Introduction to Matrix Analytic Methods in Stochastic Modelling. SIAM, Philadelphia (1999)

Melchiors, P., Dekker, R., Kleijn, M.J.: Inventory rationing in an (s, Q) inventory model with lost sales and two demand classes. J. Oper. Res. Soc. **51**(1), 111–122 (2000)

Mitrani, I., Chakka, R.: Spectral expansion solution for a class of Markov-models – application and comparison with the Matrix-Geometric method. Perform. Eval. **23**, 241–260 (1995)

Pena-Perez, A., Zipkin, P.: Dynamic scheduling rules for a multi-product make-to-stock queue. Oper. Res. **45**(6), 919–930 (1997)

Shiue, G., Altiok, T.: Two-stage, multi-product production/inventory systems. Perform. Eval. **17**(3), 225–232 (1993)

Song, D.P.: Stability and optimization of a production inventory system under prioritized base-stock control. IMA J. Manage. Math. **20**(1), 59–79 (2009)

Varga, R.: Matrix Iterative Analysis. Prentice-Hall, Englewood Cliffs (1962)

Wein, L.M.: Dynamic scheduling of a multiclass make-to-stock queue. Oper. Res. **40**(4), 724–735 (1992)

Zhao, N., Lian, Z.T.: A queueing-inventory system with two classes of customers. Int. J. Prod. Econ. **129**(1), 225–231 (2011)

Zheng, Y., Zipkin, P.: A queueing model to analyze the value of centralized inventory information. Oper. Res. **38**(2), 296–307 (1990)

Chapter 13
Optimization of Threshold Control Parameters via Numerical Methods

13.1 Introduction

In the last few chapters, threshold-type control policies have been presented in various stochastic supply chain systems, and their closeness to the optimal policies has been evidenced. Although threshold control policies are significantly simpler than the optimal policies in many cases, finding optimal threshold parameters can be a challenging issue if there are multiple threshold parameters to be determined.

The optimization problem in our context is to seek the best set of threshold parameters of the threshold-type control policy for a given supply chain system. One set of the threshold parameters is called a solution. The optimization problem can be solved by iterative search methods that converge or approximate to the optimal solution. Iterative search methods may be classified into three categories: calculus-based methods, enumerative methods, and meta-heuristic methods. The calculus-based methods often rely heavily on problem-specific knowledge such as the gradient of the objective function. Typical examples are Newton's method and gradient-based methods. This type of iterative methods is usually not applicable to our problem because the closed form of the objective function does not exist and the threshold parameters are discrete variables.

Enumerative methods are based on the simple idea by evaluating the objective function for every solution in the search space. Typical examples are branch and bound, linear programming, and dynamic programming. The biggest advantage of this type of iterative methods is that they are able to find the global optimal solution. The shortcoming is the lack of efficiency, in particular for complex systems.

Meta-heuristic methods utilize a guided random search mechanism that optimizes a problem by iteratively trying to improve candidate solutions. Typical examples include genetic algorithms/evolutionary strategy, simulated annealing, tabu search, swarm intelligences, and artificial immune system. The advantage of meta-heuristic methods is their general applicability. They do not need prerequisite

D.-P. Song, *Optimal Control and Optimization of Stochastic Supply Chain Systems*, Advances in Industrial Control, DOI 10.1007/978-1-4471-4724-4_13, © Springer-Verlag London 2013

information relating to the problem and have been evidenced to be able to tackle many complex optimization problems. However, meta-heuristics do not guarantee that an optimal solution is ever found.

This chapter will present some numerical search methods within the enumerative search category to optimize threshold control parameters for different types of supply chain systems. We will also explore the robustness of some threshold policies, namely, how well the threshold policies perform if some input data are not accurate. Optimization of threshold control parameters using meta-heuristic methods will be addressed in the next chapter.

13.2   Optimization of Threshold Parameters in Discounted-Cost Situations

Consider a Markov decision process with the state space X, the admissible control set Ω, the one-step cost function $g(\mathbf{x}, \mathbf{u})$ where $\mathbf{x} \in X$ and $\mathbf{u} \in \Omega$, and the discount factor β. The optimal discounted-cost control problem is to find the optimal policy $\mathbf{u} \in \Omega$ by minimizing the infinite horizon expected discounted cost (depending on the initial state).

$$J\left(\mathbf{x}_0\right) = \min_{\mathbf{u}} E\left[\int_0^\infty e^{-\beta t} g\left(\mathbf{x}(t), \mathbf{u}(t)\right) dt \, | \mathbf{x}(0) = \mathbf{x}_0\right]. \qquad (13.1)$$

Assuming the uniform transition rate being v and letting $\mathrm{Prob}\{\mathbf{y} \mid (\mathbf{x}, \mathbf{u})\}$ represent the one-step transition probability from state $\mathbf{x}$ into $\mathbf{y}$ under the control $\mathbf{u}$, then the Bellman optimality equation is given by

$$J\left(\mathbf{x}_0\right) = \frac{1}{\beta + v} \min \left\{g\left(\mathbf{x}, \mathbf{u}\right) + \sum \mathrm{Prob}\left\{\mathbf{y} \mid (\mathbf{x}, \mathbf{u})\right\} \cdot J\left(\mathbf{y}\right)\right\}. \qquad (13.2)$$

The standard value iteration algorithm for the discounted-cost situations is given below.

Proposition 13.1. *Specify the maximum iteration number K and the error allowance ε which is a small positive number. Let k denote the iteration number:*

Step 1: Compute the one-step transition probability $\mathrm{Prob}\{\mathbf{y} \mid (\mathbf{x}, \mathbf{u})\}$.
Step 2: Set $k = 0$ and $J_0(\mathbf{x}) \equiv 0$ for any $\mathbf{x} \in X$.
Step 3: Compute $J_{k+1}(\mathbf{x}) = \min\{g(\mathbf{x}, \mathbf{u}) + \sum \mathrm{Prob}\{\mathbf{y} \mid (\mathbf{x}, \mathbf{u})\} \cdot J_k(\mathbf{y})\}/(\beta + v)$.
Step 4: Set $\delta = |J_{k+1}(0) - J_k(0)|$. If $\delta < \varepsilon$ or $k > K$, go to Step 6.
Step 5: Replace k by $k + 1$, and go to Step 3.
Step 6: Output $J_k(0)$ and a stationary policy realizing the minimization of $\{g(\mathbf{x}, \mathbf{u}) + \sum \mathrm{Prob}\{\mathbf{y} \mid (\mathbf{x}, \mathbf{u})\} \cdot J_k(\mathbf{y})\}$.

13.2.1   Optimization of Threshold Values via Value Iteration Method

The above value iteration algorithm provides a numerical method to approximate the optimal cost function and the optimal control policy. By redefining the admissible control set being the specific threshold-type policy, for example, $\Omega = \{u_\theta\}$, where u_θ is a specific threshold policy characterized by a set of threshold parameters θ, then we can similarly compute the discounted-cost function under the policy u_θ.

Proposition 13.2. *Specify the maximum iteration number K and the error allowance ε which is a small positive number. Let k denote the iteration number:*

Step 1: Compute the one-step transition probability Prob$\{y \mid (x, u_\theta)\}$.
Step 2: Set $k = 0$ and $J_0(x) \equiv 0$ for any $x \in X$.
Step 3: Compute $J_{k+1}(x) = \{g(x, u_\theta) + \sum \text{Prob}\{y \mid (x, u_\theta)\} \cdot J_k(y)\}/(\beta + v)$.
Step 4: Set $\delta = |J_{k+1}(0) - J_k(0)|$. If $\delta < \varepsilon$ or $k > K$, go to Step 6.
Step 5: Replace k by $k + 1$, and go to Step 3.
Step 6: Output $J_k(0)$ as the discounted cost under the threshold policy u_θ.

Proposition 13.2 provides a numerical algorithm to approximate the discounted cost under a given threshold policy. The optimal threshold control policy for the discounted-cost case can be obtained from the following parameter optimization algorithm:

Proposition 13.3. *The optimal threshold policy for the discounted-cost case can be determined by $\theta^* = argmin\{J(u_\theta)\}$, where $J(u_\theta)$ can be evaluated using Proposition 13.2.*

13.2.2   Application and Computational Performance

This section discusses the application of the algorithm in Propositions 13.2 and 13.3 and its computational performance. Two sets of examples are presented. The first set is to optimize the linear switching threshold control policy in Chap. 8, and the second set is to optimize the threshold control policies for the supply chains with assembly operations in Chap. 11.

13.2.2.1   Optimization of the Linear Switching Threshold Policy in Chap. 8

The linear switching threshold control policy shown in Fig. 8.2 is characterized by six threshold parameters, that is, (i_1, j_1), (i_2, j_2), (i_3, j_3), and (i_4, j_4). From the analytical results in Chap. 2, we know that $i_3 = 1$ and $i_2 = x''_1$, which can be analytically determined by Proposition 2.7. Moreover, these threshold parameters have the following structural relationships, for example, $0 \le i_1 \le x''_1$, $j_2 \le j_1 \le N$, $1 \le i_4 \le i_2$, and $0 \le j_3 \le j_4 \le N$.

Table 13.1 CPU time in seconds of Algorithm 13.1 with different M

Case	1	2	3	4	5	6	7	8	9	10
$M = 6$	83	81	81	34	34	35	30	34	34	34
$M = 8$	81	81	81	34	35	35	32	36	35	35
$M = 10$	82	82	81	35	34	35	30	34	34	34

The key part of the threshold parameter optimization algorithm can be explained as follows (define $\theta := (i_1, j_1, j_2, j_3, i_4, j_4)$).

Algorithm 13.1: Optimization of the Linear Switching Threshold Policy in Sect. 8.2 for the Basic Supply Chain System

Set $i_3 = 1$, and calculate $i_2 = x''_1$; let $\theta = (i_1, j_1, j_2, j_3, i_4, j_4)$;
   For $i_1 \in [0, i_2]$,
    for $j_1 \in [0, N]$,
     for $j_2 \in [-N, j_1]$,
      for $i_4 \in [i_3, i_2]$,
       for $j_3 \in [0, N]$,
        for $j_4 \in [j_3, N]$,
         Evaluate the linear switching threshold policy $\mathbf{u}_\theta$.
         Update the best solution so far.
   Return the best solution and cost.

It should be pointed out that in the above algorithm, j_2 could take any negative integer in theory. However, experiments show that it rarely falls below $-N$. Note that the parameter i_2 is determined analytically. If it is optimized together with other parameters (i.e., without utilizing the analytical knowledge), the computational time of the above algorithm would be at least $(M + 1)/2$ times more since it could take any value in $[1, M]$. In the example below, we give the computational performance of the above algorithm.

Example 13.1 Consider the ten cases in Table 8.1. Suppose Algorithm 13.1 and Proposition 13.2 are applied to optimize the threshold parameters (on a PC with 1.86-GHz processor), in which the maximum iteration number in the value iteration procedure is 200. Table 13.1 gives the CPU times in seconds required for each case. The best solutions obtained from Algorithm 10.1 for each case with different raw material warehouse capacity are the same and shown in Table 13.2, in which "% above" represents the percentage of the cost under the best linear switching threshold policy above the optimal cost. In fact, Table 13.2 has the same results as Table 8.2.

The reason that the optimal threshold values are the same for M being 6, 8, and 10 is that Algorithm 13.1 is related to $i_2 = x''_1$ rather than M, while Table 13.2 shows that i_2 is the same for $M > 6$. For the same reason, we can observe that the CPU times of the same case with different M are similar. However, for different cases, the CPU times are different because the value iteration procedure may terminate with different numbers of iterations.

Table 13.2 The best solutions from Algorithm 13.1 for $M = 6$, 8, and 10

Case	% above	$(i_1, j_1) (i_2, j_2) (i_3, j_3) (i_4, j_4)$
1	0.00%	$(0, 4) (4, -4) (1, 0) (1, 0)$
2	0.11%	$(0, 3) (4, -3) (1, 1) (1, 1)$
3	0.00%	$(0, 5) (4, -3) (1, 1) (1, 1)$
4	0.12%	$(0, 1) (3, -2) (1, 0) (1, 0)$
5	0.26%	$(0, 4) (3, -2) (1, 1) (1, 1)$
6	0.00%	$(0, 4) (3, -2) (1, 1) (1, 1)$
7	0.38%	$(0, 2) (3, -4) (1, 0) (1, 0)$
8	0.03%	$(0, 3) (3, -3) (1, 1) (1, 1)$
9	0.01%	$(0, 3) (3, -3) (1, 1) (1, 1)$
10	0.00%	$(0, 4) (3, -2) (1, 2) (1, 2)$

13.2.2.2 Optimization of the Threshold Policies for an Assembly Supply Chain in Chap. 11

In Chap. 11, for the supply chain systems with assembly operations, two types of threshold control policies are presented. The first is termed Kanban threshold control (KTC) policy, and the second policy is termed as base-stock Kanban control (BKC) policy. Both of them are required to determine $n + 1$ threshold parameters, denoted as L_i, for $i = 1, 2, \ldots, n$, and H. To simplify the narrative, let $\theta = (L_1, L_2, \ldots, L_n, H)$.

Physically, L_i is regarded as the target inventory level for raw material i, and H is the target inventory level for the finished goods. Based on the discussions in Chaps. 6 and 11, the threshold parameters can be narrowed down by $L_i \geq 0$, for $i = 1, 2, \ldots, n$, and $H \geq 0$. If each raw material is constrained by the RM warehouse capacity M_i and the finished goods is constrained by the FG warehouse capacity N, the solution space could be further narrowed down to $0 \leq L_i \leq M_i$, for $i = 1, 2, \ldots, n$, and $0 \leq H \leq N$.

The key part of the threshold parameter optimization algorithm can be explained as follows ($\theta = (L_1, L_2, \ldots, L_n, H)$),

---

Algorithm 13.2: Optimization of the Threshold Control Policies in Sect. 11.2.2 for the Supply Chain Systems with Assembly Operations
for $L_1 \in [0, M_1]$,
    for $L_2 \in [0, M_2]$,
      $\cdots \cdots$
      for $H \in [0, N]$,
          evaluate the KTC or BKC threshold policy $\mathbf{u}_\theta$.
         Update the best solution so far.
    Return the best solution and cost.

Example 13.2 Consider the cases in Table 11.2. Suppose Algorithm 13.2 and Proposition 13.2 are applied to optimize the threshold parameters for KTC and BKC policies (on a PC with 1.86-GHz processor), in which the maximum iteration

Table 13.3 CPU time in seconds of Algorithm 13.2 for KTC and BKC

Case	1	2	3	4	5	6	7	8	9
Group A	2	3	2	2	2	2	2	2	3
Group B	2	2	2	2	2	2	2	2	2
Group C	67	66	57	68	61	78	61	71	92
Group D	68	68	58	69	58	68	58	69	69

number of the value iteration procedure is 100. To save the computational time, we limit the warehouse spaces by $M_i = 5$ and $N = 5$. Table 13.3 gives the CPU time in seconds required by KTC and BKC in each case.

Group A and group B have two suppliers, whereas group C and group D have three suppliers. Therefore, the required computational effort is increasing dramatically as the number of suppliers increases.

The optimization method presented in this section has two limitations. First, if there are many threshold parameters to be optimized and each of them takes values in a large interval, then the optimization procedure could be very time consuming. Secondly, if the system is complicated and has a large state space, this will make the value iteration algorithm time consuming, and the optimization procedure even more time consuming.

13.3   Optimization of Threshold Parameters in Long-Run Average Cost Situations

Consider a Markov decision process with the state space X, the admissible control set Ω, and the one-step cost function $g(\mathbf{x}, \mathbf{u})$ where $\mathbf{x} \in X$ and $\mathbf{u} \in \Omega$. The optimal long-run average cost control problem is to find the optimal policy $\mathbf{u} \in \Omega$ by minimizing the infinite horizon expected average cost (independent of the initial state):

$$J^* = \min_{\mathbf{u}} \lim_{T \to \infty} \frac{1}{T} E \int_0^T g\left(\mathbf{x}(t), \mathbf{u}(t)\right) \mathrm{d}t. \tag{13.3}$$

Assume that the uniform transition rate is v and let $\mathrm{Prob}\{\mathbf{y}|(\mathbf{x}, \mathbf{u})\}$ represent the one-step transition probability from state $\mathbf{x}$ into $\mathbf{y}$ under the control $\mathbf{u}$. We also assume that the Bellman optimality equation for the long-run average cost case holds, which can be given by

$$w(\mathbf{x}) + \frac{J^*}{v} = \frac{1}{v}\left[g(\mathbf{x}, \mathbf{u}) + \sum \mathrm{Prob}\{\mathbf{y}|(\mathbf{x}, \mathbf{u})\} \cdot w(\mathbf{y})\}\right], \tag{13.4}$$

where $w(\mathbf{x})$ is finite function defined on X. The standard value iteration algorithm for the average cost case is given below.

Proposition 13.4. *Pre-specify the maximum iteration number K and the error allowance ε which is a small positive number. Let k denote the iteration number:*

Step 1: Compute the one-step transition probability Prob $\{y \mid (x, u)\}$.
Step 2: Set $k = 0$ and $w_0(x) \equiv 0$ for any $x \in X$.
Step 3: Compute $J_{k+1}(x) = \min \{g(x, u) + \sum \text{Prob}\{y \mid (x, u)\} \cdot w_k(y)\}$.
Step 4: Set $\delta = |J_{k+1}(0) - J_k(0)|$. If $\delta < \varepsilon$ or $k > K$, go to Step 6.
Step 5: Set $w_{k+1}(x) = J_{k+1}(x) - J_{k+1}(0)$.
Step 6: Replace k by $k + 1$, and go to Step 3.
Step 7: Output $J_k(0)$ as the optimal average cost and a stationary policy realizing the right-hand side of Step 3 as the optimal control policy.

13.3.1   Optimization of Threshold Parameters via Value Iteration Method

Similar to Sect. 13.2.1, the average cost under a threshold control policy can be obtained via the value iteration procedure by redefining the admissible control set being the specific threshold-type policy, for example, $\Omega = \{u_\theta\}$, where u_θ is a specific threshold policy determined by a set of threshold parameters θ.

Proposition 13.5. *Specify the maximum iteration number K and the error allowance ε which is a small positive number. Let k denote the iteration number:*

Step 1: Compute the one-step transition probability Prob $\{y \mid (x, u_\theta)\}$.
Step 2: Set $k = 0$ and $w_0(x) \equiv 0$ for any $x \in X$.
Step 3: Compute $J_{k+1}(x) = \{g(x, u_\theta) + \sum \text{Prob}\{y \mid (x, u_\theta)\} \cdot w_k(y)\}$.
Step 4: Set $\delta = |J_{k+1}(0) - J_k(0)|$. If $\delta < \varepsilon$ or $k > K$, go to Step 6.
Step 5: Set $w_{k+1}(x) = J_{k+1}(x) - J_{k+1}(0)$.
Step 6: Replace k by $k + 1$, and go to Step 3.
Step 7: Output $J_k(0)$ as the average cost under the threshold policy u_θ.

The optimal threshold control policy for the average cost case can be obtained from the following parameter optimization algorithm.

Proposition 13.6. *The optimal threshold policy for the average cost case can be determined by $\theta^* = \text{argmin}\{J(u_\theta)\}$, where $J(u_\theta)$ can be evaluated using Proposition 13.5.*

13.3.2   Optimization of Threshold Parameters via Stationary Distribution

Assuming that the induced Markov chain of the stochastic supply chain system under a threshold control policy u_θ is stable, that is, the stationary distribution exists. Let $\pi^\theta(x) = \text{Prob}\{$the system's state stays at $x\}$ be the stationary distribution, which is independent of the initial state.

If we can obtain the stationary distribution $\{\pi^\theta(\mathbf{x})\}$ of the system state under the threshold policy $\mathbf{u}_\theta$, then the long-run average cost $J(\mathbf{u}_\theta)$ in (13.3) can be calculated as $\sum_{\mathbf{x}} \pi^\theta(\mathbf{x}) \cdot g(\mathbf{x}, \mathbf{u}_\theta)$. The optimal threshold parameters can then be found in the same way as Proposition 13.6 by replacing the $J(\mathbf{u}_\theta)$ evaluation method with the stationary distribution method.

13.3.3 Application and Computational Performance

This section discusses the application of the threshold parameters optimization via stationary distribution. A prerequisite of this method is that the stationary distribution of the system state under the threshold policy can be obtained. However, analytically deriving the stationary distribution is often difficult, particularly when the system state is infinite. Nevertheless, in some special cases, we are able to derive the stationary distribution, for example, the supply chains in Chap. 9 (Sect. 9.3), Chap. 10 (Sect. 10.3), and Chap. 12 (Sect. 12.4).

13.3.3.1 Optimization of the Threshold Parameters for a Failure-Prone Manufacturing System with Backordering Decisions in Chap. 9

In Chaps. 4 and 9, we have shown that the optimal policy for a failure-prone manufacturing supply chain with backordering decision is actually a threshold control policy, which can be characterized by three parameters, that is, l^*, m^*, n^*, and they satisfy $-\infty < l^* \le m^* \le 0 \le n^* < +\infty$.

Two methods can be used to obtain the optimal threshold parameters. The first is the value iteration method to seek the optimal control policy, in which those threshold parameters can be obtained from the optimal policy.

The second is the optimization procedure via the stationary distribution as shown in Algorithm 13.3. Let $\theta = (l, m, n)$, in which l determines whether the arriving customer demand should be accepted when the machine is up, m determines whether the arriving demand should be accepted when the machine is down, and n determines whether products should be produced. The threshold parameter optimization algorithm is given as follows ($\theta = (l, m, n)$), where the system state is limited by $-N \le x \le N$.

Algorithm 13.3: Optimization of the Threshold Policy in a Failure-Prone Manufacturing System with Backordering Decisions

For $l \in [-N, 0]$,
    for $m \in [l, 0]$,
        for $n \in [0, N]$,
            evaluate $\mathbf{u}_\theta$ via the stationary distribution.
            Update the best solution so far.
    Return the best solution and cost.

Table 13.4 CPU time in seconds to find the optimal threshold parameters in failure-prone manufacturing supply chain with backordering decisions

Case	1	2	3	4	5	6	7	8
Value iteration	0.016	0.016	0.016	0.016	0.015	0.016	0.031	0.031
Optim. via stat. distr.	3.906	3.890	3.906	3.922	3.891	3.922	3.328	4.063

Example 13.3 Consider the eight cases in Table 9.5 for the failure-prone manufacturing supply chain with backordering decisions. Let $N = 40$. The value iteration algorithm and the Algorithm 13.3 with stationary distribution are experimented on a PC with 1.86-GHz processor. Table 13.4 gives the CPU time in seconds required for each case. The best threshold parameters and the associated average costs are the same from two methods, which have been given Table 9.5.

It can be observed from Table 13.4 that the value iteration algorithm is more efficient than the Algorithm 13.3. This can be explained from two aspects. First, the Algorithm 13.3 involves an iterative searching procedure to find the best threshold parameters in a three-dimension control space. Second, the value iteration algorithm is to seek the optimal policy that is the same as the optimal threshold policy and only involves the iterations over the two-dimension state space (machine state and inventory-on-hand level). Clearly, if the optimal control policy is not of threshold structure, the optimization procedure in Algorithm 13.3 together with the value iteration evaluation would be much more time consuming than the Algorithm 13.3 with the stationary distribution.

13.3.3.2 Optimization of the Threshold Parameters for a Failure-Prone Manufacturing System with Preventive Maintenance in Chap. 10

In Chap. 10, we present two types of threshold control policies for the failure-prone supply chain systems with preventive maintenance. Both of them can be characterized by two threshold parameters, l and h with $h \geq l$. Let $\theta = (l, h)$.

The threshold parameter optimization algorithm can be described as follows ($\theta = (l, h)$), in which either the value iteration algorithm or the stationary distribution method could be used to evaluate the threshold policy $\mathbf{u}_\theta$.

Algorithm 13.4: Optimization of the Threshold Policy in a Failure-Prone Manufacturing System with Preventive Maintenance
For $h \in [0, N]$,
    for $l \in [-N, h]$,
        evaluate the two types of threshold policy $\mathbf{u}_\theta$.
        Update the best solution so far.
    Return the best solution and cost.

Table 13.5 CPU time in seconds to find the optimal threshold parameters in failure-prone manufacturing supply chain with preventive maintenance

μ	0.80	0.9	1.0	1.1	1.2
Optimization via value iteration	130.235	162.266	162.438	162.578	163.063
Optimization via stationary distr.	6.094	5.875	5.703	5.750	5.765

Example 13.4 Consider the five cases in Table 10.3 for the failure-prone manufacturing supply chain with preventive maintenance with varying demand rate. We apply Algorithm 13.4 with $N = 50$ to optimize the threshold parameters on a PC with 1.86-GHz processor, in which the maximum iteration number in the value iteration procedure is 2,000. Table 13.5 gives the CPU time in seconds required for each case. The best threshold parameters and the associated average costs are the same from two methods, which have been given Table 10.3.

This example reveals that optimizing threshold parameters using the Algorithm 13.4 via the value iteration evaluation is substantially more time consuming compared to that via the stationary distribution. In general the computational performance of different optimization methods to obtain the optimal threshold parameters in the average cost situations can be summarized as follows: (1) If the optimal control is of a threshold type, then a single value iteration algorithm may be the most efficient method to obtain the best threshold parameters; (2) if the optimal policy is not of a threshold type, but we are able to derive or approximate the stationary distribution of the induced Markov process under the threshold control policy, then a numerical optimization procedure via the stationary distribution evaluation is more efficient than via the value iteration evaluation; and (3) the value iteration algorithm is sensitive to the system stability condition. For example, as the system is close to unstable, much larger state space is required in order to evaluate the performance accurately, and the convergence rate becomes very slow.

13.4   Robustness of Threshold-Type Control Policies

Robustness of a threshold policy refers to its sensitivity to the misestimation of the system parameters. In practice, it is not always possible to obtain accurate input data. For instance, we may overestimate or underestimate the customer demand rate. Therefore, the derived best threshold policy may deviate from the actual best threshold policy. An interesting question is how far the difference will be in such cases.

Let λ denote the system input parameter, $\mathbf{u}^*_{\theta(\lambda)}$ denote the best threshold control policy obtained under the input λ, and $J(\mathbf{u}^*_{\theta(\lambda)}, \lambda)$ represent the cost under the corresponding threshold policy and parameters. Now if the actual system input value is λ', then the robustness of the threshold policy with respect to parameter λ can be measured by comparing the performance of the estimated threshold policy $\mathbf{u}^*_{\theta(\lambda)}$ and the actual best threshold policy $\mathbf{u}^*_{\theta(\lambda')}$ under the actual input parameter λ'.

Table 13.6 Robustness of threshold policies wrt the raw material lead-time rate in an assembly supply chain system

λ'_i	−20%	−10%	0%	+10%	+20%
KTC					
$\theta(\lambda)$	(2, 2, 0)	(2, 2, 0)	(2, 2, 0)	(2, 2, 0)	(2, 2, 0)
$\theta(\lambda')$	(2, 2, 0)	(2, 2, 0)	(2, 2, 0)	(2, 2, 0)	(2, 2, 0)
$J(\mathbf{u}^*_{\theta(\lambda)}, \lambda')$	42.159	40.822	39.639	38.592	37.665
$\gamma(\lambda/\lambda')$	1.000	1.000	1.000	1.000	1.000
BKC					
$\theta(\lambda)$	(3, 3, 2)	(3, 3, 2)	(3, 3, 2)	(3, 3, 2)	(3, 3, 2)
$\theta(\lambda')$	(3, 3, 2)	(3, 3, 2)	(3, 3, 2)	(3, 3, 2)	(3, 3, 2)
$J(\mathbf{u}^*_{\theta(\lambda)}, \lambda')$	41.648	40.334	39.181	38.169	37.277
$\gamma(\lambda/\lambda')$	1.000	1.000	1.000	1.000	1.000

Definition 13.1. The ratio $\gamma(\lambda/\lambda') := J(\mathbf{u}^*_{\theta(\lambda)}, \lambda') / J(\mathbf{u}^*_{\theta(\lambda')}, \lambda')$ is defined as the indicator of the robustness of the threshold policy with respect to parameter λ.

Clearly, if the ratio $\gamma(\lambda/\lambda')$ is close to one in a wider range of λ/λ', the threshold control policy is more robust with respect to λ. An example is given below to explore the robustness of some threshold policies.

Example 13.5 Consider the first scenario in Table 11.1 for the assembly supply chain. More specifically, the assembly supply chain has two types of raw materials, that is, $n = 2$, and they have the same raw material supply parameters, that is, $Q_1 = Q_2 = 3$, $\lambda_1 = \lambda_2 = 0.8$, and $c_1 = c_2 = 1$. Other system input parameters are $r = 1$, $\mu = 0.7$, $c_{n+1}^+ = 5$, $c_{n+1}^- = 20$, and $\beta = 0.5$. The system state space is limited into a finite area with $x_i \leq 10$, $x_{n+1} \in [-20, 20]$.

Table 13.6 gives the robustness indicators of KTC policy and BKC policy with respect to (wrt) the raw material lead-time rate λ, where the actual lead-time rate λ' is up to 20% below or above the estimated value.

Table 13.7 gives the robustness indicators of KTC policy and BKC policy wrt the assembly rate r, where the actual assembly rate r' is up to 20% below or above the estimated value.

Table 13.8 gives the robustness indicators of KTC policy and BKC policy wrt the demand rate μ, where the actual demand rate μ' is up to 20% below or above the estimated value.

From Tables 13.6, 13.7, and 13.8, it can be seen that both KTC and BKC threshold policies are fairly robust to the misestimation of the system parameters such as the raw material lead-time rate, the assembly rate, and the customer demand rate.

13.5 Discussion and Notes

According to the way of evaluating the performance of a given threshold policy, this chapter presents two numerical approaches to optimize threshold control parameters. The first is based on the stochastic dynamic programming approach

Table 13.7 Robustness of threshold policy wrt the assembly rate in an assembly supply chain system

r'	-20%	-10%	0%	$+10\%$	$+20\%$
KTC					
$\theta(r)$	(2, 2, 0)	(2, 2, 0)	(2, 2, 0)	(2, 2, 0)	(2, 2, 0)
$\theta(r')$	(2, 2, 1)	(2, 2, 0)	(2, 2, 0)	(2, 2, 0)	(2, 2, 0)
$J(\mathbf{u}^*_{\theta(r)}, r')$	42.057	40.773	39.639	38.634	37.741
$\gamma(r/r')$	1.0004	1.0000	1.0000	1.0000	1.0000
BKC					
$\theta(r)$	(3, 3, 2)	(3, 3, 2)	(3, 3, 2)	(3, 3, 2)	(3, 3, 2)
$\theta(r')$	(3, 3, 2)	(3, 3, 2)	(3, 3, 2)	(3, 3, 2)	(3, 3, 2)
$J(\mathbf{u}^*_{\theta(r)}, r')$	41.915	40.461	39.181	38.052	37.054
$\gamma(r/r')$	1.0000	1.0000	1.0000	1.0000	1.0000

Table 13.8 Robustness of threshold policy wrt the demand rate in an assembly supply chain system

μ'	-20%	-10%	0%	$+10\%$	$+20\%$
KTC					
$\theta(\mu)$	(2, 2, 0)	(2, 2, 0)	(2, 2, 0)	(2, 2, 0)	(2, 2, 0)
$\theta(\mu')$	(2, 2, 0)	(2, 2, 0)	(2, 2, 0)	(2, 2, 1)	(2, 2, 1)
$J(\mathbf{u}^*_{\theta(\mu)}, \mu')$	31.077	35.256	39.639	44.200	48.914
$\gamma(\lambda/\lambda')$	1.0000	1.0000	1.0000	1.0004	1.0030
BKC					
$\theta(\mu)$	(3, 3, 2)	(3, 3, 2)	(3, 3, 2)	(3, 3, 2)	(3, 3, 2)
$\theta(\mu')$	(3, 3, 1)	(3, 3, 2)	(3, 3, 2)	(3, 3, 2)	(3, 3, 2)
$J(\mathbf{u}^n_{\theta(\mu)}, \mu')$	30.953	34.967	39.181	43.576	48.129
$\gamma(\mu/s')$	1.0070	1.0000	1.0000	1.0000	1.0000

(e.g., the value iteration algorithm), and the second is based on the steady-state probability distribution. The stochastic dynamic programming can be used to tackle the optimal control problem for both discounted-cost situation and long-run average cost situation (e.g., Bertsekas 1987; Puterman 1994; Sennott 1999). It is a natural extension to use the dynamic programming approach to optimize threshold control parameters iteratively. However, there are two challenges in this optimization method.

The first challenge is the curse of dimensionality in terms of the system state space. That is, as the dimension of the state space increases (e.g., greater than 3), numerical methods such as the value iteration algorithm become computationally expensive. This is an inherent issue in standard dynamic programming. More recently, techniques such as approximate dynamic programming (Si et al. 2004; Powell et al. 2004; Topaloglu and Powell 2006; Powell 2007) and reinforcement learning (Bertsekas and Tsitsiklis 1996; Sutton and Barto 1998) have been developed to estimate the optimal cost function and approximate the optimal policy by combining the intelligence of optimization with the flexibility of simulation.

Aggregation and decomposition are common ways to reduce the size of the state space. For example, Van Houtum and Zijm (1991) decomposed a multi-echelon production system under echelon base-stock policies into one-dimensional problems so that optimal order-up-to levels for all stages can be quickly determined.

The second challenge is the optimization of multiple threshold parameters. Clearly, as the number of threshold parameters increases, the solution space is increasing exponentially, which is computational difficult for enumerative search methods. This may be called the curse of dimensionality in terms of the solution space.

Steady-state probability distribution is a piece of useful information, which can easily yield many interesting steady-state performance measures, such as average cost, resource utilization, average inventory level, customer service level, and out-of-stock probability. Based on the explicit or implicit relationships between the steady-state (stationary) distribution and the control parameters, analytical or numerical methods can then be developed to evaluate the threshold policies and optimize the corresponding threshold parameters.

There is a rich literature on deriving and estimating the stationary distributions for stochastic production/inventory systems. For example, Johansen (2005) considered a single-item inventory system with Erlangian lead times and Poisson demand under the control of a base-stock policy. The optimal single base-stock level is computed by minimizing the long-run average cost, which can be expressed from the steady-state distribution. Haji et al. (2011) considered a two-level inventory system consisting of one supplier and one retailer facing a Poisson demand. Under the base-stock policy for both supplier and retailer, the steady-state distribution is derived and analyzed to optimize the control parameters.

In the aspect of failure-prone manufacturing systems, Berg et al. (1994) considered a failure-prone production system with N identical machines producing the same type of item continuously under a single threshold control. They established the stationary distribution of the inventory process and computed various performance measures such as service level, expected inventory level, machines' utilization, and repairmen utilization. Liu and Cao (1999) considered a failure-prone manufacturing system under a two-critical-number policy to meet compound Poisson process of demand arrivals. An expression of steady-state distribution is derived and used to determine the optimal two threshold parameters. Das and Sarkar (1999) considered a production–inventory system with preventive maintenance and lost sales. A single threshold parameter is used to control production and two threshold parameters to control maintenance. Numerical search methods are used to find the optimal control parameters based on the steady-state distribution and the performance measure.

In the aspect of assembly systems, Avsar et al. (2009) considered an assembly system with Poisson demand arrivals and exponential service times under the base-stock control on the continuous review basis. Through partial aggregation, the steady-state probability distribution of this approximate model is shown to be a product-form distribution. The approximate steady-state distribution is used to optimize the base-stock levels. Topan and Avsar (2011) extended the work in Avsar et al. (2009) to a Kanban controlled two-stage assembly system.

In the aspect of multiproduct inventory systems, Zheng and Zipkin (1990) considered a single facility producing two products to meet independent Poisson demands with the same arriving rate. The production of two products is controlled by a single threshold parameter with the lower inventory level product having the priority. The steady-state distributions of the inventory levels under the given threshold parameter is obtained. Zhao and Lian (2011) considered an exponential service inventory system with two classes of customers following Poisson arriving processes. Under the (r, Q) ordering policy, the steady-state probability distribution is obtained.

Similar to the first approach, there are two challenges to use the steady-state distribution-based optimization approach. The first is the derivation of the steady-state distribution. This often tends to be difficult when the systems become more complicated. In other words, we may not be able to obtain the explicit or even implicit expression of the stationary distribution for many practical systems. The second is the same as the one of the first approach, that is, the computational complexity of optimizing multiple threshold parameters. For any given combination of threshold parameters, we have to recalculate the stationary distribution and the relevant performance measures. In the above literature using the second approach, only one, two, or at most three threshold parameters are required to be optimized.

References

Avsar, Z.M., Zijm, W.H., Rodoplu, U.: An approximate model for base-stock-controlled assembly systems. IIE Trans. **41**(3), 260–274 (2009)

Berg, M., Posner, M.J.M., Zhao, H.: Production-inventory systems with unreliable machines. Oper. Res. **42**(1), 111–118 (1994)

Bertsekas, D.P.: Dynamic Programming: Deterministic and Stochastic Models. Prentice-Hall, Englewood Cliffs (1987)

Bertsekas, D.P., Tsitsiklis, J.N.: Neuro-Dynamic Programming. Athena Scientific, Belmont (1996)

Das, T.K., Sarkar, S.: Optimal preventive maintenance in a production inventory system. IIE Trans. **31**, 537–551 (1999)

Haji, R., Haji, A., Saffari, M.: Queueing inventory system in a two-level supply chain with one-for-one ordering policy. J. Ind. Syst. Eng. **5**(1), 337–347 (2011)

Johansen, S.G.: Base-stock policies for the lost sales inventory system with Poisson demand and Erlangian lead times. Int. J. Prod. Econ. **93–94**(8), 429–437 (2005)

Liu, B., Cao, J.: Analysis of a production-inventory system with machine breakdowns and shutdowns. Comput. Oper. Res. **26**(1), 73–91 (1999)

Powell, W.B.: Approximate Dynamic Programming: Solving the Curses of Dimensionality. Wiley, Hoboken (2007)

Powell, W.B., Ruszczynski, A., Topaloglu, H.: Learning algorithms for separable approximations of stochastic optimization problem. Math. Oper. Res. **29**(4), 814–836 (2004)

Puterman, M.L.: Markov Decision Processes: Discrete Stochastic Dynamic Programming. Wiley, New York (1994)

Sennott, L.I.: Stochastic Dynamic Programming and the Control of Queueing Systems. Wiley, New York (1999)

Si, J., Barto, A., Powell, W.B., Wunsch, D.: Learning and Approximate Dynamic Programming: Scaling up to the Real World. Wiley, New York (2004)

Sutton, R.S., Barto, A.G.: Reinforcement Learning: An Introduction. The MIT Press, Cambridge (1998)

Topaloglu, H., Powell, W.B.: Dynamic programming approximations for stochastic, time-staged integer multicommodity flow problems. Informs J. Comput. **18**(1), 31–42 (2006)

Topan, E., Avsar, Z.M.: An approximation for Kanban controlled assembly systems. Ann. Oper. Res. **182**(1), 133–162 (2011)

Van Houtum, G.J., Zijm, W.H.M.: Computational procedures for stochastic multi-echelon production systems. Int. J. Prod. Econ. **23**(1–3), 223–237 (1991)

Zhao, N., Lian, Z.T.: A queueing-inventory system with two classes of customers. Int. J. Prod. Econ. **129**(1), 225–231 (2011)

Zheng, Y., Zipkin, P.: A queueing model to analyze the value of centralized inventory information. Oper. Res. **38**(2), 296–307 (1990)

Chapter 14
Optimization of Threshold Control Parameters via Simulation-Based Methods

14.1 Introduction

To optimize the parameters of threshold control policies, a key step is to evaluate the performance of the threshold policy in the underlying supply chain system. For many realistic supply chain systems, it is difficult to analytically or numerically obtain such performance. Simulation is often used as an alternative due to its powerful ability and flexibility of modeling.

In computer simulation, it can be based on discrete time and discrete event. The former implies equal-size time steps to update the dynamic states of the system. The latter models the operations of a dynamic system driven by a chronological sequence of events. Discrete event means that the time advances until the next event can occur. Therefore, time steps during which nothing happens are skipped, and the system states are only updated when an event occurs. In stochastic supply chain systems, many activities may have random durations, and the evolution of the dynamic system is naturally event driven. Therefore, we will use discrete-event simulation to model stochastic supply chain systems.

Simulation-based optimization is an emerging field that integrates optimization techniques into simulation analysis. Contemporary simulation-based optimization methods include response surface methodology, stochastic approximation, and meta-heuristic methods. In particular, simulation-based meta-heuristic methods have been developed and applied extensively in the last three decades.

Although meta-heuristic methods such as simulated annealing and evolutionary algorithms were originally formulated to address the optimization of designs/solutions in deterministic environments, they can be adapted to optimize stochastic problems (Song 2001; Fu 2002). Optimization in stochastic domains involves two important parts: generating candidate solutions and quantitatively estimating their performance. The key difference between deterministic and stochastic optimization is that the precise evaluation of the objective function is computationally more costly in the stochastic domain due to the extensive simulations required to measure the performance of a solution (Fu 2002). This difficulty has been well

D.-P. Song, *Optimal Control and Optimization of Stochastic Supply Chain Systems*, Advances in Industrial Control, DOI 10.1007/978-1-4471-4724-4_14,
© Springer-Verlag London 2013

recognized. For example, Banks et al. (2000, p. 488) stated: *"optimization via simulation adds an additional complication because the performance of a particular design cannot be evaluated exactly, but instead must be estimated. Because we have estimates, it may not be possible to conclusively determine if one design is better than another, frustrating optimization algorithms that try to move in improving directions. In principle, one can eliminate this complication by making so many replications, or such long runs, at each design point that the performance estimate has essentially no variance. In practice, this could mean that very few alternative designs will be explored due to the time required to simulate each one."*

In this chapter, we will first discuss the application of the event-driven simulation to evaluate performance measure in our context. Then, we discuss the simulation-based optimization methods focusing on two typical meta-heuristics: genetic algorithms and simulated annealing. We take the assembly supply chain system discussed in Chap. 6 and the KTC and BKC threshold policies presented in Chap. 11 as examples to demonstrate the application of these two simulation-based meta-heuristics. Afterward, we present the concept of ordinal optimization technique, which can be used to reduce the computational burden in the optimization procedure for stochastic systems. This chapter ends with some notes on relevant issues.

14.2 Performance Evaluation Through an Event-Driven Simulation Model

For the assembly supply chain shown in Chap. 6 (Fig. 6.1), the system state space can be described by $X = \{\mathbf{x} := (x_1, x_2, \ldots, x_n, x_{n+1}) \,|\, x_i \geq 0 \text{ and } x_{n+1} \in Z\}$. The manufacturer needs to make two types of decisions: the assembly rate $u(t)$ $\in \{0, r\}$ and the raw material order quantity $q_i(t) \in [0, Q_i]$ for $i = 1, 2, \ldots, n$. Under a given control policy $\mathbf{u}$, the long-run average cost (independent of the initial state) is given by

$$J(\mathbf{u}) = \lim_{T \to \infty} \frac{1}{T} E \int_0^T g(\mathbf{x}(t), \mathbf{u}(t)) \mathrm{d}t. \tag{14.1}$$

Let ω_k represent the kth sample process of the system and $J(\mathbf{u}, \omega_k)$ be the kth sample cost under the control policy $\mathbf{u}$. The long-run average cost can be rewritten as

$$J(\mathbf{u}) = \lim_{K \to \infty} \frac{1}{K} \sum_{k=1}^{K} J(\mathbf{u}, \omega_k). \tag{14.2}$$

$$J(\mathbf{u}, \omega_k) = \lim_{T \to \infty} \frac{1}{T} \int_0^T g(\mathbf{x}(t), \mathbf{u}(t), \omega_k) \mathrm{d}t. \tag{14.3}$$

In order to evaluate the sample cost under a given control policy $\mathbf{u}$, the event-driven simulation is used. Note that the planning time T should be finite in simulation. Practically, a sufficient large T is often selected in advance, which represents the planning horizon.

There are three types of events in the system, that is, demand arrivals, assembly operation completions, and arrivals of the placed order for any type of raw materials. From the supply chain entity's perspective, each raw material supplier has two states: "delivering the fulfilled order to the manufacturer" (if $q_i(t) > 0$) and "waiting for a new order from the manufacturer" (if $q_i(t) = 0$). The manufacturer has two states: "assembling the finished goods" (if $u(t) > 0$) and "stop producing finished goods" due to managerial decisions or unavailability of any raw materials ($u(t) = 0$). Alternatively, the state of the manufacturer can be represented by the following: The assembly machine is "busy" (if $u(t) > 0$) or "idle" (if $u(t) = 0$). The customer has two states: "customer demand not arrived" and "customer demand arrived."

The evolution of the system state and the active event list in the simulation is briefly described as follows. Initially, set the current event occurring epoch to be zero, that is, currentEventEpoch $= 0$. Suppose that the current decision is to order raw materials from suppliers with order size q_i and does not perform assembly operation. Then, the active event list consists of all the raw material arrival events and the customer demand arrival event. This implies that all suppliers are in a state of delivering fulfilled orders toward the manufacturer and the assembly machine is idle. The next event epoch is determined by the time corresponding to the first occurring event in the active event list, denoted by nextEventEpoch. As the system state remains the same in the period from the current event epoch to the next event epoch, the incurred cost during this period can be calculated. The next event will trigger a transition of the system state at nextEventEpoch as follows:

- If the next event is the arrival of raw material j, then the inventory-on-hand of raw material x_j will increase by q_i units. The state of supplier j becomes "waiting for a new order from the manufacturer."
- If the next event is the arrival of a customer demand, then the inventory-on-hand of finished goods x_{n+1} will decrease by one unit (representing either a finished goods is used to meet the demand or the demand is backordered). The state of the customer becomes "customer demand arrived."

Then, the currentEventEpoch is updated by nextEventEpoch, and the next event is removed from the active event list. Based on the current system states, a set of new control decisions is determined. For example, if the new decision decides to perform an assembly operations (it implies that all types of raw materials must be available), then a new event "assembly operation completion" is added to the active event list, and the assembly machine's state is changed from "idle" to "busy." The actual assembly time is generated following a known probability distribution. Check the states of suppliers and customer; if a supplier's state is "waiting for a new order from the manufacturer" and the current decision is to place a new order from this supplier, then a corresponding new raw material arrival event is added to the active event list and its state becomes "delivering the fulfilled order to the manufacturer";

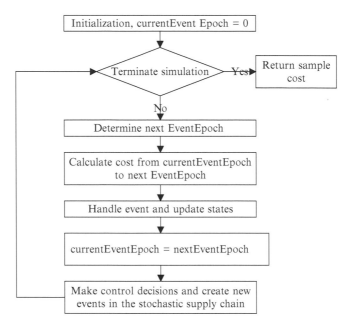

Fig. 14.1 The flow chart of the simulation model

if the customer's state is "customer demand arrived," then a new customer demand arrival event is added to the active event list and its state is changed to "customer demand not arrived." It should be pointed out that for the new events their occurring epochs are generated following the corresponding probability distributions.

The above process keeps progressing until the termination criteria are satisfied, for example, it reaches to the maximum planning time, or the sample cost has converged. The flow chart of the event-driven simulation model to evaluate a single sample cost $J(\mathbf{u}, \omega_k)$ is illustrated in Fig. 14.1.

It should be pointed out that in order to evaluate the performance of a threshold policy in the stochastic situations, multiple samples or replications are required to do averaging. By repeating the evaluation procedure in Fig. 14.1 K times with different samples, the cost function defined in (14.1) for a given threshold control policy, $J(\mathbf{u})$, can be estimated by averaging over K replications if the number of samples, K, is sufficiently large.

14.3   Simulation-Based Optimization Methods

Among all groups of simulation-based optimization methods, meta-heuristic search methods are probably the most popular methods in the last two decades. Typical examples of meta-heuristic search methods include simulated annealing (Kirkpatrick

Fig. 14.2 The flow chart
of GA optimization

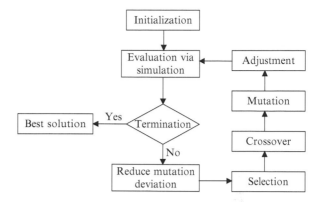

et al. 1983), tabu search (Glover 1989), genetic algorithms (Holland 1975; Goldberg 1989), ant colony optimization algorithms (Dorigo et al. 1996), particle swarm optimization algorithm (Eberhart and Kennedy 1995), and artificial immune system (Luh et al. 2003). Most of these methods have been extensively studied, in particular, on the applications to real-world combinatorial optimization problems.

In the remainder of this section, we apply two popular meta-heuristic methods, genetic algorithm (i.e., evolutionary strategy) and simulated annealing, to the assembly supply chain system to optimize threshold control parameters.

14.3.1   Genetic Algorithms (Evolutionary Strategy)

Genetic algorithms (GA) and evolutionary strategy (ES) are random search methods that take over the principle of biological evolution for the optimization of technical systems (Goldberg 1989; Back 1996; Aytug et al. 2003). Traditionally, GA uses binary string to represent a solution, whereas ES uses real variable to represent a solution. GA uses the term "crossover" while ES uses "recombination" to represent the genetic operation of swapping genes between parents. ES focuses more on the mutation operation with a mechanism to reduce mutation deviation gradually in the generation procedure. However, recent development in the last decade enables both of them to use real (or integer) variables to code solutions and suitable for parameter optimization. There is no essential difference between them. Therefore, we do not distinguish GA and ES in this book and simply use the term GA.

In our problem, a solution can be naturally represented by a vector of threshold parameters, for example, $\mathbf{S} := (S_1, S_2, \ldots, S_N)$, where N is the number of threshold parameters to be optimized. The GA procedure employed in our work is illustrated in Fig. 14.2. The GA optimization procedure involves a few control parameters, which are introduced in the following subsections.

14.3.1.1 Initialization

To conduct the GA search procedure, the following GA parameters have to be specified in the initialization step:

- GA_Gen – the maximum number of generation
- GA_μ – the number of solutions in the parent population
- GA_λ – the number of solutions in the offspring population
- $GA\_MutationProb$ – the mutation probability of each element in a solution
- GA_σ – the initial mutation deviation
- GA_N_s – the number of consecutive generations without improvement that triggers the reduction of the mutation deviation
- GA_α – the constant factor to reduce the mutation deviation
- GA_N_c – the maximum number of consecutive generations without improvement
- K – the number of samples used to estimate the expected cost
- T – the planning horizon to average the sample cost over the time

The details of the above GA parameters are explained in the following steps. Initially, a set of individuals (solutions) that is called the parent population with size GA_μ is generated randomly (or purposely). Let n denote the generation.

14.3.1.2 Selection

From the parent population, two individuals are selected randomly, which will be used to generate a descendant to be a member of the offspring population. The offspring population has a size of GA_λ, which is not smaller than GA_μ. The selection process follows the roulette wheel mechanism, for example, the parents with a higher fitness value have a higher probability to be chosen. Here the fitness of a solution can be defined as the maximum cost within the parent population minus this solution's cost. It is worth noting out that multiple parents could be selected to produce a descendant. This reflects the multi-sexual mechanism in the nature.

14.3.1.3 Crossover

A descendant is generated by randomly copying the elements from two selected parents. That is, each gene of the descendant is randomly obtained from the corresponding gene in one of its parents. The number of genes remains to be N (the number of threshold parameters). An example of global uniform crossover for two parents is shown in Fig. 14.3, where the shaded boxes represent the genes copied from parent 2.

Fig. 14.3 A global uniform
crossover in the GA
optimization

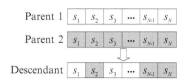

14.3.1.4   Mutation

For each descendant, its genotype may differ slightly from that of its parents. The
deviations refer to individual genes and are random and independent of each other.
For each element of a descendant, S_i, its mutation probability is pre-specified,
denoted as *GA_MutationProb*. A uniform random number $z_i \sim \sigma_n * U(-0.5, 0.5)$ is
generated, where σ_n is the mutation deviation at the nth generation and $\sigma_1 = GA\_\sigma$
is pre-specified in the initialization. The mutation operator means that with a
probability *GA_MutationProb*, the element S_i is mutated to be $S_i + z_i$ for any i.

14.3.1.5   Adjustment

After mutation, each descendant is adjusted appropriately considering the problem
context. For example, whether the threshold parameters should be nonnegative, and
whether they should satisfy specific relationships and bounds that may have been
established analytically or heuristically. The purpose of this step is to make the new
solution (descendant) feasible and also to speed up the search process by imposing
logical relationships or utilizing structural knowledge of the system.

14.3.1.6   Evaluation via Simulation

For each descendant, it is required to run K replications using the event-driven
simulation model in Fig. 14.1. The performance measure (14.1) can be approximated by averaging the costs over K replications. By the law of large numbers, the
approximated cost converges to (14.1) almost surely as K tends to infinity.

14.3.1.7   Termination Criteria

Multiple termination criteria can be defined. The most common criterion is to
specify the maximum number of generations, denoted by *GA_Gen*. Secondly, let n_c
denote the number of consecutive generations in which no improvement is achieved
for the fitness function. Let *GA_N_c* be the maximum number of n_c to be allowed
in the procedure. Thirdly, as the threshold control parameters are integers, if the
mutation deviation σ_n becomes less than 1, there is little need to further optimize
the parameters. Therefore, we assume that the optimization procedure is terminated
if $n > GA\_Gen$, or $n_c > GA\_N_c$ or $\sigma_n < 1$. The best solution up to now, denoted by
$\mathbf{S}^*$, is then returned as the "optimal" solution.

If the termination criteria are not satisfied, the best GA_μ solutions are selected from the offspring population (which has a size GA_λ) to replace the parent population in the next generation. In other words, the parent population selection operation follows the elitism principle (i.e., a policy always keeps a certain number of best solutions when each new solution is generated). When $GA\_\lambda = GA\_\mu$, we call it the standard GA; when $GA\_\lambda > GA\_\mu$, we call it the elite GA.

14.3.1.8 Reduce Mutation Deviation

The standard deviation for mutation, σ_n, is reduced by a constant factor GA_α if n_c is greater than a predetermined number (denoted by GA_N_s), and remains the same otherwise. In other words, GA_N_s is a threshold value to trigger the reduction of the standard deviation σ_n. It has been observed that this can improve the search quality significantly compared with the fixed standard deviation.

14.3.1.9 Computational Complexity

It should be pointed that the total number of runs at each generation is the product of the population size and the replication number, that is, $GA\_\lambda * K$. Take into account the fact that multiple generations, GA_Gen, are required; this gives rise to total $GA\_\lambda * K * GA\_Gen$ runs to be required in the GA procedure.

At the end of the search procedure, the "optimal" solution $\mathbf{S}^*$ provides the best threshold control policy to minimize the cost function defined in (14.1). However, it should be emphasized that multiple high-quality local optima may exist due to the combinatorial nature of the problem and GA may only find a suboptimal solution.

14.3.2 Simulated Annealing

Simulated annealing (SA) is another well-known meta-heuristic. It mimics the physical process undergone by misplaced atoms in a metal when it is heated and then slowly cooled. The heat causes the atoms to wander randomly away from their initial positions, and the slow cooling provides them more opportunities to find configurations with lower internal energy than the initial one. The structure and principle of SA can be referred to Kirkpatrick et al. (1983). Figure 14.4 illustrates the flow chart of the SA optimization procedure that is used in this chapter.

The main steps of the SA optimization procedure include the following: initialization, inner neighborhood search, terminate inner neighborhood search, cooling temperature and reduce step size, and terminate outer loop search, which are explained in the following subsections.

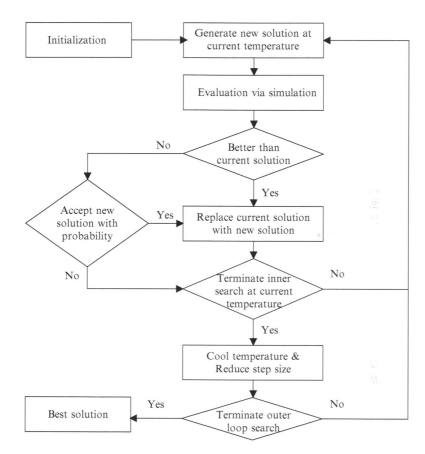

Fig. 14.4 The flow chart of the SA optimization

14.3.2.1 Initialization

In the SA search procedure, the following parameters are pre-specified in the initialization step:

- *SA_OuterNum* – the maximum number of the outer loop search
- *SA_InnerNum* – the maximum number of the inner neighborhood search
- *SA_T* – the initial temperature
- *SA_σ* – the initial neighborhood step size
- *SA_α* – the constant factor to cool the temperature
- *SA_β* – the constant factor to reduce the neighborhood step size
- *SA_N_s* – the number of the consecutive outer loops without improvement that triggers the reduction of the neighborhood step size
- *SA_N_I* – the maximum number of the consecutive inner search loops without improvement

- SA_N_O – the maximum number of the consecutive outer loops without improvement
- K – the number of samples used to estimate the expected cost
- T – the planning horizon to average the sample cost

Let $\mathbf{S}_c$ denote the current feasible solution generated randomly and $J(\mathbf{S}_c)$ be the corresponding costs evaluated via simulation.

14.3.2.2 Inner Neighborhood Search

The inner neighborhood search loop includes the following steps in Fig. 14.4: Generate new solution at current temperature; evaluate via simulation, better than current solution; accept new solution with probability; replace current solution with new solution; and terminate inner search at current temperature.

More specifically, for the current solution $\mathbf{S}_c$, we randomly generate a vector $\mathbf{u}$ with the same size of $\mathbf{S}_c$, whose each element follows a uniform distribution $U(-0.5, 0.5)*SA\_\sigma$. The new neighborhood solution is given by $\mathbf{S} = \mathbf{S}_c + \mathbf{u}$. Make appropriate adjustments to ensure that the new solution $\mathbf{S}$ is feasible. Evaluate the new solution via simulation to obtain the cost $J(\mathbf{S})$. If $J(\mathbf{S}) < J(\mathbf{S}_c)$, then accept $\mathbf{S}$ as the new solution, that is, $\mathbf{S}_c = \mathbf{S}$ and $J(\mathbf{S}_c) = J(\mathbf{S})$. On the other hand, if $J(\mathbf{S}) > J(\mathbf{S}_c)$, then accept $\mathbf{S}$ with a probability $Prob := \exp((J(\mathbf{S}) - J(\mathbf{S}_c))/SA\_T)$.

The inner neighbor search is terminated if no improvement is found after SA_N_I consecutive trials or the total number of inner trials at the current temperature exceeds $SA\_InnerNum$.

14.3.2.3 Cool Temperature and Reduce Step Size

The temperature, SA_T, is reduced by a factor SA_α whenever the inner neighborhood search is terminated. Meanwhile, the neighborhood search step size, SA_σ, will be reduced by a factor SA_β if there are more than SA_N_s consecutive outer loops without improvement. The purpose is to search new solutions with a smaller step size at a lower temperature.

14.3.2.4 Terminate Outer Loop Search

The outer loop will be terminated if one of the following termination criteria is satisfied. Firstly, the number of the outer search loop exceeds the pre-specified maximum number, $SA\_OuterNum$. Secondly, the number of the consecutive outer loops without improvement exceeds the maximum number SA_N_O. Thirdly, the neighborhood search step size becomes less than 1. After the termination of the outer loop, the best solution up to now, denoted by $\mathbf{S}^*$, is then returned as the "optimal" solution.

Table 14.1 Optimization of threshold policies for an assembly supply chain $(n = 5)$

	Threshold policy	Cost	CPU(s)	Threshold values
Standard GA	KTC	58.5949	156	(6, 6, 6, 7, 6, 4)
	BKC	53.3371	138	(5, 12, 9, 15, 14, 7)
Elite GA	KTC	57.6150	164	(6, 6, 6, 8, 6, 3)
	BKC	47.9256	165	(6, 16, 13, 13, 10, 6)
SA	KTC	61.2842	130	(6, 6, 8, 8, 6, 5)
	BKC	48.8789	113	(6, 16, 14, 18, 18, 6)

14.3.3 Numerical Examples

In this section, we present a few examples to demonstrate the applications of GA and SA to optimize threshold control parameters. We take the assembly supply chain in Chap. 6 as an example. The threshold control policies are KTC and BKC policies proposed in Chap. 11.

For the assembly supply chain in Chap. 6, when the number of raw material types is greater than 3, that is, $n > 3$, the numerical methods presented in Chap. 13 become computational difficult. For example, the value iteration-based optimization methods in Sect. 13.2 become computationally intractable due to the curse of dimensionality and the complexity of combinatorial optimization, whereas the stationary distribution-based methods become inapplicable due to the difficulty of deriving the stationary distribution and the complexity of combinatorial optimization. The simulation-based meta-heuristics are probably the best alternatives to tackle such complex combinatorial optimization problems.

For both GA and SA, there is a set of control parameters that are required to be pre-specified. A common way to select these control parameters is through pilot experiments. In our context, the GA control parameters are set as follows: $GA\_Gen = 200$; $GA\_\lambda = 40$; $GA\_\mu = 20$ (for the elite GA); $GA\_MutationProb = 0.5$; $GA\_\sigma = 10$; $GA\_N_s = 5$; $GA\_\alpha = 0.95$; $GA\_N_c = 100$; and $K = 100$. The SA control parameters are set as follows: $SA\_OuterNum = 200$; $SA\_InnerNum = 1{,}000$; $SA\_T = 10$; $SA\_\sigma = 10$; $SA\_\alpha = 0.98$; $SA\_\beta = 0.95$; $SA\_N_I = 30$; $SA\_N_O = 100$; and $K = 100$. The simulation-based optimization experiments are performed on a PC with 3.00-GHz processor.

Example 14.1. Consider an assembly supply chain in Fig. 7.1 with five types of raw materials, that is, $n = 5$. Other system parameters are set as follows: The demand rate $\mu = 0.7$; the inventory unit cost for finished goods $c_{n+1}^+ = 5$; the backlog cost $c_{n+1}^- = 20$; the maximum raw material order quantity $Q_i = 3$ for $i = 1, 2, \ldots, 5$; the lead-time rate for replenishing raw materials λ_i is randomly generated in the interval $[0.6, 1.0]$; and the raw material inventory unit cost c_i is randomly generated in the interval $[0, 2.0]$. The sample average cost is calculated over a period $T = 100$. The results of the standard GA, the elite GA, and the SA methods with computational time (CPU time in seconds) are summarized in Table 14.1.

Table 14.2 Optimization of threshold policies for an assembly supply chain ($n = 10$)

	Threshold policy	Cost	CPU(s)	Threshold values
Standard	KTC	81.5039	261	(8, 6, 7, 8, 5, 8, 6, 8, 10, 8, 4)
GA	BKC	71.9091	290	(10, 5, 8, 15, 6, 9, 18, 7, 18, 13, 7)
Elite	KTC	77.2103	390	(6, 5, 7, 8, 7, 8, 5, 10, 10, 6, 5)
GA	BKC	65.7504	399	(12, 10, 16, 13, 11, 20, 17, 14, 7, 5, 6)
SA	KTC	82.8386	234	(7, 6, 8, 6, 5, 9, 10, 10, 8, 5, 4)
	BKC	73.6383	215	(16, 11, 8, 20, 10, 19, 20, 17, 7, 10, 8)

It can be observed that three optimization methods achieve reasonably close "optimal" costs with the elite GA finding the best solutions for both KTC and BKC. This indicates that the use of the elitism selection policy for parent generation does improve the search quality. The found "optimal" BKC policy outperforms the "optimal" KTC policy quite significantly for all three optimization methods. This is in agreement with the analytical and numerical results in Chap. 11. In terms of the computational time, SA uses slightly less CPU time compared to GA, while the elite GA uses a bit more CPU time than the standard GA.

Example 14.2. Consider an assembly supply chain in Fig. 7.1 with ten types of raw materials, that is, $n = 10$. Other system parameters are set in the same way as Example 14.1. The results of the standard GA, the elite GA, and the SA methods are summarized in Table 14.2.

Generally, the similar findings to that from Table 14.1 can be observed with respect to the relative performance of three search methods, two threshold policies, and the computational time. Although the dimension of the solution space in this example almost doubles compared to Example 14.1, the required CPU time only increases by 67–142%. The reason is that the computational time of GA and AS is mainly determined by the total number of solution evaluations, which may not change dramatically as the complexity of the system increases, even though the solution space is increasing exponentially. However, it should also be pointed out that meta-heuristics such as GA and SA are often ending with suboptimal solutions.

In practice, a good balance between the solution quality and the computational time should be achieved when using meta-heuristic optimization methods such as GA and SA. However, this may be problem specific and is also related to the setting of the control parameters in GA and SA. How to select appropriate control parameters in GA and SA has been an ongoing issue. Nevertheless, it is fair to say that the simulation-based meta-heuristics do provide with an alternative to be able to find reasonably good solutions within affordable computational time for many complex optimization problems.

14.4 Ordinal Optimization Technique

This section introduces another optimization approach, called ordinal optimization (OO). Ordinal optimization was first proposed by Ho et al. (1992) and has been successfully applied to many stochastic optimization problems.

14.4.1 The Concept of Ordinal Optimization

The basic idea of ordinal optimization is simple. Instead of trying to find the best solution, ordinal optimization (OO) concentrates on finding a "good enough" solution requiring a significantly reduced computation time. By doing that, OO can give a measurement of the quality of the best solution that has been found.

Let Θ denote the entire set of feasible solutions and assume that the solutions have been ordered with respect to their performance measure from best to worst. Let G denote the set of good feasible solutions, and any solution in G is a satisfactory solution. For example, G can consist of the top α percent solutions in Θ, that is,

$$\alpha = \frac{|G|}{|\Theta|} \tag{14.4}$$

where $|.|$ denotes the size of the set. If we randomly sample m solutions from the entire solution space, the probability that none of these solutions is in G is $(1 - \alpha)^m$. Therefore, the probability that at least one of the m solutions in G is given by $1 - (1 - \alpha)^m$. If this probability is desired to be no less than P_{sat}, then we have

$$m \geq \ln (1 - P_{sat}) / \ln(1 - \alpha) \tag{14.5}$$

By (14.5), we can create Table 14.3, which shows the minimum required values of m for different values of P_{sat} and α. For example, if at least one of the m solutions is desired to be in the top 0.1% of the solution space with no less than 99% probability, then m should be at least 4,603. Note this number is independent of the size of the entire solution space Θ.

More importantly, the numbers in Table 14.3 are substantially smaller than the size of the solution space in most optimization problems. Therefore, OO enables us to quickly find a good solution with a high probability even with a blind sampling scheme. In many practices, we may have some knowledge of the direction of

Table 14.3 Required number of solution samples for applying OO

P_{sat}, α	Top 0.1%	Top 0.01%
99	4,603	46,050
99.9	6,905	69,075
99.99	9,206	92,099
99.999	11,508	115,124

sampling. Such knowledge can improve the sampling process and facilitate us to obtain better results with the same number of solution samples. However, it should be pointed out that the set of "good" solutions, for example, top 0.1% solutions, could still be a huge set, and some of its solutions may be far away from the optimal.

14.4.2 An OO-Based Elite GA

It has been noticed that by applying meta-heuristic methods to stochastic situations, the iterative search procedure means that simulations are done repeatedly to obtain a measure of performance. It is this multiplicative effect of the number of iterations in the search procedure and the number of replications in each evaluation that incurs an expensive computational burden. This section explains that by integrating OO with the elite GA appropriately, more efficient algorithms can be developed, which are able to find "optimal" solution with significantly reduced computational time in stochastic situations (Song et al. 2006).

One of the key tenets in ordinal optimization is that "it is much easier to determine whether or not alternative A is better than B than to determine $A - B = ?$" (Ho 1999, p. 173). In other words, it is much easier to estimate approximate relative order than precise (absolute) value. This has been termed the *order versus value* tenet. More details and explanations of this tenet can be found in Ho (1999) and references therein.

In the elite GA procedure, a key step is to select the elite group of solutions from the offspring population to serve as the parent population in next generation. This relies on the evaluation through simulation of all solutions in the offspring population. The order versus value tenet speeds up the procedure to identify the elite group of solutions. For example, instead of evaluating all solutions using K samples, their performance is only estimated approximately, by using a very small number k (much less than K) of samples. Based on the approximate performance, a number of "best" solutions can then be selected and further evaluated using K samples. It is postulated that such a selection scheme will keep a high proportion of the actual elite group of solutions, but with significantly less computational time.

More specifically, define G as a set of the "good" solutions, called the elite set which would be generated by using K samples and S as a set of the selected "good" solutions based on approximate estimation by averaging over k samples.

To illustrate the application of the order versus value tenet in this context, a simple numerical experiment is performed following Ho (1999). Consider a population (with size 100) of solutions ($\theta_1, \theta_2, \ldots, \theta_{100}$) ordered by performance $J(\theta_i)$. Assume that the true performance values of $J(\theta)$ are linearly increasing from 1 through 100, that is, $J(\theta_i) = i$ for $i = 1, 2, \ldots, 100$. Let G be the best ten solutions, that is, $G = \{\theta_1, \theta_2, \ldots, \theta_{10}\}$. If $\hat{J}(\theta)$.is an estimation of $J(\theta)$ by averaging over a sample size of k, then the estimation $\hat{J}(\theta)$ can be represented by

$$\hat{J}(\theta) = J(\theta) + (w_1 + w_2 + \cdots + w_k)/k \tag{14.6}$$

Fig. 14.5 Alignment percentage versus the size of the selected set

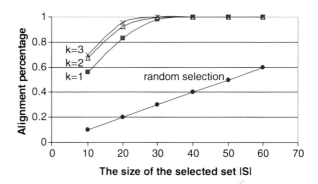

where w_i is the sample of a random noise w. It is assumed that the random noise w has a large range and follows a uniform distribution $U(-25, 25)$.

Ho (1999) shows that if θ_1 and θ_2 are two solutions with expected performances $J(\theta_1)$ and $J(\theta_2)$ and $\hat{J}(\theta_1)$ and $\hat{J}(\theta_2)$ are the estimates observed from a single sample, then

$$\text{Prob}\left\{J(\theta_2) < J(\theta_1) \mid \hat{J}(\theta_2) < \hat{J}(\theta_1)\right\} > \text{Prob}\left\{J(\theta_2) < J(\theta_1) \mid \hat{J}(\theta_1) < \hat{J}(\theta_2)\right\}. \tag{14.7}$$

Namely, if one sample is observed or estimated to have better in performance than another, then it is likely to be actually better.

An alignment percentage $|G \cap S| / |G|$, where $|G|$ refers to the size of the set G, reflects the fraction of good solutions that have been chosen in the selected set S. The set S uses the top $|S|$ solutions from the 100 small sample observations, that is, $S = \{\theta_i \mid \hat{J}(\theta_i)$ is one of the best ten observed performances$\}$. Figure 14.5 shows the alignment percentage versus the size of the selected set S for $k = 1$, 2, and 3. The results in Fig. 14.5 use (1) the alignment percentage obtained by averaging over 1,000 experiments or replications, and (2) in each experiment, Eq. (14.6) yields estimate values $\{\hat{J}(\theta_i)\}$ using $k = 1$, 2, or 3; the set S is selected based on $\{\hat{J}(\theta_i)\}$; and the corresponding alignment percentage is computed by $|G \cap S| / |G|$. To make a comparison, the random selection method is also shown in Fig. 14.5, in which the set S is selected randomly from the total 100 candidates.

Figure 14.5 shows that if just a single sample ($k = 1$) is used to estimate the performance of solutions and the "best" ten are selected, then over 50% of the good solutions in G are actually in the set S. As the size of the selected set S increases to 30, more than 90% of the good solutions are in the set S. Compared with the random selection method, the alignment percentage is greatly improved. Furthermore, if two or three samples are used to approximate the performance of the solutions and make selections, then the alignment percentage becomes over 90% when the size of S increases to 20. This example indicates that by estimating solutions' performance very approximately, good solutions can be identified quickly.

The elite GA in Sect. 14.3.1 can be developed into an OO-based elite GA procedure by replacing the solution evaluation and reproduction steps with the followings "*OO-based evaluation and reproduction*," which consists of three steps:

1. Coarse selection: For each descendant in the offspring population, its performance is estimated through a very small number of samples (denoted by k). Based on these approximate estimations, the "best" $|S|$ individuals are selected from the offspring population and form the selected set S, where $|S|$ should be not less than the parent population size GA_μ.
2. Fine selection: For each individual in the set S, its performance is fully evaluated through a large number of samples (i.e., K samples). The current best solution is updated if the performance of a solution in S is better.
3. Reproduction: Based on the fully evaluated performance measures, the best GA_μ individuals are selected from the set S, and they will form the parent population at the next generation.

The above three-step selection process combined with the elite GA can be justified because (1) it can identify a high proportion of high-quality solutions, and (2) even if a poor solution has been selected to be a parent, the roulette wheel selection scheme within the GA guarantees that its genes have a very low probability of being passed on to the next generation.

In terms of computational complexity, with the assumption that the numbers of generations are the same, the elite GA in Sect. 14.3.1 involves total $GA\_\lambda \cdot K \cdot GA\_Gen$ runs, where GA_λ is the offspring population size, K is the number of samples to fully evaluate a solution, and GA_Gen is the maximum number of generations. On the other hand, the OO-based elite GA involves total $GA\_\lambda \cdot k \cdot GA\_Gen + |S| \cdot K \cdot GA\_Gen$ runs, where k is the number of samples used to very approximately estimate a solution's performance in the coarse selection step, and $|S| = GA\_\mu$. Normally, $K \gg k$, which means that the OO-based elite GA requires about $|S|/GA\_\lambda$ of the computational time required in the basic GA. For example, if $GA\_\mu = GA\_\lambda/2$, then the OO-based elite GA can save about 50% of the computational time compared to the elite GA.

14.5   Notes

Simulation is often used for three purposes in the literature: (1) to confirm and verify the analytical or heuristic results, (2) to demonstrate the effectiveness of a specific policy and make a comparison between different control policies, and (3) to optimize the control policies through "what-if" scenario analysis or meta-heuristic search methods. Simulation is particularly useful in situations where analytical methods are not applicable and efficient heuristics are not available.

For example, in the first aspect, simulation has been used to validate the approximate procedure for determining the (r, Q) policy parameters (Moinzadeh and Nahmias 1988), to confirm the accuracy of an analytical model to estimate performance measures (Alfredsson and Verrijdt 1999), to validate the approximation of a queuing network-based analytical method that is used to calculate the performance of assembly Kanban systems (Matta et al. 2005), to assess the accuracy of an approximate method that finds the optimal reorder points and the optimal backorder number (Thangam and Uthayakumar 2008), to test the accuracy of the steady-state probability distribution of an approximate model for a base-stock controlled assembly system (Avsar et al. 2009), and to validate the estimates of throughput and queue length obtained using an aggregation procedure for a synchronization station with multiple inputs (Ramakrishnan and Krishnamurthy 2012).

In the second aspect, simulation has been used to illustrate the effectiveness of a heuristic rule in Graves (1980) and a parametric scheduling policy in Wein (1992), to compare a pre-allocation policy based on a decomposition method with several commonly used allocation policies in general assembly systems subject to service level constraints in De Kok and Visschers (1999), and to compare a heuristic decision rule with three traditional ordering policies in Ismail et al. (2011).

In the third aspect, Ahmed and Alkhamis (2002) presented a simulation-based optimization approach using simulated annealing to optimize (s, S) policy in an inventory system. Shahabudeen et al. (2002) applied the simulated annealing to optimize the number of Kanbans and lot size in a three-stage stochastic production line. Alabas et al. (2002) compared the performance of different meta-heuristic search methods such as genetic algorithm, simulated annealing, and tabu search to optimize the number of Kanbans. A rule-based genetic algorithm together with a simulation model is proposed to find the optimal control parameters for the multi-loop CONWIP policy in an actual lamp assembly production line (Ip et al. 2007). A simulation-based optimization procedure is presented to calculate the optimal inventory parameters in a dual-index policy (Veeraraghavan and Scheller-Wolf 2008). Simulation and genetic algorithms are combined to optimize maintenance and inventory control policies in a failure-prone production line consisting of multiple machines without intermediate buffers (Rezga et al. 2004). A genetic algorithm and a discrete-event simulation model are used to optimize a hybrid push/pull policy in an assemble-to-order manufacturing system (Ghrayeb et al. 2009). Koulouriotis et al. (2010) utilized discrete-event simulation and a genetic algorithm to optimize a few pull-type control policies including base-stock, Kanban, CONWIP, and hybrid policies in serial manufacturing lines and assembly systems.

More literature on the application of simulation and simulation-based optimization methods in supply chain management can be referred to Aytug et al. (2003), Terzi and Cavalieri (2004), and Yoo et al. (2010).

However, applying simulation-based meta-heuristics to stochastic systems also faces computational challenges due to the multiplication factor of the large number of samples (required to estimate the performance measure in stochastic situations) and the large number search iterations (required by the meta-heuristic mechanism to find a good solution). Ordinal optimization is one of the techniques to reduce the computational burden in optimization problems (Ho et al. 2007). There have been some studies to combine ordinal optimization with meta-heuristics, for example, Luo et al. (2001) and Yen et al. (2004).

References

Ahmed, M.A., Alkhamis, T.M.: Simulation-based optimization using simulated annealing with ranking and selection. Comput. Oper. Res. **29**(4), 387–402 (2002)

Alabas, C., Altiparmak, F., Dengiz, B.: A comparison of the performance of artificial intelligence techniques for optimizing the number of Kanbans. J. Oper. Res. Soc. **53**, 907–914 (2002)

Alfredsson, P., Verrijdt, J.: Modeling emergency supply flexibility in a two-echelon inventory system. Manag. Sci. **45**(10), 1416–1431 (1999)

Avsar, Z.M., Zijm, W.H., Rodoplu, U.: An approximate model for base-stock-controlled assembly systems. IIE Trans. **41**(3), 260–274 (2009)

Aytug, H., Khouja, M., Vergara, F.E.: Use of genetic algorithms to solve production and operations management problems: a review. Int. J. Prod. Res. **41**(17), 3955–4009 (2003)

Back, T.: Evolutionary Algorithms in Theory and Practice: Evolution Strategies, Evolutionary Programming, Genetic Algorithms. Oxford University Press, Oxford (1996)

Banks, J., Carson, J.S., Nelson, B.L., Nicol, D.M.: Discrete Event Systems Simulation, 3rd edn. Prentice Hall, Englewood Cliffs (2000)

De Kok, A.G., Visschers, J.W.C.H.: Analysis of assembly systems with service level constraints. Int. J. Prod. Econ. **59**(1–3), 313–326 (1999)

Dorigo, M., Maniezzo, V., Colorni, A.: The ant system: optimization by a colony of cooperating agents. IEEE Trans. Syst. Man Cybern. Part-B **26**, 29–41 (1996)

Eberhart, R., Kennedy, J.: A new optimizer using particle swarm theory. In: Proceedings of the Sixth International Symposium on Micro Machine and Human Science, pp. 39–43. Nagoya, Japan (1995)

Fu, M.C.: Optimization for simulation: theory vs. practice. INFORMS J. Comput. **14**(3), 192–215 (2002)

Ghrayeb, O., Phojanamongkolkij, N., Tan, B.A.: A hybrid push/pull system in assemble-to-order manufacturing environment. J. Intell. Manuf. **20**(4), 379–387 (2009)

Glover, F.: Tabu search – part I. ORSA J. Comput. **1**(3), 190–206 (1989)

Goldberg, D.E.: Genetic Algorithms in Search, Optimization and Machine Learning. Addison-Wesley, Reading (1989)

Graves, S.C.: The multi-product production cycling problem. AIIE Trans. **12**(3), 233–240 (1980)

Ho, Y.C.: An explanation of ordinal optimization: soft computing for hard problems. Inf. Sci. **113**, 169–192 (1999)

Ho, Y.C., Sreenivas, R., Vakili, P.: Ordinal optimization of DEDS. Discret. Event Dyn. Syst. Theory Appl. **2**, 61–88 (1992)

Ho, Y.C., Zhao, Q.C., Jia, Q.S.: Ordinal Optimization: Soft Optimization for Hard Problems. Springer, New York (2007)

Holland, J.: Adaptation in Natural and Artificial Systems: An Introductory Analysis with Applications to Biology, Control, and Artificial Intelligence. University of Michigan Press, Ann Arbor (1975)

Ip, W.H., Huang, M., Yung, K.L., Wang, D.W., Wang, X.W.: CONWIP based control of a lamp assembly production line. J. Intell. Manuf. **18**(2), 261–271 (2007)

Ismail, C., Burak, E., Joseph, G.: A decision rule for coordination of inventory and transportation in a two-stage supply chain with alternative supply sources. Comput. Oper. Res. **38**(12), 1696–1704 (2011)

Kirkpatrick, S., Gelatt, C.D., Vecchi, M.P.: Optimization by simulated annealing. Science **220**(4598), 671–679 (1983)

Koulouriotis, D.E., Xanthopoulos, A.S., Tourassis, V.D.: Simulation optimisation of pull control policies for serial manufacturing lines and assembly manufacturing systems using genetic algorithms. Int. J. Prod. Res. **48**(10), 2887–2912 (2010)

Luh, G.C., Chueh, C.H., Liu, W.W.: MOIA: multi-objective immune algorithm. Eng. Optim. **35**(2), 143–164 (2003)

Luo, Y.C., Guignard, M., Chen, C.H.: A hybrid approach for integer programming combining genetic algorithms, linear programming and ordinal optimization. J. Intell. Manuf. **12**(5–6), 509–519 (2001)

Matta, A., Dallery, Y., Di Mascolo, M.: Analysis of assembly systems controlled with Kanbans. Eur. J. Oper. Res. **166**(2), 310–336 (2005)

Moinzadeh, K., Nahmias, S.: A continuous review model for an inventory model with two supply modes. Manag. Sci. **34**(6), 761–773 (1988)

Ramakrishnan, R., Krishnamurthy, A.: Performance evaluation of a synchronization station with multiple inputs and population constraints. Comput. Oper. Res. **39**(3), 560–570 (2012)

Rezga, N., Xie, X., Mati, Y.: Joint optimization of preventive maintenance and inventory control in a production line using simulation. Int. J. Prod. Res. **42**(10), 2029–2046 (2004)

Shahabudeen, P., Gopinath, R., Krishnaiah, K.: Design of bi-criteria Kanban system using simulated annealing technique. Comput. Ind. Eng. **41**(4), 355–370 (2002)

Song, D.P.: Stochastic models in planning complex engineer-to-order products. Ph.D. thesis, University of Newcastle upon Tyne (2001)

Song, D.P., Hicks, C., Earl, C.F.: An ordinal optimization based evolution strategy to schedule complex make-to-order products. Int. J. Prod. Res. **44**(22), 4877–4895 (2006)

Terzi, S., Cavalieri, S.: Simulation in the supply chain context: a survey. Comput. Ind. **53**(1), 3–16 (2004)

Thangam, A., Uthayakumar, R.: A two-level supply chain with partial backordering and approximated Poisson demand. Eur. J. Oper. Res. **187**(1), 228–242 (2008)

Veeraraghavan, S., Scheller-Wolf, A.: Now or later: a simple policy for effective dual sourcing in capacitated systems. Oper. Res. **56**(4), 850–864 (2008)

Wein, L.M.: Dynamic scheduling of a multiclass make-to-stock queue. Oper. Res. **40**(4), 724–735 (1992)

Yen, C.H., Wong, D.S.H., Jang, S.S.: Solution of trim-loss problem by an integrated simulated annealing and ordinal optimization approach. J. Intell. Manuf. **15**(5), 701–709 (2004)

Yoo, T., Cho, H., Yucesan, E.: Hybrid algorithm for discrete event simulation based supply chain optimization. Expert Syst. Appl. **37**(3), 2354–2361 (2010)

Chapter 15
Conclusions

15.1 Conclusions and Managerial Insights

Supply chain management (SCM) aims to achieve efficient and effective flow and storage of materials and information across the entire supply chain system to meet customer requirements. This book treats supply chain management problem as stochastic optimal control and optimization problems by minimizing total system-wide costs subject to a set of constraints.

To tackle the challenge of pursuing global optimization for the supply chains in the presence of multiple uncertainties such as random demands, stochastic processing times, machine failure and repair, and stochastic order replenishment lead times, this book has taken the following approach:

- Several relatively simple but typical stochastic supply chain systems are studied comprehensively, including the basic supplier–manufacturer–customer supply chains, multistage serial supply chains, supply chains with backordering decisions, supply chains with preventive maintenance decisions, supply chains with assembly operations, and supply chains with multiple products. On the one hand, these supply chains are good representatives; on the other hand, the results can serve as reference points for more complicated supply chains.

- For the selected stochastic supply chain systems, analytical models are formulated using the stochastic dynamic programming approach, which leads to the optimal integrated control policies. Useful structural characteristics of the optimal integrated policies are established rigorously and/or numerically. For example, the optimal policies are often characterized by a set of monotonic switching curves (in two-dimension state space) or switching manifolds (in multi-dimension state space) with good asymptotic behaviors.

- Based on the established structural knowledge of the optimal policies, easy-to-operate suboptimal threshold-type control policies are constructed for both the discounted cost case and the long-run average cost case. For example, linear switching threshold policies are proposed for the basic

D.-P. Song, *Optimal Control and Optimization of Stochastic Supply Chain Systems*, Advances in Industrial Control, DOI 10.1007/978-1-4471-4724-4_15, © Springer-Verlag London 2013

supplier–manufacturer–customer supply chains and the multistage serial supply chains; Kanban threshold control policies and base-stock Kanban control policies are presented for the supply chains with assembly operations. Based on the matrix analytic method and Markov chain theory, the system stability of the underlying supply chains has been addressed, which is a more fundamental issue than the control problem when steady-state performance measures are considered. The impact of storage capacity and ordering/production capacity on the system stability is explicitly established. Steady-state probability distributions are obtained for several types of manufacturing supply chain systems.

- Several methods are presented to implement the parameterized threshold control policies. First, numerical methods based on value iteration algorithm and stationary distribution are presented to optimize the threshold parameters. Second, simulation-based optimization methods including genetic algorithms, simulated annealing, and ordinal optimization are introduced to find the optimal threshold parameters.
- Throughout the book, a large number of numerical examples are provided to illustrate the results and compare the performance of policies and algorithms.

Based upon the research in this book and the results in the literature, the following managerial insights can be summarized:

1. In order to achieve the optimal performance of stochastic supply chains, it is necessary to integrate the supply chains through information sharing and management coordination. Under certain assumptions on the stochastic factors, it is possible to formulate the optimal control and optimization problem analytically. The resulting optimal policies may have good structural characteristics.
2. The optimal integrated ordering and production policies rely on all the items of information in the supply chain systems including demand rate, production rate, machine failure and repair rates, replenishment lead times, storage capacity, and relevant cost coefficients. However, the values that these system parameters take may not influence the structure of the optimal policies. Namely, the structural characteristics of the optimal policies (such as the number of switching manifolds, their monotonicity, and asymptotic behaviors) generally hold for the system parameters and distribution types to a large extent. On the other hand, the system parameters do affect the position, shape, and asymptotes of the switching manifolds.
3. Different items of information have different degrees of impact on the optimal policies. The analytical models and numerical methods enable us to evaluate and quantify their impact. It has been observed that dynamic information of customer demand and downstream inventory plays a significant role to manage upstream entities in the optimal policies, whereas the dynamic information of upstream inventory has less impact on the management of the downstream entities. In addition, the reliability, lead times, and capacity at downstream entities have more significant impact on the system performance than those at upstream entities.

4. The optimal policies can be reasonably approximated by parameterized threshold-type control policies. For example, for the basic supplier–manufacturer–customer supply chain, a set of piece-wise linear switching curves can closely approximate the optimal policy. For multistage serial supply chains and assembly supply chains, the combined echelon base-stock and Kanban policies can be good approximations to the optimal policies. However, the quality of a threshold-type control policy depends on the choice of its threshold parameters.

5. Echelon base-stock control policies rely on the information of downstream entities' inventory and final customer's demands. Although such information is utilized in a rather simple format, that is, accumulated into the echelon base-stock position, it does reflect a certain degree of information sharing and management coordination. In that sense, the echelon base-stock policies may be regarded as a type of vendor managed inventory (VMI) policy (because upstream entities make ordering decisions utilizing the demand and inventory information at downstream entities). On the other hand, if threshold-type control policies, for example, base-stock policies, Kanban policies, CONWIP policies, and their combination forms, are implemented after the threshold parameters have been optimized within the underlying systems, it is equivalent to saying that the supply chain members are collaboratively designing the threshold control parameters by considering all relevant items of information (not just demand and inventory). In that sense, the optimized threshold-type control policies can be regarded as a further development of VMI policies. The above discussions indicate that there is sufficient theoretical supports to the good performance of VMI policies and optimized threshold policies in supply chain management, and the VMI polices could be further developed through better collaboration and optimization.

6. Different ways are available to design/optimize threshold control parameters, for example, numerically, heuristically, or simulation-based. Numerical and heuristic methods are often more computationally efficient, whereas simulation-based optimization methods are more widely applicable. The key point is the trade-off between solution quality and computational effort.

15.2   Limitations and Further Research

This book presents an approach to the optimal control and optimization of stochastic supply chain systems, which aims to bridge the gap between the theoretical analysis and the practical applicability for stochastic supply chain management. The main contents are based on the author's previous and present research work. Relevant literatures and their contributions are noted in the "discussion and notes" sections at the end of each chapter.

This book has several limitations. Firstly, we focused on several relatively simple stochastic supply chain systems to demonstrate the approach. In practice, supply chain systems are likely to be more complicated in both chain infrastructure (that describes the network structure) and probability distributions (that describe the stochastic factors). However, the structural characteristics of the optimal policies tend to be preserved for more complicated supply chains. For example, Benjaafar et al. (2011) considered a multistage assembly system with variable batch sizes. Under the lost-sale assumption, they showed that the optimal production policy for each stage is a state-dependent base-stock policy with the base-stock level decreasing in the inventory level of items that are downstream and increasing in the inventory level of all other items. Chapter 3 showed that the main analytical results carry over to more general supply chains after relaxing the key assumptions. In addition, a number of studies have shown that some complicated supply chains can be decomposed into a series of simple ones (De Kok and Visschers 1999; Muharremoglu and Tsitsiklis 2008; Janakiraman and Muckstadt 2009).

Secondly, we consider the optimal control problem in stochastic supply chains under the assumption that the supply chain is integrated, and therefore, information is shared and the decisions are coordinated. However, supply chain members often have contractual relationship that defines the level of integration, in which information sharing and management coordination may be only partially achievable (Arshinder et al. 2008). In addition, strategic decisions such as supply chain network design and relationship management are also important issues in supply chain optimization (Min and Zhou 2002). This book did not address these issues in very detail, although some of them have been mentioned and discussed in the "discussion" sections, for example, Chap. 2.

Thirdly, this book has used the stochastic dynamic programming and the induction approach to investigate the structure of the optimal policies; applied the matrix analytic method and queuing theory to study the system stability; used the flow balance equations, the characteristic equations, and the spectral expansion method to derive the steady-state probability distribution; and developed the simulation-based meta-heuristics to optimize the threshold parameters. However, there are other approaches that could be used to tackle the supply chain management problems (e.g., Sarimveis et al. 2008).

Further research and development could be pursued in the following directions:

- Supply chain integration issues: Although there has been a trend that supply chain systems are moving towards integration, at present many supply chains are still not well integrated. It would be interesting to determine an appropriate integration level for a given supply chain, and investigate the optimal control problem for stochastic supply chains when certain types of information are not shared and certain types of decisions are not coordinated.
- Multi-objective supply chain optimization: The approach of this book mainly takes the system-wide cost as the supply chain performance measure to optimize. There are other measures that are of interest, for example, customer service level, asset utilization, and safety. Recently, the environmental impact of supply chains has attracted much attention, for example, greener supply chain, by reducing carbon footprint (Elhedhli and Merrick 2012). One way to consider multiple

performance measures is to consolidate them into a single objective or treat some measures as constraints. A more natural way is to treat them as multi-objective optimization problems (Wang et al. 2011).

- Analytical and approximate methods: Analytical methods such as approximate dynamic programming (Si et al. 2004; Powell 2007), reinforcement learning (Bertsekas and Tsitsiklis 1996; Sutton and Barto 1998), and stochastic stability-based solutions (Arruda and Do Val 2008) could be applied to circumvent the computational burden of standard dynamic programming. Approximation procedures via aggregation and decomposition could be developed to reduce the size of the state space but still capturing the main interacting effect between subsystems.
- Simulation-based methods: Simulation-based optimization is often performed in a top-down structure that is more suitable for centralized decision-making in supply chain systems. Multi-objective meta-heuristics (e.g. multi-objective GA) can be used to tackle multi-objective supply chain optimization problems. On the other hand, agent-based simulation takes the bottom-up structure that is more suitable for decentralized decision-making systems (Chaib-draa and Muller 2006).

References

Arruda, E.F., Do Val, J.B.R.: Stability and optimality of a multi-product production and storage system under demand uncertainty. Eur. J. Oper. Res. **188**(2), 406–427 (2008)

Arshinder, K., Kanda, A., Deshmukh, S.G.: Supply chain coordination: perspectives, empirical studies and research directions. Int. J. Prod. Econ. **115**(2), 316–335 (2008)

Benjaafar, S., ElHafsi, M., Lee, C.Y., Zhou, W.H.: Optimal control of an assembly system with multiple stages and multiple demand classes. Oper. Res. **59**(2), 522–529 (2011)

Bertsekas, D.P., Tsitsiklis, J.N.: Neuro-Dynamic Programming. Athena Scientific, Belmont (1996)

Chaib-draa, B., Muller, J.P.: Multiagent-Based Supply Chain Management. Springer, Berlin (2006)

De Kok, A.G., Visschers, J.W.C.H.: Analysis of assembly systems with service level constraints. Int. J. Prod. Econ. **59**(1–3), 313–326 (1999)

Elhedhli, S., Merrick, R.: Green supply chain network design to reduce carbon emissions. Transp. Res. Part D **17**(5), 370–379 (2012)

Janakiraman, G., Muckstadt, J.A.: A decomposition approach for a class of capacitated serial systems. Oper. Res. **57**(6), 1384–1393 (2009)

Min, H., Zhou, G.: Supply chain modeling: past, present and future. Comput. Ind. Eng. **43**(1–2), 231–249 (2002)

Muharremoglu, A., Tsitsiklis, J.N.: A single-unit decomposition approach to multiechelon inventory systems. Oper. Res. **56**, 1089–1103 (2008)

Powell, W.B.: Approximate Dynamic Programming: Solving the Curses of Dimensionality. Wiley, Hoboken (2007)

Sarimveis, H., Patrinos, P., Tarantilis, C., Kiranoudis, C.: Dynamic modeling and control of supply chain systems: a review. Comput. Oper. Res. **35**(11), 3530–3561 (2008)

Si, J., Barto, A., Powell, W.B., Wunsch, D.: Learning and Approximate Dynamic Programming: Scaling up to the Real World. Wiley, New York (2004)

Sutton, R.S., Barto, A.G.: Reinforcement Learning: An Introduction. The MIT Press, Cambridge (1998)

Wang, F., Lai, X., Shi, N.: A multi-objective optimization for green supply chain network design. Decis. Support Syst. **51**(2), 262–269 (2011)

Index

D.-P. Song, *Optimal Control and Optimization of Stochastic Supply Chain Systems*, Advances in Industrial Control, DOI 10.1007/978-1-4471-4724-4, © Springer-Verlag London 2013